Thermometry

Author
James F. Schooley, Ph.D.
Staff Physicist
National Bureau of Standards
Gaithersburg, Maryland

CRC Press, Inc.
Boca Raton, Florida

Library of Congress Cataloging-in-Publication Data

Schooley, James F.
   Thermometry.

   Bibliography: p.
   Includes index.
   1. Thermometers and thermometry.   I. Title.
QC271.S36   1986   536'.5'028   86-6142
ISBN 0-8493-5833-7

This book represents information obtained from authentic and highly regarded sources. Reprinted material is quoted with permission, and sources are indicated. A wide variety of references are listed. Every reasonable effort has been made to give reliable data and information, but the author and the publisher cannot assume responsibility for the validity of all materials or for the consequences of their use.

The work herein is the official writing of an employee of the U.S. Government. As such, it resides in the public domain and is not subject to copyright.

Direct all inquiries to CRC Press, Inc., 2000 Corporate Blvd., N.W., Boca Raton, Florida, 33431.

International Standard Book Number 0-8493-5833-7

Library of Congress Card Number 86-6142
Printed in the United States

PREFACE

This book has been written with two major groups of people in mind — students and professional thermometrists. During some 10 years of service as chief of the temperature measurement group at the National Bureau of Standards, the author has seen ample evidence of the needs of both groups with respect to an understanding of thermometry and a comprehensive reference to current work in the field.

For students, the book can best serve as an auxiliary text in the study of chemistry, physics, and mechanical engineering. The subject matter touches most closely upon thermodynamics and statistical mechanics, heat transfer, and phase equilibria, although it should be self-evident that a study of thermometry will benefit the students of other topics as well. The treatment here is neither complete nor comprehensive on any topic, save perhaps thermometry itself; however, it should serve well to promote a fuller understanding of the areas mentioned above, in addition to providing much material of use for examples and problems.

Professional thermometrists rarely seem to have been trained for their duties in coherent academic programs. Rather, they generally have come to their present employment from various scientific and engineering backgrounds that leave large gaps in one's understanding of the bases for the practice of thermometry. Moreover, their sources of information on present activities in thermometry are widespread and of uneven quality. For this group, the material included here should provide a useful springboard to refresher study in neglected scientific topics. During the period of time when the material is current, it should serve the additional valuable purpose of providing a single source book with comprehensive discussions of temperature scales, recommended practices in thermometry, available thermometers, and areas of active research in this field. Approximately 500 references to other publications on thermometry are included.

In preparing this book, the author has drawn heavily on the work of colleagues at the National Bureau of Standards, which for two generations has provided leadership to both the U.S. and the world in the field of thermometry. Many of the figures used and much of the data quoted originated at the NBS.

The assistance given to the author in preparing this book has been most generous and gratifying. Several colleagues, including G. T. Furukawa, L. A. Guildner, B. W. Mangum, R. J. Soulen, Jr., and C. T. Van Degrift, reviewed the entire book in draft form; others, including G. W. Burns, R. E. Edsinger, J. P. Evans, W. E. Fogle, H. Marshak, M. R. Moldover, M. L. Reilly, and G. J. Rosasco, offered discussions and reviews of various sections. Although inconsistencies and omissions no doubt remain, the book is very much the better as a result of the helpfulness of these talented people. The author is most grateful, too, to S. C. Ramboz, who diligently typed the manuscript more than once in bringing it to final form.

James F. Schooley
Gaithersburg, Md.

THE AUTHOR

James F. Schooley, Ph.D., is a staff physicist in the Center for Basic Standards at the National Bureau of Standards, Gaithersburg, Maryland. Dr. Schooley received the A.B. degree from Indiana University and M.S. and Ph.D. degrees from the University of California at Berkeley. For about 10 years he headed the thermometry group at the National Bureau of Standards. He is the author or co-author of more than 50 scientific publications, including 3 book chapters. His research interests are concentrated in thermometry, particularly at low temperature, and in superconductivity. He has received the U.S. Department of Commerce Silver and Gold Medal Awards for his research in cryogenics.

DEDICATION

To Michael

TABLE OF CONTENTS

Chapter 1
Introduction 1
I.   A Short Story 1
II. Strange Zeros and Other Thermometry Problems 2
References 5

Chapter 2
The Concept of Temperature 7
I.   Introduction 7
II. Temperature Simplified. Early Thermometers and Temperature Scales 7
     A. Definitions of Basic Thermometry Terms 7
     B. The First Thermometers and Temperature Scales 8
III. Temperature and Thermodynamics 10
     A. The Ideal Gas and an Absolute Zero of Temperature 10
     B. Heat and Work. The Carnot Cycle 16
     C. The Laws of Thermodynamics 18
IV. Temperature and Statistical Mechanics 26
     A. Kinetic Theory and the Ideal Gas 27
     B. Quantization of Light and Heat 29
References 31

Chapter 3
Temperature Fixed Points 33
I.   Introduction 33
II.   Phases, Phase Equilibria, and Phase Transitions 33
III. Types of Temperature Fixed Points 35
IV.   Water, a Prototypical Fixed-Point System 37
      A. Ice Point and Triple Point of Water 37
         1. Phase Diagram of Ice 37
         2. Effect of Pressure on Melting. The Clapeyron Equation 37
         3. Triple Points of Water 38
         4. The "Triple-Point" Cell 39
      B. Isotopic Effects on the Water Triple Point 43
      C. Impurity Effects in the H_2O System 44
      D. Dilute Solutions. Raoult's Law 45
      E. The Steam Point 48
V.   High-Pressure Cells for Low-Temperature Triple Points 49
VI.   Melting, Freezing, and Triple Points of Metals 55
VII. Fixed Points for the Life Sciences 63
VIII. Solid-Solid Equilibria as Temperature Reference Points 65
      A. Solid-Solid Equilibria in Oxygen 67
      B. Superconductive-Normal Phase Equilibria in Metals 68
References 72

Chapter 4
International Scales of Temperature 77
I.   Introduction 77
II. Elements of Temperature Scales and Thermometry 77
     A. Characteristics of a Temperature Scale 77
     B. Temperature-Scale Thermometry 81

III.		Conference on the Meter and its Consequences	86
IV.		International Temperature Scales	87
	A.	The Normal Hydrogen Scale	88
	B.	The International Temperature Scale of 1927 (ITS-27)	88
		1. The Range −190 to +660°C	89
		2. The Range 660 to 1063°C	90
		3. The Range Above 1063°C	90
	C.	The International Temperature Scale of 1948 (ITS-48) and the International Practical Temperature Scale of 1948 (IPTS-48)	91
		1. The Range −183 to +630°C	91
		2. The Range 630 to 1063°C	92
		3. The Range Above 1063°C	92
		4. A "Practical" ITS-48	93
	D.	The International Practical Temperature Scale of 1968 (IPTS-68) and the IPTS-68(75)	94
		1. The Range 13.81 to 273.15 K	96
		2. The Range 273 to 903 K	105
		3. The Range 903 to 1337 K	106
		4. The Range Above 1337 K	107
		5. Supplementary Information	107
	E.	1976 Provisional 0.5 to 30 K Temperature Scale (EPT-76)	108
		1. Development of EPT-76	108
		2. Features of the EPT-76	108
V.		International Temperature Scales of the Future	109
References			111

Chapter 5
The Measurement of Thermodynamic Temperatures 115

I.		Introduction	115
II.		Gas Thermometry	115
	A.	Uncertainty of the Ideal Gas Constant	116
	B.	Deviations from the Ideal Gas Law	117
	C.	Dead Space Corrections	120
	D.	Corrections for Thermal Expansion	122
	E.	Hydrostatic Pressure Head Corrections	123
	F.	Thermomolecular Pressure Corrections	123
	G.	Effects of Adsorbed Impurities	124
	H.	Current Gas Thermometry Experiments	125
		1. Guildner and Edsinger's Measurements from 273 to 730 K	125
		2. Berry's Measurements from 2.6 to 27.1 K	131
		a. Absolute P-V Isotherm Thermometry	132
		b. Relative P-V Isotherm Thermometry	133
		c. Constant-Volume Gas Thermometry	134
		3. Acoustic Thermometry	136
		4. Dielectric-Constant Gas Thermometry	136
		5. Constant-Volume Gas Thermometry of Kemp, Besley, and Kemp	137
		6. Constant-Volume Gas Thermometry of Steur et al.	138
III.		Noise Thermometry	138
	A.	Thermometry Using Standard Noise Circuitry	139
	B.	Cross-Correlation Methods	140
	C.	Noise-Voltage Detection by Superconductive Josephson Junctions	141

	IV.	Nuclear Orientation Thermometry 143
	V.	Kelvin Thermodynamic Temperatures Below 273 K 147
	VI.	Radiation Thermometry 148
		A. Total Radiation Thermometry 149
		B. Spectral Radiation Thermometry 151
		1. Measurements by Coates and Andrews 152
		2. Measurements by Jung.. 152
		3. Spectral Radiation Measurements Above 630°C 153
	VII.	Kelvin Thermodynamic Temperatures Above 273 K 156
References 157

Chapter 6
Modern Thermometers 163
I. Introduction 163
II. Thermal Expansion Thermometers 164
  A. Filled-System Thermometers 164
  B. Bimetallic Thermometers 165
  C. Liquid-in-Glass Thermometers 166
III. Thermocouple Thermometry 172
  A. Seebeck and Peltier Effects 172
  B. The Laws of Thermocouple Circuits 176
    1. Law of Homogeneous Metals 176
    2. Law of Intermediate Metals at a Single Temperature 176
    3. Law of Intermediate Metals at Different Temperatures 178
    4. Law of Intermediate Temperatures 178
  C. Standard Thermocouple Thermometers and Their Calibration 179
    1. Annealing 180
    2. Test Junction Assembly 181
    3. emf Measurement 181
    4. Construction of Reference Tables 182
  D. Special Problems in Thermocouple Thermometry 182
  E. Thermocouples for Special Applications in Thermometry 183
    1. Thermocouples for Use in Cryogenics 183
    2. Thermocouples for High Temperatures 183
IV. Resistance Thermometers 186
  A. Resistance Thermometer Measurement 186
    1. Potentiometer Measurements 187
    2. Bridge Measurements 189
    3. AC Measurements 190
  B. Pure Metal Resistance Thermometers 192
    1. Standard Platinum Resistance Thermometers 192
    2. High-Temperature Platinum Resistance Thermometers 194
    3. Resistance Temperature Detectors (RTDs) 199
  C. Thermistor Thermometers 202
  D. Rhodium-Iron Resistance Thermometers 204
  E. Semiconducting Thermometers in Cryogenics 206
V. Radiation Thermometers 208
  A. Radiation Properties of a Black Body 209
  B. Radiation Properties of Real Materials 212
  C. General Features of Radiation Thermometers 213
    1. Radiation Thermometers for Thermometry Research 214
    2. Commercial Radiation Thermometers 215

	D.	Temperature Measurement in Gases, Flames, Plasmas, and Stars 219
VI.	Other High-Precision Thermometers 222
	A.	Vapor-Pressure Thermometry..222
	B.	Nuclear Quadrupole Resonance Thermometry 223
	C.	Quartz Resonance Thermometers 225
	D.	Paramagnetic Thermometers..225
	E.	Industrial Johnson Noise Thermometry 227

References..228

Index...239

Chapter 1

INTRODUCTION

I. A SHORT STORY

Michael was twelve and he wanted to impress his grandfather, so he said, "Did you know that it's zero outside, Gramps?"

The old man looked quickly at the boy. Then with a quizzical smile, he said, "No, I didn't, son. It doesn't seem that cold outside, to me."

Michael grinned. "It isn't, really. I mean in Fahrenheit. Our thermometer says 32. But our teacher told us that freezing is zero on the scientific Celsius scale."

"Well, Michael, I had no idea that you were such a scientist. How did you come to hear about the Celsius scale?" asked his grandfather.

"Mr. Yoxtheimer said that a new temperature scale was just announced in 1968 and that we all should learn about it. But Gramps," said Michael, hesitating.

"What is it, son?" asked the old man.

"Well, it seems funny that zero isn't the *beginning* of temperature. Either in Celsius or in Fahrenheit. Isn't temperature just how hot it is? How can things have a minus hotness?"

The old man's eyes twinkled. Michael could tell that something good was going to happen.

"The trouble is, people keep putting zeros in the middle of things instead of at the beginning," said Gramps.

"What do you mean, Gramps?" asked Michael.

"Well, for example, time," said Gramps. "When were you born?"

"May fourteenth, 1959," chirped Michael.

"All right," said Gramps, "you were born in 1959 and it is 1971 now. Do you think that time started one thousand nine hundred seventy-one years ago?"

"No, no. Zero is when Christ was born. But I see what you mean — there's a zero in the middle of time, too," said Michael. Seeing his grandfather's smile, Michael pursued the question. "Are there any other zeros, Gramps?"

"There might be," was the response. "Tell me about your bicycle tire when you repaired it today — do you remember anything about its air pressure?"

"I guess so." Michael pondered a moment. "The pressure gauge said zero when the tire was flat, but after I pumped it tight, it said sixty pounds." Michael wasn't sure where this topic might lead.

Gramps sighed. "I'm not sure that this will be clear to you, Michael, but the pressure in the air around us really isn't zero, either. Have you ever heard the weatherman mention 'atmospheric pressure'?"

Michael brightened. "Sure," he said, "he says, 'The pressure is thirty-point-oh-six and rising'." Then his face clouded. "What does that mean?" he asked.

"See if you can't get your teacher to help you find out about temperature, time, and pressure, to see whether real scientists use zeros in the middle," said Gramps. "He might have an encyclopedia or a regular science book. Who knows, you might be onto something," he said, with that same twinkle in his eye.

Michael left for school the next morning with a determined air. Every day for a week he was late getting home. Finally, on a crisp, late spring evening after the table was cleared, he spread a much-folded paper across it and sought out his grandfather. "Mr. Yoxtheimer said that you and I both will get extra credit in science this time, Gramps. It took some research, but we found out that you were right."

Grandfather surveyed the page. It looked like this:

2 Thermometry

	Scientific scale	**Common scale**
Temperature	Kelvin (K) 0 K means no heat at all	Degrees Celsius (°C) or degrees Fahrenheit (°F) 0°C means 273 K; and 0°F means 255 K
Time	Sidereal year 0 year means the beginning of the universe	Year (BC or AD) on the Gregorian calendar 0 AD may mean 20,000,000,000 years after the beginning of the universe
Pressure	Pascal (Pa) 0 Pa means no pressure at all	Gage pressure (lb/in^2) or vacuum (inches of mercury) 0 psig means 101,325 Pa 0 in Hg means 101,325 Pa

Grandfather stroked his chin thoughtfully. "Well, Michael, it certainly seems that there are different ways of looking at everyday measurements, doesn't it?"

"It sure does," said Michael. Then he grinned mischievously. "By the way, Gramps, did you know that the temperature outside is 276? Kelvins, that is."

A few days later, Michael was relaxing in the big easy chair in front of the fire. "Gramps," he asked, "does temperature have some really important meaning other than just how hot it is?"

"I think it does, Michael, if you use the right scale," the old man responded. "Scientists use it to understand how things work."

"Well, I wonder how you know when you are using just the right scale."

The older man was quiet for a bit, and Michael looked at him to see if he was still listening. Finally Gramps said, "Maybe the temperature scale is closest to being right when the most things work right for the scientists."

II. STRANGE ZEROS AND OTHER THERMOMETRY PROBLEMS

Michael and his grandfather have touched upon the single problem that, more than any other, keeps even well-educated people from understanding scientific thermometry. It is necessary to know the significance of a scale before its numbers can have meaning.

In order to make very plain the distinction between scientific scales of measurement and everyday scales — the scales that often are the ones used in engineering, for example — we present in Figures 1, 2, and 3 comparisons of scientific and everyday scales for temperature, time, and pressure/vacuum, respectively. The fact that common substances such as water, nitrogen, and mercury have melting or boiling temperatures that are given different numbers on different scales really should be no more surprising than the fact that atmospheric pressure on earth possesses several different identifying numbers or the fact that the birth of the solar system — uncertain as that event might be — is given different time values on different scales of time.

In Table 1, we present readings on the Fahrenheit, Celsius, Rankine, and Kelvin temperature scales that correspond to several physical transitions. We also provide equations that relate the secondary scales to the Kelvin scale. Note that, to avoid adding numbers that refer to different units, each number in the equations of Table 1 is divided by its unit so as to be made dimensionless.

The Rankine and Fahrenheit scales have been used in engineering work in the U.S. for many years. Recently, a trend toward increasing use of the metric system and the "Système Internationale" units has increased the use of the Celsius and Kelvin scales in engineering. Use of the Fahrenheit scale still is common throughout the U.S. for commercial products and for meteorology. In most other countries of the world, the Celsius scale is preferred for these purposes.

We spend a little time in the next chapter discussing the reasons why more than one scale of temperature has come into common use in the world. At that point, we hope that the

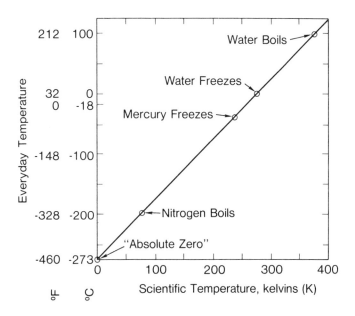

FIGURE 1. Relations between the everyday Fahrenheit (°F) and Celsius (°C) temperature scales and the scientific temperature scale in kelvins (K). Several "fixed points" of temperature are included on the curve for reference.

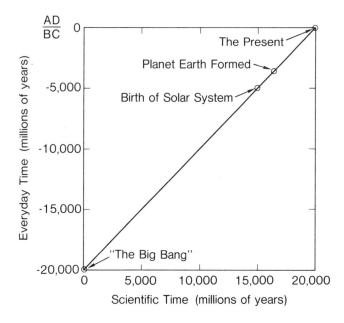

FIGURE 2. Relation between the everyday time scale and the scientific time scale. Approximate times of several epochal events in the history of the universe are included on the curve for reference.

reader will find "strange zeros" to be only a nuisance which cause some annoyance in converting numbers given on one scale to the analogous numbers on another; no conceptual difficulty should remain.

There is a lot more to the study of thermometry than simply relating the everyday scales to the scientific ones. We use the second chapter also to discuss the scientific basis for

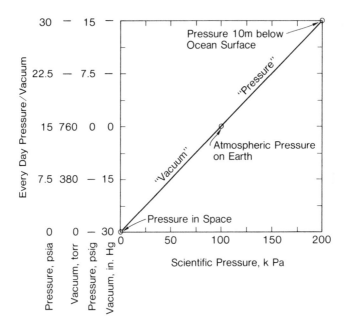

FIGURE 3. Relations comparing everyday pressure and vacuum scales, including pounds per square inch absolute (psia), torr, pounds per square inch gage (psig), and inches of mercury (in. Hg), with the scientific scale in pascals (kPa). Three reference points are included on the curve.

Table 1
TEMPERATURES OF PHYSICAL TRANSITIONS ON FOUR
INTERRELATED TEMPERATURE SCALES

Physical transition	Scale reading			
	Kelvin (K)	Rankine (°R)	Celsius (°C)	Fahrenheit (°F)
Melting of ice	273.150	491.670	0.000	32.000
Boiling of water	373.150	671.670	100.000	212.000
Boiling of nitrogen	77.344	139.219	−195.806	−320.451
Melting of tin	505.118	909.212	231.968	449.542
"Absolute zero"	0.000	0.000	−273.150	−459.670

$W/(K) = [5/9] \, X/(°R)$
$W/(K) = [Y/(°C)] + 273.150$
$W/(K) = [(Z/(°F)) - 32.000][5/9] + 273.150$

Note: W = Temperature expressed in kelvins (K); X = Temperature expressed in degrees Rankine (°R); Y = Temperature expressed in degrees Celsius (°C); Z = Temperature expressed in degrees Fahrenheit (°F). The units (°R) = (°F) are only 5/9 as large as the units (K) = (°C).

thermometry. By the time they have finished studying Chapter 2, the readers should have gained an appreciation of the vital role that thermometry has played in the development of the science of thermodynamics, as well as the fundamental interrelationships of temperature and temperature measurement with the sciences of thermodynamics and statistical mechanics.

Chapter 2 also should convince the reader of the important role in thermometry that is played by reference temperatures — the so-called "fixed points" of the science. The third chapter contains a discussion of the scientific origins of temperature fixed points as well as detailed descriptions of the most important ones.

Having established the central position in thermometry occupied by reference points of temperature, we turn, in Chapter 4, to a discussion of scales of temperature that are recommended internationally for scientific use. Emphasis is placed upon the Kelvin thermodynamic temperature scale and the International Practical Temperature Scale of 1968, which provide scientific guidance to all present-day thermometrists.

In Chapter 5, we present synopses of the methods currently in use to provide a better understanding of the relationship of the international temperature scales to the laws of physical science. The reader will be able to see that accurate determinations of thermodynamic temperatures present a continuing challenge to the best efforts that science can offer in temperature metrology.

We have taken the viewpoint that many of the useful types of thermometers can be described in a meaningful way even in a volume as slim as this one. Chapter 6 contains these descriptions, begun in each case with a brief analysis of the principles involved and including some of the major uses of each type. It should come as no surprise to the thoughtful reader that certain thermometry problems or applications have "created their own thermometers." These also have been included, so far as possible, in Chapter 6.

Each chapter contains general references to the topics at hand in order to provide guidance for the acquisition of a more detailed understanding than can be given in a short space, as well as specific references to current research in the area under discussion. It is hoped that, in this way, this text will serve both as an introduction to the many facets of thermometry and as a point of departure for pursuit of special topics in current temperature research.

REFERENCES

General Temperature References
Fairchild, C. O., Hardy, J. D., Sosman, R. B., and Wensel, H. T., Eds., *Temperature, Its Measurement and Control in Science and Industry,* Reinhold, New York, 1941.
Wolfe, H. D., Ed., *Temperature, Its Measurement and Control in Science and Industry,* Vol. 2, Reinhold, New York, 1955.
Herzfeld, C. M., Ed.-in-Chief, *Temperature, Its Measurement and Control in Science and Industry,* Vol. 3 (Part 1, Brickwedde, F. G., Ed., Part 2, Dahl, A. I., Ed., Part 3, Hardy, J. D., Ed.), Reinhold, New York, 1962.
Hall, J. A., *The Measurement of Temperature,* Chapman and Hall, London, 1966.
Plumb, H. H., Ed.-in-Chief, *Temperature, Its Measurement and Control in Science and Industry,* Vol. 4, Instrument Society of America, Pittsburgh, 1972.
Barber, D. R., Ed., *Temperature Measurement at the National Physical Laboratory; Collected Papers 1934—1970,* Her Majesty's Stationery Office, London, 1973.
Billing, B. F. and Quinn, T. J., Eds., *Temperature Measurement, 1975,* Conference Series Number 26, The Institute of Physics, London, 1975.
Benedict, R. B., *Fundamentals of Temperature, Pressure and Flow Measurements,* 2nd ed., John Wiley & Sons, New York, 1977.
Schooley, J. F., Ed.-in-Chief, *Temperature, Its Measurement and Control in Science and Industry,* Vol. 5, American Institute of Physics, New York, 1982.
Quinn, T. J., *Temperature,* Academic Press, London, 1983.

Units and Metric References
Deming, R., *Metric Power,* Thomas Nelson, Nashville, 1974.
Liptai, R. G. and Pearson, J. W., Eds., *Metrication — Managing the Industrial Transition,* ASTM Special Tech. Publ. 574, American Society for Testing and Materials, Philadelphia, 1975.
Page, C. H. and Vigoureux, P., Eds., *SI The International System of Units,* Her Majesty's Stationery Office, London, 1977; National Bureau of Standards Special Publ. 330, 1981.
Jespersen, J. and Fitz-Randolph, J., *From Sundials to Atomic Clocks: Understanding Time and Frequency,* National Bureau of Standards Monogr. 155, December 1977.

Chapter 2

THE CONCEPT OF TEMPERATURE

I. INTRODUCTION

It is easy to compose superficial definitions of temperature, but difficult to convey its full conceptual meaning. Temperature is "understood" nowadays even by very young children. They have seen and used thermometers and their senses permit them to make reasonable estimates of the temperatures of objects about them in terms of the local temperature scale. Regarding the significance of that scale of temperature for physical and chemical properties of the world around them, however, neither children nor most adults possess any real knowledge.

II. TEMPERATURE SIMPLIFIED. EARLY THERMOMETERS AND TEMPERATURE SCALES

A. Definitions of Basic Thermometry Terms

We begin our discussion of temperature by providing definitions of a few fundamental terms; others will be given as they arise in the discussion.

Temperature — The "hotness" of a given object or system as described by a numerical value expressed on a definite scale of temperatures. For example, we could contemplate a tinkling glass of ice water and decide that its *temperature* should be 0, 32, or 273, depending upon the scale we choose to use.

Fixed point and reference temperature — A *fixed point* is a physical phenomenon that occurs reproducibly at the same temperature. This very useful property of matter was one of the first discovered by thermometrists. A *reference temperature* can be achieved by properly achieving a *fixed point*. The melting of ice mentioned above, the bodies of healthy humans, and boiling water all were shown to be reasonably constant-temperature fixed points for reference work in thermometry. Other phenomena were thought to be fixed in temperature, but eventually were found wanting as reference points. These include the temperatures of caves and cellars as well as the freezing points of various wines and salt solutions.

Temperature scale — A numerical reference standard for temperature, continuous over a specified range, by means of which unknown temperatures may be evaluated. A temperature scale may be derived strictly from thermodynamic considerations, in which case it is known as a "thermodynamic" temperature scale; or it may be constructed by specifying temperature values for various fixed points, with intervening values to be determined by the response of a temperature-dependent instrument, in which case it is known as a "practical" or "laboratory" temperature scale. Our instinctive sense of hot and cold can judge the relative temperatures of two objects, but for most practical purposes (and for *all* scientific aims) we need a continuous scale of temperature. With a temperature scale, we can compare measurements that are separate in time or place and we can derive expressions for the temperature dependences of material properties so as to understand them better. A scale of temperature can serve the simple purpose of providing a number that we can attach to any given level of hotness, or it can perform a more fundamental function. The earliest scales were constructed to monitor simple physical and biological phenomena — such goings-on as a royal fever, changes in weather, and the freezing and boiling of water. The makeup of these scales also reflected the nature of newly created thermometers.

Thermometer — An instrument by means of which one can measure the temperature of an object or system with respect to a particular scale of temperatures.

FIGURE 1. The air thermoscope.[2] The air in the bulb B contracts or expands in proportion to its temperature, moving the meniscus O of the liquid up or down in capillary E. The cup D provides a reservoir for the liquid. (Adapted from Middleton, W. E. K., *A History of the Thermometer and Its Uses in Meteorology*, The Johns Hopkins Press, Baltimore, 1966.)

B. The First Thermometers and Temperature Scales

It is well to review the instruments used in early thermometry, since they affect our practice of temperature measurement even today.

The question of who made the first thermometer is a difficult one to answer; the 17th-century scientific literature is a haphazard collection of personal diaries, letters to friends, and sketchy reports to learned bodies. Middleton, in a well-documented history of thermometry,[1] has argued plausibly that one Santorio Santorre, a physician practicing in Venice, may have been the first (in 1612) to use an air thermoscope, such as the one shown in Figure 1, to measure temperature both in medicine and in meteorology.

The air thermoscope is a simple device, but it is not very accurate by present standards. Its construction is shown in Figure 1. It is made by connecting a glass tube E to a bulb B, partly filling the tube with water, alcohol, or light oil, and inverting the tube so that the open end penetrates the liquid surface of the reservoir D. As the temperature increases or decreases, the air in the bulb expands or contracts accordingly, moving the liquid meniscus O. The temperature is measured in terms of the position of the meniscus within the length of the tube. Not only is the air thermoscope awkward to use, but also it responds to changes in the local air pressure as well as to changes in the temperature, and thus its accuracy is compromised.

The well-known astronomer Galileo, often credited with the invention of the thermometer, certainly was among the first to use the air thermoscope.

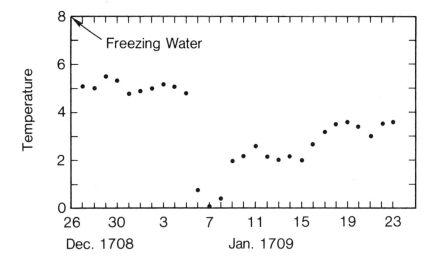

FIGURE 2. Air temperatures in Copenhagen as recorded by Rømer during an exceptionally cold winter. These data probably provided part of the basis for the present-day Fahrenheit temperature scale.

Daniel Gabriel Fahrenheit, born in 1686 in Danzig, was one of the first of many 17th- and 18th-century thermometrists to make thermometers of glass tubing filled variously with mercury or with an alcohol solution. He used at least three different schemes at various times for calibrating thermometers, thereby providing the same level of hotness with at least three different numerical values of temperature. The last of his scales, triply calibrated in an ice/salt-water solution, in an ice-water bath, and " . . . under the armpit, of a living man in good health . . . ",[3] closely corresponds to the present-day Fahrenheit scale. The existence of multiple scales was not unusual at that time; Middleton calls attention to the fact that thermometers occasionally were fitted with a backboard bearing a dozen or more different contemporary scales. Thermometers made from different kinds of glass or incorporating different liquids generally did not agree in all their measurements even when calibrated in the same way.

All early thermometers were based upon the principles of the thermal expansion of a gas or of a liquid — electrical and radiation thermometers came much later.

The question of how to choose a temperature scale was a difficult one for scientists of the 17th and 18th centuries. Logic would impel one to encompass all possible temperatures within the set of positive real numbers, since both negative and imaginary temperatures would pose problems in interpretation. As early as the Iron Age, humans knew that fires could burn or melt virtually any earthly object; thus it was natural to suppose that temperatures could increase without an upper limit. A zero, or lower limit, for temperatures was another matter — without access to mechanical refrigeration, scientists were restricted to their local outdoor environments as sources of low temperatures.

Rømer, a Copenhagen thermometrist, calibrated his thermometers so that they did not register temperatures below zero even during the winter of 1708 to 1709, one of the coldest recorded in European history.[4] The appropriate data from Rømer's notebook are plotted in Figure 2. On Rømer's scale, the temperature of the human body would have been about 24 degrees and that of boiling water 50 to 55 degrees. Shortly after Rømer developed this scale, Fahrenheit visited his laboratory. It appears that Fahrenheit accepted Rømer's calibration principle, but divided each of Rømer's degrees into four in order to gain more resolution from his thermometers. In the absence of lower calibration temperatures, a simple procedure such as Rømer's provided a workable zero for early temperature scales.

The Fahrenheit scale that persists today is approximately the one described by him in 1724 for the Royal Society of London.[3] As we have mentioned already, Fahrenheit's zero was not really a fixed point, but only reflected the outdoor temperature on a very cold day in Copenhagen; he describes its calibration in the laboratory as resulting from an indefinite mixture of ice, water, and sea salt or sal ammoniac. Fahrenheit's two real fixed points were derived from an ice-water mixture (32°F) and — as mentioned above — from the blood heat of a healthy person (apparently assigned by Fahrenheit as 96°F). As we now realize, both of the latter temperatures are reproducible within 0.1°F.

Anders Celsius, a Swedish astronomer, also created a temperature scale about the year 1742. In his mercury-in-glass thermometer scale, the two fixed points, the ice point and the water boiling point, were separated by 100 degrees. The remarkable feature of the scale was that Celsius selected 100° for the ice point and 0° for the boiling point![5] In 1744, Celsius died; his successor as laboratory astronomer was a good friend, Märten Strømer. Middleton finds the evidence convincing that Strømer favored the use of larger numbers for hotter temperatures and that, after Celsius' death, Strømer quietly substituted his own scale for that of Celsius. If this is true, Celsius himself might be both flattered and surprised at seeing today's Celsius scale.

We have noted the fact that all the first thermometers were gas- or liquid-expansion devices. Not all were made alike, however. A rich variety of thermometers was created almost from the first, including air-expansion, water-expansion, mercury-expansion, and alcohol- and alcohol-water-solution-expansion instruments. The thermometric fluids at first were enclosed by various kinds of glass in many different geometric forms. Eventually, in order to measure temperatures above the solid range of glass, air thermometers contained in metal bulbs came into use. Prinsep,[6] for example, used a gold-bulb air thermometer in 1828 for measurements up to 1000°C, and Pouillet developed a similar thermometer with a bulb of platinum.[7]

We could describe many other temperature scales that were developed in various parts of the world during the 17th and early 18th centuries. These scales were highly individual and often short-lived. For more than a century in this era, thermometrists struggled to find a rational basis for thermometry. It is now time to abandon the simplistic approach to temperature that characterized the earliest thermometry, however. We now briefly describe the successful search for the logical basis of temperature that culminated in the work of Kelvin and of many others in the 19th and 20th centuries.

III. TEMPERATURE AND THERMODYNAMICS

An adequate discussion of the scientific basis for temperature requires the introduction of the principles of thermodynamics and those of statistical mechanics. Temperature plays an essential part in both of these sciences. Simply put, the science of thermodynamics, discussed in this section, explains the behavior of systems of gases, liquids, and solids with respect to what are called "state variables" — temperature, pressure, volume, gravitational, magnetic, or electric fields, and the like. In considering strictly thermodynamic relationships, no attention is paid to the internal structure of the substances involved. On the other hand, in the science of statistical mechanics, discussed in Section IV, macroscopic properties are derived from the microscopic characteristics of the substances involved, including their atomic or molecular constitution, energy level schemes, and atomic and nuclear interactions.

A. The Ideal Gas and an Absolute Zero of Temperature

Central to the historical development of the idea that there exists a scientifically rational zero of temperature is the experimental demonstration of the *ideal gas law*. In this section we attempt to show how these two basic features of thermometry — an absolute zero and the ideal gas — came to be understood and how we understand them today.

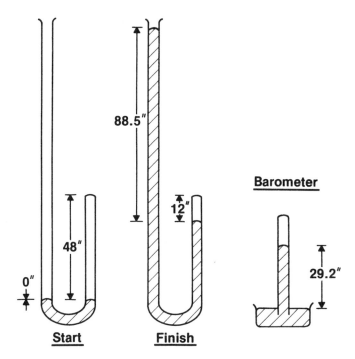

FIGURE 3. Principle of Boyle's P-v apparatus, showing the condition of the U-tube at the start of the experiment and at the finish, in addition to his use of a barometer to establish the base pressure at the start of the experiment.

The fact that a volume of gas responds to changes in pressure as well as temperature made the first air thermometer (see Figure 1) difficult to use for accurate measurements. Air pressure varies hourly during periods of changeable weather. Day-to-day changes can amount to several percent of atmospheric pressure. Such variations had to be overcome before reliable gas thermometry could be achieved.

Robert Boyle and his colleagues prepared the way in the mid-17th century for an understanding of the ideal gas law; his landmark experiments on pressure-volume relations of air are well worth noting.[8]

Scientists of that time understood the fundamentals of air pressure and barometry — this history is well documented by Conant,[9] among others. Using the apparatus sketched in Figure 3, Boyle established very clearly the relation between the pressure P exerted by a gas and its volume v as

$$Pv = \text{constant} \qquad (1)$$

This relation is now called "Boyle's law". It describes a fundamental property of an ideal gas.

Boyle's pressure-volume apparatus was nearly 3 m high; he recorded the length of a column of air trapped in the closed arm of a glass U-tube as an assistant upstairs added quantities of mercury to the open arm. As we have intimated, mercury barometers already were in use in the 17th century; Boyle's account indicates that measurements of the local air pressure were a part of his experimental procedure. Boyle properly determined the total pressure on the trapped air as a quantity proportional to the difference in height of the mercury columns in his U-tube, augmented by the height of the mercury column in his barometer. Boyle's data are plotted in Figure 4. Note that the representation of the volume

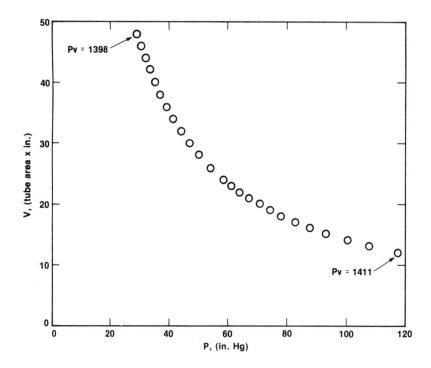

FIGURE 4. Pressure-volume data for air at constant temperature as observed by Boyle.[8] Note the constancy of the product Pv.

of the trapped air column by its length implies that the closed side of the U-tube was made with a uniform cross-sectional area.

The reader easily can verify that the hyperbolic relation dictated by Boyle's law is satisfied by his data. We note in Figure 4 the close agreement of the Pv products at the two extrema of the curve.

Two aspects of Boyle's experiment are especially noteworthy. First, Boyle properly took account of atmospheric pressure in correlating his results, adding the measured barometric pressure to that of the open end of his U-tube. Second, Boyle was quite aware of the necessity to maintain the temperature of the gas at a fixed value during his experiment, using a wet cloth to cool the air after adding more mercury.

Besides Boyle's law, the remaining ingredient needed to establish the ideal gas law empirically is the law we now attribute to Charles or Gay-Lussac, whose well-known experiments were performed at the beginning of the 19th century. It is remarkable that 100 years before these famous men performed their fundamental experiments, Guillaume Amontons wrote to the Paris Academy of Science in 1699 of his own ideas on the "spring" of the air, and the way that different amounts of air expanded by the same proportion over the same range of temperature if the pressure on the air was allowed to remain the same. He believed that the spring or elasticity of air might disappear if all heat were removed from it. Amontons measured the ratio of pressures required to keep a quantity of air confined to the same volume at different temperatures;[10] converted to the Celsius scale, P(100)/P(0) amounted to 1.404. Here, P(100) refers to the pressure measured at a temperature of 100°C. This observation by Amontons constitutes a striking precedent for all of the later work on the pressure-volume-temperature relations of gases. Eighty years after the report of Amontons, J. H. Lambert engaged in similar studies, commenting that "in absolute cold the air is packed so tightly together that its particles quite touch each other".[11] Again we see the clear realization that the absence of heat is equivalent to a contraction in the gas volume. Lambert's work gave 1.370 for the same pressure ratio measured by Amontons.

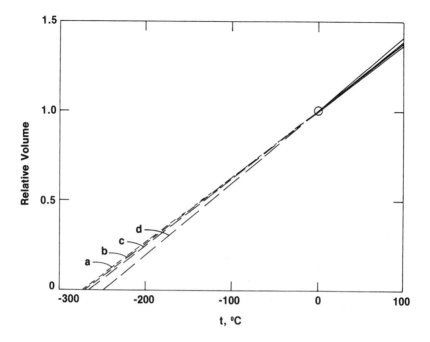

FIGURE 5. Comparison of v-T data of early thermometrists. Both scales have been normalized for the sake of clarity. The curves are identified as follows: (a) Regnault ($t_0 = -272.8°C$); (b) Lambert ($t_0 = -270°C$); (c) Gay-Lussac ($t_0 = -266°C$); (d) Amontons ($t_0 = -248°C$).

Two decades after Lambert's report, Gay-Lussac reported values of 1.375 to 1.380 for the ratio V(100)/V(0) measured at constant pressure for air. Significantly, Gay-Lussac obtained nearly the same ratio for other gases tested, including hydrogen, oxygen, and nitrogen.[12]

In writing of his own work, Gay-Lussac gave credit for similar (and previous) results to J. A. C. Charles. Evidently Charles never published his results in a form that could persist until the present time; nevertheless, because of the generous acknowledgment of Gay-Lussac, the law embodied in Equation 2 often is ascribed to Charles. It is also known as Gay-Lussac's law, however, and sometimes is given the jolly title of the Charles-Gay-Lussac law. Perhaps it should be called the law of Amontons.

All of the experimental results on the temperature-volume relations of gases can be correlated by the equation

$$(v_1/v_2) = (t_1 - t_0)/(t_2 - t_0) \qquad (2)$$

where t_0 connects the zero of the scale of temperature used in the experiments with the zero of a more fundamental temperature scale.

In Figure 5, we portray the work of Amontons, Gay-Lussac, and Lambert, along with similar measurements reported in 1847 by Victor Regnault;[13] all the data are normalized to unit volume at $t = 0°C$. In this figure we see 150 years of evidence for an absolute zero of temperature based empirically upon the "shrinking to zero" of a given volume of gas as its temperature is reduced while its pressure is held constant. One could equally well describe the same experiments in terms of the complete loss of "spring", or pressure, by a fixed volume of gas as its temperature is reduced continually. Inescapably, either method predicts the value of that absolute zero as approximately $-270°C$.

There was by no means a general appreciation of the significance of the data of Figure 5, even by the time of Regnault. However, the laws of Boyle and Charles provide the necessary basis for the creation of an ideal gas law

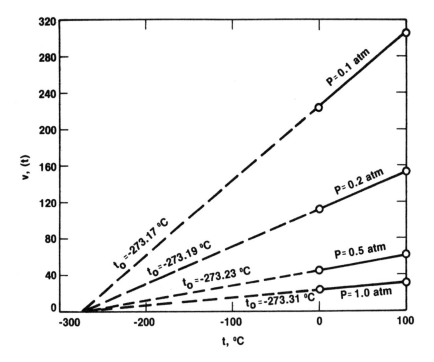

FIGURE 6. Extrapolation of Charles-Law (isobaric) measurements to obtain the constant t_0. The curves were constructed using present-day thermodynamic data for ^{4}He gas, so that agreement is bound to occur between the value of t_0 defined by extrapolation of these data to zero pressure and the value set by international agreement, $-273.15°C$.

$$Pv = a(t - t_0) \qquad (3)$$

where a and t_0 are constants to be determined by experiment. Note that Equation 3 reduces to Boyle's law if the temperature is held constant and that it reduces to Charles' law if the pressure is held constant.

Beattie[14] summarized the results of virtually all of the determinations of the thermodynamic temperature of the ice point from the time of Chappius in 1888 to Beattie's own work in 1938. The technique used was not different from that of Charles or Gay-Lussac; one assumes that exactly 100°C separate the freezing and boiling temperatures of water. Many different values of t_0 were obtained in the twenty-odd experiments that Beattie summarized. Based upon measurements on several different gases, the various determinations of t_0 ranged as low as $-273.25°C$ and as high as $-273.08°C$. The "Grand Average" value quoted by Beattie was $-273.165°C$. From the work of Guildner and Edsinger described in Chapter 5, we now believe that a more accurate value for the ice point on such a "centigrade" thermodynamic temperature scale — one that has exactly 100 degrees in the range between the ice and steam points — is 273.22°C above the absolute zero. We discuss this point further in Chapter 4.

The value $-273.15°C$ has been adopted officially to relate the Celsius scale to the thermodynamic scale in temperature. This choice created the scale that has come to be known as the Kelvin thermodynamic temperature scale (KTTS). The KTTS is now known to be inconsistent with the "centigrade" feature of the earlier "centigrade" thermodynamic temperature scale, although the inconsistency was masked for many years by the inaccuracy of thermodynamic temperature determinations. Figure 6 shows data that could be used to obtain this value of t_0 by means of Charles' law. The method of extrapolation used in Figure 6 differs only in precision from that used by the early thermometrists. The curves show that

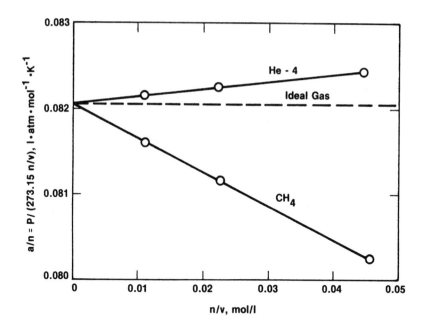

FIGURE 7. Extrapolation of the ideal gas law Pv = aT to provide a limiting value of the constant a/n for the gases ^{4}He and CH_4.

real gases are more nearly ideal as their working pressures are reduced, since the lower-pressure data better reflect the defined value of t_0.

Once a value for t_0 has been established, similar experiments can be used to study the behavior of the constant factor "a" in Equation 3. In examining the values of the constant for different gases, many scientists came to realize that the factor "a" approached a value that was constant for different gases only when multiplied by the ratio (M/m) = 1/n, where M is the molecular weight of the gas under study, m is the mass of the measured sample in grams, and n, consequently, is the number of moles of gas in the sample. To evaluate the constant a = r(m/M) = rn, we examine Equation 3 for smaller and smaller gas pressures, or equivalently, for smaller and smaller gas densities.

Examples of this type of analysis are shown in Figure 7 for the gases ^{4}He and CH_4 measured at 0°C. Here the value of a/n

$$a/n = p/(273.15 \, n/v) \, \ell\text{-atm/(mol-K)} \qquad (4)$$

is plotted as a function of gas density for the two gases. It is evident that neither gas behaves ideally except in the limit of negligible density.

By experimenting with many gases in this type of measurement, one can discover that, for any real gas, at the low-density limit

$$(a/n) = Pv/(273.15 \, n) = 0.0820568 \, (\ell\text{-atm})/(\text{mol-K}) \qquad (5)$$

or in SI units

$$(a/n) = 8.31441 \, J/(\text{mol-K})$$

The constant a/n has come to be known as the *gas constant,* usually given the designation R. The ideal gas law thus can be written

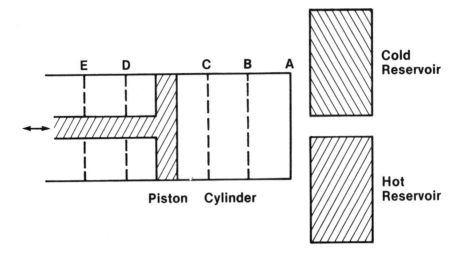

FIGURE 8. Idealization of Carnot's heat engine. The cold reservoir and the hot reservoir can be placed at will in thermal contact with the cylinder base A. Work can be done or absorbed by the external system at the end of the connecting rod.

$$Pv = nRT \quad \text{or} \quad PV = RT \tag{6}$$

where the use of the capital V often indicates the presence of exactly 1 mol of gas.

The laws of Boyle and Charles have led us to an empirical understanding of an absolute zero of temperature, to an equation of state for an idealized gas, and to an evaluation of a constant factor relating the pressure-volume product to an adjusted temperature for gases at low pressures. Now we should discuss the introduction of the science of thermodynamics.

B. Heat and Work. The Carnot Cycle

Study of the mechanical properties of gases was paralleled in 1769 by James Watt's perfection of the steam engine.[15] It became clear immediately that this device provided man with a prime mover of considerable value, and its principles of operation have received the attention of scientists and engineers ever since its invention.

Sadi Carnot idealized the operation of Watt's steam engine in order to determine the maximum efficiency achievable by an engine utilizing a heat cycle.[16] Carnot's idealized engine is shown in schematic form in Figure 8.

The stages of the ideal cycle are indicated by the position of the (frictionless) piston:

1. Thermal contact is made between the bottom of the cylinder A and the hot reservoir.
2. The gas absorbs an amount of heat energy Q_1 from the hot reservoir and expands, driving the piston leftward to position D and performing external work, such as turning a wheel.
3. When the piston reaches position D, the hot reservoir is disconnected. The gas continues to expand, driving the piston to position E, but cooling to the temperature of the cold reservoir in the process.
4. Thermal contact is made between the bottom of the cylinder A and the cold reservoir. The piston is driven by an external force back to position C while rejecting a quantity of heat Q_2 to the cold reservoir to maintain the gas temperature at a constant value.
5. The cold reservoir is disconnected when the piston reaches position C, but the piston continues to compress the gas, raising its temperature to that of the hot reservoir by the time that the piston reaches position B.

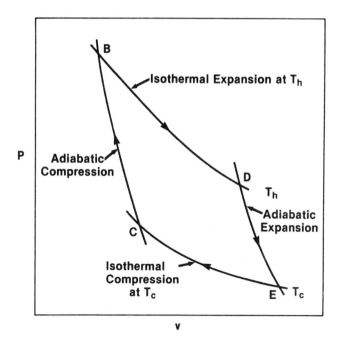

FIGURE 9. The pressure-volume diagram of a working gas in use in an ideal Carnot heat engine. The stages correspond to those in Figure 8.

6. Again the hot reservoir is connected thermally to surface A and the gas expands, driving the piston leftward once again to position D and again performing work on the external system.

Stages 3, 4, 5, and 6 are repeated, constituting each time a cycle of the ideal heat engine. The system constitutes an "engine" in the sense of providing a net source of mechanical energy, equal to the difference ($Q_1 - Q_2$), to the piston at the expense of the extra heat drawn from the hot reservoir. The reason for the excess of external power delivered at stage 6 over that demanded at stage 4 is that the temperature of the gas is higher during the expansion stages than it is during the compression stages. Thus the engine utilizes the working gas to perform external work by moving heat energy from a hot reservoir to a cold reservoir.

The process we have described is reversible, that is to say, an external force could equally well be used to move heat energy from the cold reservoir to the hot one. (This is the principle of the "heat pump".)

The Carnot cycle can be shown as a closed loop in a P-v diagram such as the one in Figure 9.

Note that Work, defined as Force × Distance, can be found from the area in a P-v diagram like that in Figure 9, since Force = Pressure × Area (PA) and Distance = (Volume/Area) = Length (L):

$$W = PAL = Pv$$
$$= Q_1 - Q_2 \qquad (7)$$

The efficiency of Carnot's ideal reversible heat engine

$$\frac{W}{Q_1} = \frac{Q_1 - Q_2}{Q_1} \qquad (8)$$

can be shown to depend only upon the temperatures of the hot and cold reservoirs and not upon the nature of the working gas.

Kelvin realized that a universal scale of temperatures could be devised by defining ratios of its temperature values (T_1/T_2) to be equal to the ratios of the ideal Carnot engine heats (Q_1/Q_2), where the latter ratios are used with the meaning given above. Thus the Carnot efficiency becomes $(W/Q) = (T_h - T_c)/(T_h)$.

Several consequences of Kelvin's choice are apparent immediately. First, the temperature scale so defined is independent of material properties. Second, the scale clearly interrelates thermal properties of the working substances. Third, a natural zero of the new scale appears as that temperature at which no further work can be accomplished by a Carnot engine; moreover, referring again to Equation 8, one sees that a Carnot engine becomes 100% efficient whenever the temperature of the cold reservoir reaches 0°.

As they contemplated the consequences of these and other thermal studies, Kelvin and his colleagues published papers that initiated the science of thermodynamics and laid the basis for a thermodynamically rigorous scale of temperature.[17-17b]

Kelvin might well have suggested that the thermodynamic scale of temperature be based upon the ideal gas law, since such a scale would possess properties similar to the ones given above for the Carnot-efficiency derivation. However, the latter can be considered to be more general because, in principle, it applies not only to gases, but to liquids and solids too.

Kelvin implicitly acknowledged the applicability of the thermodynamic temperature scale to the ideal gas in specifying a means for deriving scale numbers. His suggestion involved adding 273°C — the value of t_0 derived from the work of Charles, Gay-Lussac, and Regnault — to the Celsius scale. In this way, all contemporary thermodynamic measurements were brought into harmony.

The absolute temperature scale so constructed is known as the Kelvin thermodynamic temperature scale (KTTS). Efforts were made to compare temperatures based upon Carnot-cycle measurements of real gases with those derived from the ideal gas law, but the very limited accuracy of the former prevented these efforts from producing very useful results.

In this short space we have only directed a fleeting glance at the beginning of an epoch Helmholtz, Gibbs, and others provided the logical basis for an absolute scale of temperature that had eluded thermometrists for more than two centuries. A natural zero of temperature was identified, along with an idealized method for building a scale that is independent of material properties.

C. The Laws of Thermodynamics

The Kelvin temperature scale is the backbone of thermodynamics. We present here a brief discussion of some of the principles of thermodynamics needed to understand the Kelvin scale in relation to the temperature-dependent properties of materials including the following (all assumed to be in molar units unless otherwise specified):

- P, absolute pressure
- V, volume
- C, heat capacity
- E, internal energy
- H, enthalpy or total heat content
- S, entropy
- G, Gibbs free energy

The *zeroth law of thermodynamics* — so named by Sir Ralph Fowler[18] — codifies the experimental result that heat always flows from the hotter of two systems to the cooler until

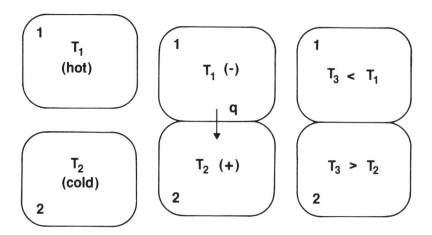

FIGURE 10. Illustration of the zeroth law of thermodynamics. If two bodies, one at a hotter temperature T_1 and the other at a colder temperature T_2, are placed in thermal contact, heat energy flows from the hotter to the colder until the temperatures reach the same level, T_3 — in general, a temperature that is intermediate between T_1 and T_2.

thermal equilibrium is achieved or until the systems are separated thermally. This well-established principle is illustrated in Figure 10. Like the other laws of thermodynamics yet to be discussed, the zeroth law has been found to be true in every case tested and thus is accepted as a reliable guide to the behavior of matter.

The internal energy E of a substance or system may comprise kinetic energy (possessed in consequence of its motion), potential energy (possessed in consequence of its position), and still further energy in consequence of its electrical or magnetic condition. The science of thermodynamics takes no account at all of the sources of a system's internal energy but only of changes in its amount. When a system is changed from state 1 to state 2, the internal energy changes by

$$\Delta E = E_2 - E_1 \tag{9}$$

The *first law of thermodynamics* states simply that energy is conserved in thermodynamic systems. If energy is gained or lost by one system, a like amount is lost or gained by a contiguous system. (Note, however, that mass and energy can be interchanged through the processes of nuclear fission and nuclear fusion, according to the principles of quantum mechanics.) The very major interest of thermodynamics is to elucidate the relations between two types of energy — work and heat. Accurate measurements of these quantities require careful attention to thermometry. The internal energy of a system can be modified by the exchange of either form of energy across its boundary

$$\Delta E = E_2 - E_1 = \Delta Q - \Delta W \tag{10}$$

Here, ΔQ denotes the magnitude of any gain in heat energy and ΔW the magnitude of any work done by the system.

Heat energy can be exchanged across a boundary by radiation, conduction, or convection. The equations governing these mechanisms are, respectively

$$\Delta Q \text{ (rad)} = \epsilon \sigma T^4 A \, \Delta(\text{time}) \tag{11}$$

where ϵ is the total emissivity, σ is the Stefan-Boltzmann constant, and A is the area;

Table 1
ENERGY UNITS[a]

"15°C Calorie" (energy to raise the temperature of 1 g of water from 14.5°C to 15.5°C)	=	4.1855
"20°C Calorie"	=	4.1816
"Btu" (energy to raise the temperature of a 1-lb mass of water from 59.5°F to 60.5°F)	=	1055.06
"Thermochemical calorie" (a defined quantity)	=	4.1840
"International Steam Table calorie" (also defined)	=	4.1868

[a] Data are in joules.

$$\Delta Q \text{ (solid cond)} = \kappa(A/L) \Delta T \Delta(\text{time}) \qquad (12)$$

where κ is the thermal conductivity, A is the area and L the length of the solid connecting two systems and ΔT their temperature difference; and

$$\Delta Q \text{ (gas convection)} = hA \Delta T \qquad (13)$$

where h is the convective heat transfer coefficient, A is the effective area of the boundary, and ΔT the temperature difference between the gas and the system.[19]

Work also can be exchanged between systems in several ways. Some of these are

$$\text{Gas expansion} \qquad \Delta W = P \Delta V \qquad (14)$$

$$\text{Transport of matter} \qquad \Delta W = F \Delta X \qquad (15)$$

$$\text{Magnetization} \qquad \Delta W = v H \Delta B \qquad (16)$$

where v is the volume, H is the magnetic intensity and ΔB the change in magnetic flux density.[20]

For many purposes, it is more convenient to discuss the enthalpy or total heat content H of a system, defined by the equation

$$H = E + PV \qquad (17)$$

The change of energy of a system measured by calorimetry commonly measures the change in enthalpy:

$$\Delta H = E_2 + (PV)_2 - E_1 - (PV)_1 \qquad (18)$$

When a system is heated, its temperature generally rises. The ratio of the heat energy thus expanded to the temperature rise is called the heat capacity C:

$$(\Delta Q/\Delta T) = C \qquad (19)$$

The heat capacity is expressed in units of energy per degree of temperature per unit quantity of substance examined. The energy unit initially was the calorie, a variable unit based upon the mechanical equivalent of heat as measured by a temperature rise in a given quantity of water. The unit of energy in the International System of Units (SI), mentioned in Chapter 1, is the joule (J). A common engineering unit of energy is the British thermal unit (Btu). The relationships among these units are given in Table 1.

Since the Celsius degree is defined to be the same size as the kelvin, these two units are interchangeable with respect to heat capacity data. The Fahrenheit degree, defined as 5/9

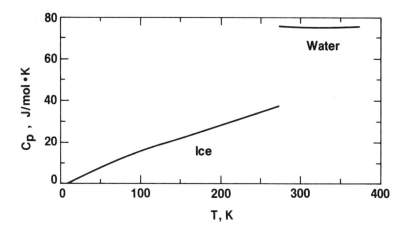

FIGURE 11. The heat capacity of ice[22] and water.[23]

the size of the Celsius degree, commonly is used in engineering work along with the Btu energy unit.

Thermal capacity data occasionally are reported in terms of "specific heat". This term denotes the ratio of the thermal capacity of a gram of a particular substance to that of a gram of water at 15°C (which is exactly equal to one "15°C calorie" per gram per °C). Thus the specific heat is a dimensionless quantity. The use of "heat capacity" in units of joules per mole per °C is now the standard practice for reporting thermal data.

The heat capacity of the ice-water system is presented in Figure 11. Interesting features of the data include the fact that the heat capacity approaches zero as the absolute temperature is reduced to zero; that there is a discontinuous change in the heat capacity at the melting point of ice; and that the heat capacity of water is rather constant over its liquid range (the actual variation, about 1%, accounts for the variability of the calorie).

We can distinguish between two situations with regard to heat capacity measurements. If the volume does not change in the process of heating, then $P\Delta V = 0$ (Equation 14) and the constant-volume heat capacity can be written

$$C_V = \left(\frac{\partial E}{\partial T}\right)_V \tag{20}$$

(The notation used in Equation 20 is that of partial differentiation. The equation may be stated in words as follows: "The heat capacity at constant volume is equal to the small change in the internal energy of the system, brought about by an infinitesimal change in the system temperature at constant volume, divided by the temperature change." C_V thus becomes the slope, or tangent, of the curve of E vs. T with V held constant, at a particular temperature.[21])

If the pressure is constant during the heating process, but the volume changes, the heat capacity is more efficiently described in terms of the enthalpy

$$C_P = \left(\frac{\partial E}{\partial T}\right)_P + P\left(\frac{\partial V}{\partial T}\right)_P$$

$$= \left(\frac{\partial H}{\partial T}\right)_P \tag{21}$$

We note here that the ideal gas has the property that, at constant temperature, its internal energy is constant. That is, for the ideal gas,

$$\left(\frac{\partial E}{\partial V}\right)_T = 0 = \left(\frac{\partial E}{\partial P}\right)_T \quad (22)$$

and any work done on or by an ideal gas in contraction or expansion cannot change its internal energy. Thus the work must be accompanied by the emission or absorption of an equivalent amount of heat. Similarly, for an ideal gas,[24]

$$\left(\frac{\partial H}{\partial V}\right)_T = 0 = \left(\frac{\partial H}{\partial P}\right)_T \quad (23)$$

One also can show that the difference between the heat capacity of an ideal gas at constant pressure and its heat capacity at constant volume is equal to the gas constant R. For the ideal gas

$$Pv = nRT \quad \text{or} \quad PV = RT \quad (6)$$

We have defined C_P and C_V already as

$$C_P = \left(\frac{\partial H}{\partial T}\right)_P$$

$$= \left(\frac{\partial E}{\partial T}\right)_P + P\left(\frac{\partial V}{\partial T}\right)_P$$

$$C_V = \left(\frac{\partial E}{\partial T}\right)_V$$

We can rewrite the term $(\partial E/\partial T)_P$ by noting that it represents the change in internal energy of the ideal gas as it is heated at constant pressure. Because the change in the internal energy is independent of the particular path used between the initial and final states, we can write

$$\left(\frac{\partial E}{\partial T}\right)_P = \left(\frac{\partial E}{\partial T}\right)_V + \left(\frac{\partial E}{\partial P}\right)_T \quad (24)$$

That is, the change in internal energy is the same if first the temperature is changed at constant volume and then the pressure is changed isothermally to its final value. Then

$$C_P - C_V = \left(\frac{\partial E}{\partial T}\right)_P + P\left(\frac{\partial V}{\partial T}\right)_P - \left(\frac{\partial E}{\partial T}\right)_V$$

$$= \left(\frac{\partial E}{\partial T}\right)_V + \left(\frac{\partial E}{\partial P}\right)_T + P\left(\frac{\partial V}{\partial T}\right)_P - \left(\frac{\partial E}{\partial T}\right)_V$$

$$= \left(\frac{\partial E}{\partial P}\right)_T + P\left(\frac{\partial V}{\partial T}\right)_P \quad (25)$$

But from Equation 22, $(\partial E/\partial P)_T = 0$ for the ideal gas. Furthermore, $(\partial V/\partial T)_P = R/P$ for an ideal gas. Therefore

Table 2
HEATS OF FUSION AND VAPORIZATION[a]

Substance	ΔH_f (J/mol)	ΔH_v (J/mol)
H_2O	6,010	40,680
Hg	2,295	59,290
H_2	117	904
Au	13,000	—
NaCl	30,220	—

[a] Where available, data were taken from Chase, M. W., *Bulletin of Alloy Phase Diagrams*, 4, 124, 1983. Other data taken from the *Handbook of Chemistry and Physics*, Weast, R. C., Ed., CRC Press, Boca Raton, Fla., 1985.

$$C_P - C_V = P(R/P) = R \qquad (26)$$

Thus, in principle, careful measurements of gas heat capacities can be used to evaluate the gas constant.

We return to the subject of heat capacities in the next section.

If a system is held at its melting or boiling temperature, heat may be added to it with no change in its temperature. Experimentally the substance is observed to undergo a phase change, from its solid to its liquid state or from the liquid to the vapor. The heat energy required to melt or evaporate the substance is called its latent heat of fusion or vaporization, respectively. Since most substances change volume upon melting or vaporizing, both processes are most efficiently described in terms of enthalpy. If we define the measured heat per mole as L, then

$$L_p \text{ (fusion)} = \Delta Q_P = [E \text{ (melted state)} + PV \text{ (melted state)}]$$
$$- [E \text{ (solid state)} + PV \text{ (solid state)}] = \Delta H_{\text{fusion}} \qquad (27)$$

Some typical values of heats of fusion and vaporization are given in Table 2.

The *second law of thermodynamics*, like the zeroth and the first, also derives from experience. It can be stated in the following form: "Any system, left undisturbed, will approach its most probable configuration." It was developed at about the same time as the first law, but it was not at all so readily accepted. The reason for this hesitation was that a new concept, the concept of entropy of a system, had to be developed in order to account for the direction that physical or chemical changes take.

By the zeroth law of thermodynamics, a system composed initially of ice and warm water at a temperature T_1 will change spontaneously as a result of flow of heat from the water to the ice. If there is a sufficient quantity of heat stored in the aggregated heat capacity of the warm water to overcome the aggregated heat of fusion of the ice, then the final state of the system will consist only of water at some temperature below T_1. The final state of the system will possess the same total energy as the initial state.

Under no circumstances will the water in the final state described above spontaneously change back into the mixture of ice and warmer water that existed in the initial state, even though no change in total energy would thus result. It is important to understand the basis for this unidirectional characteristic of thermal systems.

Many other systems could be described with respect to the second law of thermodynamics. A gas, initially confined at a certain pressure P_1 in one compartment of a large, otherwise empty tank will leak through a hole in the compartment into the remainder of the tank until the gas pressure in all available parts of the system reaches the same value, a pressure less than P_1. The gas never will spontaneously leak back into the compartment, emptying the rest of the tank and reattaining its former higher pressure P_1, even though both states of the system possess identical total energies. Similarly, a suitable mixture of hydrogen and oxygen gases at a particular temperature, when ignited with a spark, will react chemically to form water vapor at a higher temperature but with the same total energy. The reverse reaction never occurs.

The thermodynamic quantity that has been invoked to differentiate between the two possible directions of the processes such as we describe above is entropy. The introduction of this quantity allows the second law of thermodynamics to be expressed as follows: "Any isolated system will change spontaneously from a state of lower entropy to one of higher entropy." The first and second laws were stated succinctly by Clausius, "Die Energie der Welt ist konstant; die Entropie der Welt streibt einen Maximum zu." ("The energy of the world is constant; the entropy of the world tends toward a maximum.")

We stress here that the second law does not forbid the manufacture of ice, of high-pressure quantities of gas, or of hydrogen-oxygen mixtures. It does demand, however, that the external agencies that accomplish such manufacturing processes increase in their own entropies by at least as much as the decreases in entropy accomplished in preparing those systems.

We now show how to identify and measure the entropy of a system. Remember that all systems move toward equilibrium when left undisturbed. Whether we speak of dropping a hot stone into water, perforating the membrane separating two unlike gases, releasing a wound-up toy to dance its jig, or removing the battery from an electrical capacitor, we know that the system so released will seek equilibrium with its surroundings. Temperature gradients will disappear, differences in concentration will go away, and potential energies will dissipate. The systems will approach predictable final states of rest from which they will not move without external stimulus. Any effort to disturb the state of rest of a system at equilibrium necessitates the intervention of a second system which thereby dissipates some or all of *its* stored energy, proceeding unidirectionally towards a joint equilibrium with the system thus disturbed.

In principle, there are actions that are entirely reversible — the compression by a frictionless piston of a gas within a cylinder that has thermally opaque walls could be reversed to recover exactly the work of compression. Such a process would involve no change in entropy. Alas, in reality, dissipative mechanisms abound. The ubiquity of both internal and external friction, finite electrical and thermal resistances and other energy-loss characteristics force us to regard reversible processes as limiting cases that real actions cannot quite achieve. Therefore all real processes involve increases in entropy. These increases can be measured in terms of the departures from reversibility of the individual processes.

As with internal energy and enthalpy, we do not need at present to evaluate the magnitude of the entropy of any system, but only the changes in entropy, ΔS, that accompany particular processes. The change in entropy can be written

$$\Delta S = \frac{\Delta Q}{T} \tag{28}$$

The entropy change involved in the isothermal expansion of an ideal gas, for example, may be obtained by integrating the work of expansion (identical to the heat absorbed by the expanding gas from its environment)

$$\Delta W = \int_{v_1}^{v_2} P dv = \Delta Q \tag{29}$$

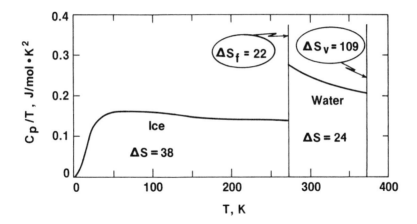

FIGURE 12. Entropy changes in ice, water, and the phase transitions of melting and boiling, as derived from the heat capacity data of Figure 11. The entropy units are J/(mol · K).

Since $P = nRT/v$, $\Delta W = nRT \ln(v_2/v_1)$, and

$$\Delta S_{exp} = nR \ln(v_2/v_1) \qquad (30)$$

As a second example, consider a substance of heat capacity $C(T)$; its entropy change during warming is

$$\Delta S = \int_{T_1}^{T_2} \frac{C(T)}{T} dT \qquad (31)$$

Finally, the entropy change resulting from a change of state can be written

$$\Delta S = \Delta H/T \qquad (32)$$

where ΔH might be the latent heat of fusion or of vaporization.

Using the data of Figure 11 and Table 2, we illustrate in Figure 12 the evaluation of the entropy changes occurring in the ice-water system as the Kelvin temperature is raised from zero to the steam point. By Equation 31, the entropy change on heating ice from 0 K to its melting temperature is 38 J/mol · K. By Equation 32, the entropy of the system increases by 22 J/mol · K on melting at 273.15 K (at 1 atm pressure). Continued heating to the steam point increases the entropy by another 24 J/mol · K, and evaporating a mole of water at 373.15 K increases the entropy by 109 J/mol · K. The second law of thermodynamics shows why a quantity of water at a particular temperature cannot spontaneously change into a mixture of ice and warmer water that possesses the same total energy — the process would involve a net decrease of total entropy.

It is not possible to derive *a priori* a description of the state of a substance for which the entropy is zero. Considering the equivalence of entropy with disorder, one is tempted to define as possessing zero entropy an ideal single crystal of a particular substance at $T = 0$ K; all molecules in a perfect array and resting at the minima of their energies. Such a definition is arbitrary, but it has been made, for example, by Lewis and Gibson:[25] "If the entropy of each element in some crystalline state be taken as zero at the absolute zero of temperature: every substance has a finite positive entropy, but at the absolute zero of temperature the entropy may become zero, and does so become in the case of perfect crystalline substances."

The above statement often is amplified by the postulate that it is not possible in any way to reduce the temperature of any substance to the zero of the thermodynamic scale of temperature. The unattainability of absolute zero is one of the most common expressions of the *third law of thermodynamics*. A simple intuitive picture of the third law can be garnered if the reader understands that, to reduce the temperature of a substance, it is necessary to remove heat energy from the substance; in turn, that process requires placing it in thermal contact with another substance whose temperature is yet lower. Since no substance can possess a temperature less than zero, it therefore is not possible to cool any substance to 0 K.

In order to describe the behavior of systems as they approach equilibrium at a particular temperature, which we consider at length in the next chapter, it is necessary to introduce the notion of "free energy". Two expressions have been defined for free energy:

$$A = E - TS \tag{33}$$

defines the "Helmholtz function" or the "Helmholtz free energy", which usually proves to be most useful for systems held at constant volume and constant temperature; and

$$G = H - TS \tag{34}$$

defines the "Gibbs function" or the "Gibbs free energy", which is most useful for systems at constant pressure and constant temperature. These functions depend only upon the state of a given system.

In the next chapter, we discuss the attainment of equilibrium in systems that consist of two or more forms of a given substance. Such equilibria are best discussed in terms of the Gibbs function, since they are, in general, constant-pressure, constant-temperature systems. In any small, reversible process that occurs in the system described above

$$dG = dH - TdS - SdT \tag{35}$$

However, dH can be shown to be equal to (TdS + VdP); therefore

$$dG = VdP - SdT \tag{36}$$

So that, for any small, reversible, isobaric, and isothermal process, dG = 0. Thus, the criterion for the attainment of equilibrium in a constant-pressure, isothermal system is simply that the Gibbs free energy remain unchanged during the period of observation. The "isothermal system" in this criterion provides us with the environment for accurate thermometry.

IV. TEMPERATURE AND STATISTICAL MECHANICS

All of thermometry, including the ideal gas law, can now be discussed by statistical mechanical techniques. These techniques involve accounting for differing properties and populations in various molecular, atomic, or nuclear quantum levels. As our understanding of these levels has improved, the power of statistical methods has grown in a corresponding way.

Whereas the science of thermodynamics offers conclusions about the empirically-observed behavior of bulk properties of matter without regard to the makeup or characteristics of individual atoms or molecules, the science of statistical mechanics attempts to predict the behavior of mechanical systems composed of very numerous, similar particles by the employment of statistical methods. Full advantage is taken of any details known about the composition of the particles and about the laws of mechanics that govern their actions.

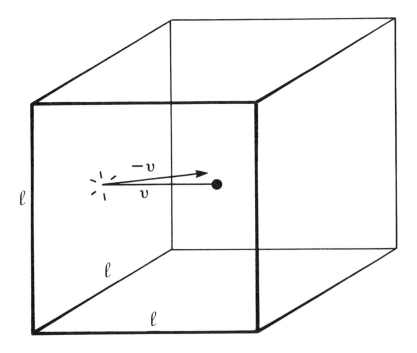

FIGURE 13. Idealized drawing of an atom of velocity v striking the wall of a perfectly elastic container in the form of a cube of side ℓ.

A. Kinetic Theory and the Ideal Gas

By the use of classical Newtonian mechanics, one can describe an ideal gas in terms of simple atomic properties; the kinetic-theory ideal gas law gives new significance to the notion of the temperature of an ideal gas.

Consider a vessel such as the cubical box of Figure 13 that contains large numbers N_i of like atoms of masses m_i of a gas. Each atom is in constant motion with an average velocity of magnitude $|v_i|$. Frequent collisions take place between atoms of the gas and the walls of the vessel. We consider further that

1. The pressure exerted upon the vessel by the gas is the sum result of all the atom-wall collisions.
2. The atoms are perfectly elastic, so that there is no net loss of energy as a result of collisions.
3. The atoms have essentially no attraction for each other or for the walls of the vessel.
4. The atoms occupy only a negligible portion of the volume of the vessel.
5. The mean kinetic energy of the atoms is the same and is constant at any temperature.

Then, referring to Figure 13, we can assume that, because of the extremely large number of atoms in the gas, the average velocity $|v_i|$ of all atoms of a given mass m_i is the same and constant, and is directed with equal probability toward any one of the faces of the vessel. The acceleration "a" of an atom — its time rate of change of velocity — during each collision that it makes with any face that is normal to its motion is

$$a = \frac{\Delta v_i}{\Delta t} = \frac{2|v_i|}{\ell |v|} = \frac{2v_i^2}{\ell} \qquad (37)$$

The force f that an atom exerts on a face during a collision is

$$f = ma = 2m_i v_i^2/\ell \tag{38}$$

The total force on each of the six faces is then

$$F = \Sigma (2N_i m_i v_i^2)/(6\ell) = \Sigma (N_i m_i v_i^2)/(3\ell) \tag{39}$$

and the pressure on each face is

$$P = F/\ell^2 = (\Sigma N_i m_i v_i^2)/(3\ell^3) = (\Sigma N_i m_i v_i^2)/(3V) \tag{40}$$

Since we have assumed that the velocity $|v_i|$ is constant at any temperature, the product PV is constant;

$$PV = (1/3) \Sigma N_i m_i v_i^2 = \text{constant} \tag{41}$$

Equation 41 is the kinetic theory expression of Boyle's law.

Equation 41 also contains the basis for Dalton's law. The total pressure P of a mixture of gases

$$P = \frac{1}{3V} [N_1 m_1 v_1^2 + N_2 m_2 v_2^2 + \ldots] \tag{42}$$

can be seen to amount simply to the sum of the pressures that each gas would exert if it occupied the container by itself.

Furthermore, we see that Avogadro's law can be derived by considering that the average kinetic energy KE of each molecule is the same at any temperature, irrespective of its mass:

$$KE = (1/2) m_1 v_1^2 = (1/2) m_2 v_2^2 = \ldots \tag{43}$$

If we now equate the kinetic-theory pressures of two gases of masses m_1 and m_2

$$P = [1/(3V)] N_1 m_1 v_1^2 = (1/3V) N_2 m_2 v_2^2 \tag{44}$$

and divide Equation 44 by Equation 43, then $N_1 = N_2$, showing that equal volumes of different gases are composed of equal numbers of molecules if their temperatures and pressures are the same. This is Avogadro's hypothesis.

Finally, considering the equivalence of Equations 44 and 6

$$PV = (1/3) Nmv^2 = nRT \tag{45}$$

we see that temperature enters kinetic theory as a quantity that is proportional to the sum of the kinetic energy of the molecules of an ideal gas. The constant of proportionality, Boltzmann's constant k, can be obtained by setting Equation 45 equal to NkT. If n is taken to be 1 mol of an ideal gas, then N becomes Avogadro's number N_o and k becomes[26]

$$k = \frac{R}{N_0} \frac{8.31441 \pm 0.00026 \text{ J/(mol} \cdot \text{K)}}{6.022045 \pm 0.000031 \times 10^{23}/\text{mol}}$$

$$= 1.380\,662 \pm 0.000\,44 \times 10^{-23} \text{ J/K} \tag{46}$$

Reviewing our efforts so far, we see that the ideal gas law can be derived from kinetic theory. We see also that kinetic theory applied to the ideal gas offers an alternative definition of the absolute temperature, that of measuring — instead of its pressure and volume — the average velocity of the molecules of an ideal gas of known molecular weight. In addition, we have obtained relations illustrating Dalton's and Avogadro's laws and derived an equivalence between the gas constant and Boltzmann's constant.

By continuing in this fashion, we could derive meaningful kinetic-theory expressions for other ideal-gas properties and for the equations of state of real gases.[27] At this point, however, it is useful to discuss statistical mechanics in a more general form.

B. Quantization of Light and Heat

Upon considering the observed spectral distribution of radiation from a black body, Planck was forced to conclude that radiant energy is not emitted as a continuum, but instead as packets, or quanta, of energy.[28] Einstein applied this same notion to explain the observed fact that only light of a certain minimum frequency would cause electrons to be emitted (the "photoelectric effect") from a metal surface.[29]

In 1913, Niels Bohr was able to explain the existence of sharp lines in atomic spectroscopy by introducing a model of the atom in which only certain energy states could be occupied by atoms.[30] This picture of the atom has been generalized to incorporate a central, massive nucleus composed of protons and neutrons — themselves constrained to quantized energy levels — surrounded by electrons inhabiting specific quantum levels.

Visible light — indeed all electromagnetic radiation from low-energy radio waves to the very high energy gamma rays — appears in the quantum theory as quanta of energy emitted by atoms and subatomic particles as they move from a quantum state of a certain energy to one of lower energy. We discuss later several techniques in thermometry that depend directly upon the quantization of matter.

The fact that diffraction always occurs in light is of little importance when the observer is recording the motion of billiard balls in mechanical experiments. When the experimenter's interest is focused upon the motion of atoms, however, diffraction effects are so large as to prevent him from knowing with certainty their location.

Other features cloud the efforts of scientists to track exactly the actions and reactions that take place in the microscopic world of molecules, atoms, and their components. One of these features is the problem caused by the origin of light, which after all is the scientist's favorite probe. By shining his beacon upon his miniature subjects, the experimenter actually changes the state of many of them — some of the quanta of light are absorbed by the atomic targets, boosting them into more energetic quantum levels. Still another difficulty lies in the fact that both light and subatomic particles show a dual wave-particle nature. Electrons in their atomic orbits, for example, are most easily described by wave-like functions that lead to a probabilistic accounting of their positions in space.

In fact, the notion of statistical probability is central to modern thermodynamics. Already implicit in the prequantum Newtonian mechanics was the idea that a system of many particles could not be followed in detail through any mechanical action; instead, the "equilibrium state" simply became the one of greatest probability. Fluctuations of the system away from that state could be expressed in a mathematically satisfactory way, too. The impossibility of following the motions of *any* particle as small as an atom and the uncertainties raised by the wave-particle nature of subatomic entities reinforced statistical methods as the most fruitful means of describing the bulk properties of matter.

From the discussion of the last section, we saw that the average kinetic energy of a free collection of gaseous atoms is $3kT/2$. If one applies the same ideas to a collection of particles that interact with harmonic forces — such as atoms in a crystalline lattice — one finds that the total energy $E = 3kT$. This energy is composed of equal parts of kinetic and potential energy.

If now one calculates the molar heat capacity of such a solid, one finds

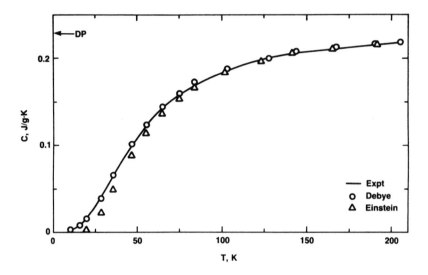

FIGURE 14. Experimental heat capacity of silver[34] showing the approach to the Dulong-Petit value at high temperatures (DP), the imperfect fitting of the one-frequency Einstein heat capacity calculation, and the improved fitting of the multiple-frequency Debye calculation.

$$C_V = dE/dT = d(3NkT)/dT = 3\,Nk = 3R \tag{47}$$

Dulong and Petit[31] recognized that many solids exhibited very similar values of molar heat capacity; the value $C_V = 3R$ is known as the Dulong-Petit value.

If one examines heat capacities at low temperatures, however, one finds strong deviations from the value of 3R. The heat capacity of silver, shown in Figure 14, is one of many such examples.

In attempting to explain low-temperature heat capacities, Einstein invoked the idea of quantized oscillators. Supposing that each particle vibrates with the same frequency v, Einstein derived an expression for E_1, the internal energy of the system in one dimension[32]

$$E_1 = hv/[\exp(hv/kT) - 1] \tag{48}$$

At high temperatures, that is, for $kT \gg hv$ Equation 48 simplifies to $E_1 \cong kT$; then, for all three dimensions, $E(tot) = 3R/mol$, the Dulong-Petit value. At low temperatures, the exponential term dominates and

$$E_1 = hv\,\exp(-hv/kT) \tag{49}$$

The molar heat capacity in three dimensions can be written

$$C_V \cong \frac{3R\,(hv/kT)^2\,\exp(hv/kT)}{[\exp(hv/kT) - 1]^2}$$

or

$$C_V = \frac{3R\,a^2 e^a}{(e^a - 1)^2} \tag{50}$$

where a is θ_E/T. Here $\theta_E = hv/k$ is called the "Einstein" temperature.

The dominating effect of the exponential term at very low temperatures results in the rapid fall of the Einstein model of the heat capacity shown in Figure 14.

The difficulty inherent in Einstein's model was the use of only one frequency of crystalline vibration. In fact, there are many modes of vibration in solids, some of them with quite low frequencies so that the approximation $h\nu \gg kT$ cannot be used.

Debye[33] proposed a different model for lattice vibrations, treating the solid as a homogeneous elastic continuum with a great many vibrational frequencies up to a limiting frequency ν_{max}. This treatment avoided the difficulty of calculating the actual vibrational spectrum without ignoring its nature. Debye's equation for the heat capacity is

$$C_V = 9Nk \left(\frac{T}{\theta_D}\right)^3 \int_0^{x_m} \frac{e^x x^4 dx}{(e^x - 1)^2} \tag{51}$$

where $x = h\nu/kT$ and $\theta_D = h\nu_{max}/k$. Note the similarity between the characteristic Einstein temperature θ_E and Debye temperature θ_D.

Figure 14 shows the improved agreement with experimental heat capacities[34] provided by the Debye model of the lattice.

GENERAL REFERENCES

Middleton, W. E. K., *A History of the Thermometer and Its Uses in Meteorology*, The Johns Hopkins Press, Baltimore, 1966.

Berger, R. L., Clem, T. R., Harden, V. A., and Mangum, B. W., Historical development and newer means of temperature measurement in biochemistry, in *Methods of Biochemical Analysis*, Vol. 30, Glick, D., Ed., John Wiley & Sons, New York, 1984, 269—331.

Schooley, J. F., Temperature measurements, in *Food Analysis: Principles and Techniques*, Vol. 2, Gruenwedel, D. W. and Whitaker, J. R., Eds., Marcel Dekker, New York, 1984, 1—40.

Zemansky, M. W., *Heat and Thermodynamics, 5th Edition*, McGraw-Hill, New York, 1968.

Moore, W. J., *Physical Chemistry*, 4th ed., Prentice-Hall, Englewood Cliffs, N.J., 1972.

Bolton, H. C., *Evolution of the Thermometer — 1592—1743*, Chemical Publishing, Easton, Pa., 1900.

Lewis, G. N. and Randall, M., *Thermodynamics and the Free Energy of Chemical Substances*, McGraw-Hill New York, 1961.

Sears, F. W., *An Introduction to Thermodynamics, the Kinetic Theory of Gases, and Statistical Mechanics, 2nd Edition*, Addison-Wesley, Reading, Mass., 1953.

Rossini, F. D., *Chemical Thermodynamics*, John Wiley & Sons, New York, 1950.

Kennard, E. H., *Kinetic Theory of Gases* McGraw-Hill, New York, 1938.

Beattie, J. A. and Oppenheim, I., *Principles of Thermodynamics*, Elsevier Scientific, New York, 1979.

REFERENCES

1. **Middleton, W. E. K.**, *A History of the Thermometer and Its Uses in Meteorology*, The Johns Hopkins Press, Baltimore, 1966.
2. **Biancani**, "Sphaera mundi, seu cosmographia demonstrativa . . . ", Bologna, 1620, 111; quoted in Middleton,[1] p. 10.
3. **Fahrenheit, D. G.**, *Philos. Trans. R. Soc. London*, 33, 78, 1724.
4. **Rømer, Ole**, "Adversaria . . . (Notebook)". Published in 1910 in Copenhagen through the efforts of Kirstin Meyer; see *Nature, (London)*, 82, 296—298, 1910.
5. **Celsius, A.**, *Kongl. Svensk. Wet. Akad.*, Handlinger, 3, 171—180, 1742; quoted in Middleton,[1] p. 97.
6. **Prinsep, J.**, *Ann. Chim. Phys.*, 2nd Series, 41, 247, 1829.
7. **Pouillet, C. S. M.**, *Comptes Rendus*, 3, 782, 1836.
8. **Boyle, R.**, *On the Spring of the Air*, 2nd ed., Oxford, 1662.
9. **Conant, J. B.**, *On Understanding Science*, Yale University Press, New Haven, 1947.
10. **Amontons, G.**, *Mem. Acad. R. Sci. Paris*, 1699, pp. 112—126; quoted in Middleton,[1] p. 111.
11. **Lambert, J. H.**, "Pyrometrie oder vom Maase des Febers und der Wärme", Berlin, 1779.
12. **Gay-Lussac, J. L.**, *Ann. Chim.*, 1st ser., 43, 137, 1802.
13. **Regnault, V.**, *Mem. Acad. R. Sci. Paris*, 21, 1—748, 1847.

14. **Beattie, J. A.**, The thermodynamic temperature of the ice point, in *Temperature, Its Measurement and Control in Science and Industry*, Fairchild, C. O., Hardy, J. D., Sosman, R. B., and Wensel, H. T., Eds., Reinhold, New York, 1941, 74—88.
15. **Sandfort, J. F.**, *Heat Engines*, Doubleday & Co., Garden City, N.Y., 1962, chap. 2.
16. **Carnot, S.**, *Reflections on the Motive Power of Fire*, transl. by Thurston, R. H., Dover Publications, New York, 1960.
17. **Thomson, W.**, On an absolute thermometric scale founded on Carnot's theory of the motive power of heat, and calculated from Regnault's observations, *Cambridge Phil. Soc. Proc.*, June 5, 1848.
17a. **Thomson, W.**, On the dynamical theory of heat, with numerical results deduced from Mr. Joule's equivalent of a thermal unit, and M. Regnault's observations in steam, Pt. 2, *Trans. R. Soc. Edinburgh*, March 1851.
17b. **Joule, J. P. and Thomson, W.**, On the thermal effects of fluids in motion, Pt. 2, *Philos. Trans. R. Soc. London*, 144, June 15, 1854.
18. **Fowler, R. C. and Gugenheim, E. A.**, *Statistical Thermodynamics*, Cambridge University Press, 1940.
19. See, for example, **Daynes, H. A.**, *Gas Analysis by Measurements of Thermal Conductivity*, Cambridge University Press, 1933; also see Dushman, S. and Lafferty, J. M., *Scientific Foundations of Vacuum Technique*, 2nd ed., John Wiley & Sons, New York, 1962, chap. 1.
20. A discussion of the thermodynamics of magnetic systems can be found, for example, in **Callen, H. B.**, *Thermodynamics*, John Wiley & Sons, 1963, chap. 14; see also **Hudson, R. P.**, *Principles and Application of Magnetic Cooling*, North-Holland, Amsterdam, 1972, chap. 2.
21. **Margenau, H. and Murphy, G. M.**, *The Mathematics of Physics and Chemistry*, 2nd ed., Van Nostrand, Princeton, N.J., 1956, chap. 1.
22. **Giauque, W. F. and Stout, J. W.**, *J. Am. Chem. Soc.*, 58, 1144, 1936.
23. **Osborne, N. S., Stimson, H. F., and Ginnings, D. C.**, *J. Res. Natl. Bur. Stand.* 23, 238, 1939.
24. **Lewis, G. N. and Randall, M.**, *Thermodynamics and the Free Energy of Chemical Substances*, McGraw-Hill, New York, 1923, 65.
25. **Lewis, G. N. and Gibson, G. E.**, *J. Am. Chem. Soc.*, 42, 1529, 1920.
26. **Cohen, E. R. and Taylor, B. N.**, *J. Phys. Chem. Ref. Data*, 2(4), 663, 1973. This compilation will be updated during 1985; see also **Pipkin, F. M. and Ritter, R. C.**, Precision measurements and fundamental constants, *Science*, 219, 913, 1983.
27. See, for example, **Halliday, D. and Resnick, R.**, *Physics for Students of Science and Engineering*, John Wiley & Sons, New York, 1962, chap. 24.
28. **Planck, M.**, *Ann. Physik*, 4, 553, 1901.
29. **Einstein, A.**, *Ann. Physik*, 17, 891, 1905.
30. **Bohr, N.**, *Philos. Mag.*, 26(1), 476, 1913.
31. **Dulong, P. L. and Petit, A. T.**, *Ann. Chim. Phys.*, 10, 395, 1819.
32. **Einstein, A.**, *Ann. Physik*, 22, 180, 1907.
33. **Debye, P.**, *Ann. Physik*, 39(4), 789, 1912.
34. **Corak, W. S., Garfunkel, M. P., Satterthwaite, C. B., and Wexler, A.**, *Phys. Rev.*, 98, 1699, 1955.

Chapter 3

TEMPERATURE FIXED POINTS

I. INTRODUCTION

We saw in Chapter 2 that temperature fixed points played an essential part in the development of the thermometer as a measuring instrument. The melting and boiling of water and the normal body temperature of humans seemed always to occur at the same levels of "hotness." Thus, these systems provided ready references for use in the study and calibration of thermometers by the pioneers in thermometry. Besides these most common examples, there are dozens of other possible choices for fixed points in thermometry, covering the range of temperature from the deepest levels of cryogenics to beyond the melting temperature of gold.

All of the fixed points used in thermometry are derived from phenomena referred to as *phase equilibria* and *phase transitions;* therefore, it is important to understand what is implied by these names. In this chapter, we begin with a discussion of phases, phase equilibria, and phase transitions. Then we describe the major types of phase equilibria that give rise to fixed points, we discuss the properties that we require of fixed points, and we examine in some detail the question "How *fixed* in temperature is a fixed point?" for some of the most important of these phenomena.

II. PHASES, PHASE EQUILIBRIA, AND PHASE TRANSITIONS

The sciences of thermodynamics and statistical mechanics provide the means to predict the behavior of physical and chemical systems by allowing one to make calculations of the energetic relationships among the possible configurations that the systems can take. The concepts of phase and phase equilibrium are of central importance to thermodynamics, as well as to the study of temperature.[1]

It will be helpful at this point to define several terms that recur in the study of phase relationships. Among these are *states of matter, phase, component, phase equilibrium,* and *phase transition.*

Ordinarily, there are considered to be three possible *states of matter* — gases, liquids, and solids. Gases, for example, possess no long-lasting intermolecular arrangement; any attractive force between their molecules is much smaller than the disruptive force of thermal energy. Thus gases have no unique density — we saw in Chapter 2 that the density of a gas is given approximately by the ideal gas law, $(n/v) = (P/RT)$ — and no resistance to shape deformation. A given quantity of gas will take the shape of any container that one might select to hold it, occupying all of the available space with a uniform density.

Liquids constitute a state that is intermediate between gases and solids. The molecules form an aggregate of fairly uniform density. Consequently, a given liquid will only partially fill a container of arbitrarily large size, generally occupying its lower portions and leaving the higher parts to be filled with its own vapor or with other available gases. The densities of liquids range from about 0.1 g/cm^3 to 13 g/cm^3; they depend somewhat upon both temperature and pressure.

Liquids, unlike gases, possess some resistance to shape deformation. This property is easily demonstrated, for example by pouring a liquid from a container and observing that it tends to agglomerate into spherical droplets while it falls free of all forces but gravity and its own surface tension. The fact that the molecules of a liquid slide past one another (i.e., exhibit "viscous flow") in response to a force so weak as gravity, however, distinguishes the liquid state from the solid state.

A solid tends to retain its shape indefinitely, unless it is exposed to forces much larger than that of gravity. If the atomic or molecular arrangement of a solid forms a regular pattern, then the solid is said to be "crystalline" and it will be found to exhibit a number of special properties. A solid without atomic or molecular regularity is said to be "amorphous."

In most cases, one can determine the state of a given substance rather easily. However, there are several types of substance that do not fit well into one of these three states. These include aerosols, colloids, emulsions, gels, and suspensions, which generally feature extremely small (but identifiable) particles of liquid or solid substances dispersed throughout a gaseous or liquid medium; rubbers and fibers, which exhibit anisotropic properties with respect to deformation; liquid crystals, which, while in most respects corresponding to liquids, yet exhibit the regular structures characteristic of the crystalline state; plasmas, which generally appear as a gas composed of ions, electrons, and photons; and surfaces, which generally provide an interface between two different states of matter.

A given quantity of matter that is uniform throughout, both in chemical composition and in physical state, and is in an equilibrium condition was described by Gibbs as consisting of a single *phase*.[2] Examples of single-phase systems include all gases, both pure and mixtures; many liquids, including liquid compounds, completely miscible liquid mixtures, and solutions; and solids that contain only one chemical and physical composition, such as common table salt or a metallic alloy.

Most systems in one's everyday experience consist of more than one phase. Often they contain two, three, or more phases. For example, a cup of coffee usually contains a gas phase above the single liquid phase composed of a solution of water, coffee extract, sugar, and milk, and it often contains one or two solid phases as well, composed perhaps of coffee grounds and undissolved sugar. In general, systems consisting of one phase are said to be homogeneous, and systems with two or more phases are called heterogeneous.

A *component* of a phase is an independent chemical constituent that must be specified as to its relative abundance in order to describe the composition of the phase. An open flask of melting ice made from pure water, for example, consists of one gaseous phase of several components (all the separate chemical species normally found in air — N_2, O_2, CO_2, H_2O, etc.); one liquid phase containing principally the component water, but in addition comprised of all of the water-soluble components of air; and one solid phase containing as its major component solid water or ice, but embracing as well those components of air that are soluble in ice. By exercising considerable care, one can reduce each of the phases in this flask to only one component, H_2O. Even though the flask would then contain three phases, it would properly be described as a one-component system.

In discussing the phase relationships of matter, it is necessary to notice the physical characteristics of the substances as well as their chemical compositions. This is true because the thermodynamic properties of a particular substance may be quite different in different physical states. The differences in properties among the three principal phases of H_2O (ice, water, and water vapor) are so well known as to need no further discussion. Less well known is the fact that ice occurs in no fewer than eight different forms, distinguishable by their differing crystal structures. Each of these structures defines a distinct phase in the ice system, and each is stable only under certain conditions of temperature and pressure. For another example, diamonds and soot, while both containing only carbon atoms bonded to each other, manifest very different physical and thermodynamic properties, including internal energy and chemical reactivity, and so constitute separate phases, with individual conditions for stability. Curiously, it is the soot that is stable at ordinary pressures; the diamond phase actually is stable only at very high pressures.

In any heterogeneous system it is possible for all of the substances to exist in thermodynamic equilibrium. For this *phase equilibrium* to exist, the Gibbs free energies of all the phases (see Chapter 2) G_l, G_s, and G_g must be equal. In turn, the condition of free-energy equality implies that the temperature, pressure, and possibly other properties are uniform throughout the system, that the concentrations of all solutions have reached the values that are characteristic of that temperature and pressure, and that the partial pressures of all gaseous species represent equilibrium values. We shall describe many phase-equilibrium systems throughout this chapter, because the state of equilibrium involving two or more phases of a given substance often occurs only within a very narrow range of temperature.

The transformation of a particular homogeneous substance from one thermodynamically distinct phase state to another is called a *phase transition*. As we noted in Chapter 2, most phase transitions are accompanied by an exchange of energy between the substance and an external source or sink of heat. This energy is known as the latent heat of transition, and it is a well-defined quantity for a particular transition. If one only very slowly adds heat to or removes heat from a system that is undergoing a phase transition, then the system can remain nearly in phase equilibrium. In that case, the system temperature will change but little during the transition.

Ideally, a system should not be in transition at all while it is serving as a reference temperature; instead, it should reside in a condition of equilibrium among the relevant phases. However, the operator often must induce the phase transition in order to "locate" the phase equilibrium condition. The achievement of phase equilibrium with its constant-temperature condition may require specific conditions also to be met with respect to other physical parameters such as pressure, magnetic field, and electric field, as well as allowable rates of thermal exchange across the system's boundaries.

Table 1 contains a representative list of phase equilibrium states that are used to provide reference temperatures. For each state, the approximate temperature on the Kelvin thermodynamic temperature scale is listed along with the latent heat of transition.

Much of the information contained in Table 1 will be discussed later in this chapter; however, some features of the equilibrium states that are listed in the table should be examined now. One outstanding feature of the list was mentioned earlier — phase equilibria occur at many different temperatures covering the range from deep cryogenics to beyond the melting temperature of gold. Indeed, the full temperature range spanned by phase equilibria is substantially wider than is indicated even by the list given above.

Another noticeable feature of Table 1 is that the latent heat associated with the phase transitions vary from zero (for superconductive and superfluid transitions) to very large values for the refractory materials. Transitions that are characterized thermodynamically as first-order transitions invariably are accompanied by a measurable latent heat which, as we showed in Chapter 2, can be expressed as the product $T\Delta S$. The superconductive and superfluid transitions, on the other hand, involve no entropy change and thus occur without any detectable latent heat;[3] furthermore, both of these transitions bring the materials involved into states that show spectacular new properties. The superconductive and superfluid states are so unusual in their properties that — like other examples given earlier — they are thought to constitute a wholly new state of matter in which quantum-mechanical features of the materials entirely dominate their responses to physical stimuli — to the flow of electrical and thermal energies, for example.[4-6]

Certain of the equilibria listed in Table 1 provide less reproducible reference temperatures than others. For example, gas-liquid equilibrium temperatures, and to a lesser extent, liquid-solid equilibrium temperatures, are noticeably dependent upon the system pressure.

In the remaining pages of this chapter, we discuss in some detail the phase equilibria that are significant from the point of view of thermometry.

III. TYPES OF TEMPERATURE FIXED POINTS

By convention, the condition of phase equilibrium that characterizes a temperature fixed point often is referred to in terms of the corresponding phase transition that is induced by

Table 1
TEMPERATURES AND LATENT HEATS OF SELECTED PHASE EQUILIBRIA

Equilibrium state[a]	Approximate Kelvin thermodynamic temperature (K)[b]	Latent heat of transition (J/mol)[c]
Liquid and solid tungsten (W)	3694	46,000
Liquid and solid niobium (Nb)	2746	(26,900)
Liquid and solid rhodium (Rh)	2236	(21,490)
Liquid and solid platinum (Pt)	2042	(19,650)
Liquid and solid iron (Fe)	1811	13,800
Liquid and solid gold (Au)	1337.58	13,000
Liquid and solid silver (Ag)	1235.08	11,300
Liquid and solid copper-silver eutectic	1053	—
Liquid and solid aluminum (Al)	933.61	10,700
Liquid and solid zinc (Zn)	692.73	7,320
Liquid and solid tin (Sn)	505.12	7,195
Gaseous and liquid water (H_2O)	373.15	40,680
Gaseous, liquid, and solid gallium (Ga)	302.924	5,565
Liquid and solid water (H_2O)	273.150	6,010
Gaseous, liquid, and solid mercury (Hg)	234.314	2,295
Gaseous and solid carbon dioxide (CO_2)	194.67	—
Gaseous, liquid, and solid xenon (Xe)	161.39	2,300
Gaseous, liquid, and solid argon (Ar)	83.798	1,190
Gaseous, liquid, and solid oxygen (O_2)	54.361	446
Beta phase and gamma phase in solid oxygen (O_2)	43.802	371
Gaseous, liquid, and solid neon (Ne)	24.562	332
Gaseous and liquid hydrogen (H_2)	20.28	904
Gaseous, liquid, and solid hydrogen (H_2)	13.81	117
Superconductive and normal solid niobium (Nb)	9.3	0
Superconductive and normal solid lead (Pb)	7.200	0
Gaseous and liquid helium-4 (^{4}He)	4.222	83
Superfluid and normal liquid helium-4 (^{4}He)	2.177	0
Superconductive and normal solid tungsten (W)	0.01	0

[a] Liquid-solid equilibria are to be prepared at a pressure of 101 kPa. Superconductive-normal equilibria are to be prepared in magnetic fields smaller than 1 μT.

[b] Temperature values obtained, where possible, from Crovini, L., Bedford, R. E., and Moser, A., Extended list of secondary reference points, *Metrologia*, 13, 197, 1977; and from Bedford, R. E., Bonnier, G., Maas, H., and Pavese, F., Recommended values of temperature for a selected set of secondary reference points, *Metrologia*, 20, 145, 1984.

[c] Enthalpies of transition obtained, where possible, from Chase, M. W., Heats of transition of the elements, *Bull. Alloy Phase Diagrams*, 4, 124, 1983.

the operator in order to detect the limits of the equilibrium. For example, one might discuss the "freezing point of tin" instead of "the temperature of the liquid-solid phase equilibrium of tin, attained by slowly withdrawing heat energy from the sample." This substitution should cause no difficulties if the reader remembers that the desired reference temperature occurs only during an equilibrium that may be noticeably disturbed by the flow of heat, by variation in system pressure, or by the existence of extraneous fields of several kinds.

Table 1 lists several different kinds of phase equilibria characteristic of elements, alloys, and compounds. These include liquid-solid (melting or freezing) equilibrium temperatures, gas-liquid (condensation or boiling) temperatures, gas-liquid-solid (triple-point) temperatures, gas-solid (sublimation) temperatures, temperatures incorporating two forms of solid crystal structure, temperatures reflecting the transformation from the normal resistive state of a metal to the superconductive state, and the temperature of transition from the normal state of liquid ^{4}He to its superfluid state. In addition, reference books such as *The Handbook* of *Chemistry and Physics* list other types of phase transitions — for example, critical temperatures, temperatures of decomposition, and temperatures of dehydration — whose phase equilibrium states can be considered for use as reference temperatures.

In addition to these types of fixed points, there are other phenomena that can provide reference temperatures, too. Most of them involve transitions from one physical state to another; ferroelectric transitions, magnetic transitions, and nematic transitions in liquid crystals are examples of such physical changes.

Let us now examine some of the more useful types of phenomena in order to see how they can be incorporated into temperature reference devices.

IV. WATER, A PROTOTYPICAL FIXED-POINT SYSTEM

The freezing and boiling equilibria of water were among the first fixed points studied by thermometrists. The integral part played by water in human life, coupled with its ready availability, made it a natural substance for examination. The importance of the water fixed points is attested by the initial choice of the ice point and the eventual choice of the water triple point as the fundamental defining point of the Kelvin thermodynamic temperature scale, as we saw in Chapter 2. We shall see in Chapter 4 that the ice and steam points were selected as the fundamental defining points of the international centigrade scale.

Because of its central role in thermometry, we use the water system as a model for fixed points in this section; many useful features common to all fixed points can be illustrated in this way.

A. Ice Point and Triple Point of Water

We mentioned earlier that there are several forms of ice. In discussing briefly the relationships among these different ice phases, we hope to clarify the meaning of the terms *ice point* and *triple point of water* as well as to introduce the useful device of phase diagrams.

1. Phase Diagram of Ice

A phase diagram is a drawing that portrays in graphical form some of the limiting conditions for the existence of the phases of a particular system as functions of two or more of the variables affecting that system. Typically the variables, or *degrees of freedom* as they are often called, include pressure, temperature, and concentration. Other variables that sometimes must be specified in order to describe the state of a system include magnetic field, electric field, and gravitational strength. A pressure-temperature phase diagram for ice and water is given[7] in Figure 1. This figure shows the limits of many of the phases of the one-component H_2O system.

The diagram shows by its uppermost curve that water, indicated by the phase field identified as L, can coexist with a particular form of ice, ice I, from low pressures up to about 0.2 GPa (a gigapascal, or GPa, equal to 10^9 Pa, is the pressure equivalent to about 10,000 times ordinary atmospheric pressure — see Figure 3 in Chapter 1). From 0.2 GPa to perhaps 0.35 GPa, water freezes into the form denoted as ice III. Two forms of ice, IV and V, share the equilibrium solid phase from 0.35 GPa to about 0.6 GPa. Yet another distinct structure, ice VI, coexists with water from 0.6 to 2.2 GPa, yielding in turn to ice VII. Each of the ice phases delineated in Figure 1 is a distinct form of ice characterized by a separate structure.

2. Effect of Pressure on Melting. The Clapeyron Equation

The common form of ice is ice I. Note that the equilibrium between water and ice I is not an ideal constant-temperature equilibrium; instead it varies with pressure with a dependence of about $-100°C/GPa$, or about $-0.01°C$ as the pressure is increased from zero

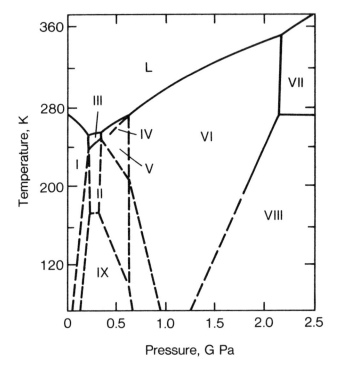

FIGURE 1. The phase diagram of ice from 0 to 2.5 GPa. The dashed lines indicate metastable or uncertain phase boundaries. Data taken from Reference 7.

to normal atmospheric pressure. The pressure variation of the equilibrium temperature between a pure solid and its melt is given by the inverse of the Clapeyron equation,

$$dT/dP = \Delta V/\Delta S = T\Delta V/L \tag{1}$$

In this equation ΔV represents the change of volume upon melting, ΔS is the entropy change upon melting, and L is the latent heat of melting. A derivation of the Clapeyron equation can be found in standard texts on thermodynamics.[3] Using the experimentally determined values for water,[8] $\Delta V = -0.0906$ cm^3/g, L = 79.8 cal/g, T = 273.15 K, and the conversion factor 0.0242 cal = 1 ℓ-atm, one obtains the pressure dependence of melting ice as $-0.0075°C$/atm, very close to the value estimated above from the phase diagram. The qualitative nature of the effect of pressure on the liquid-solid equilibrium temperature of ice can be deduced easily if one remembers that the molar volume of ice is larger than that of water. The exertion of pressure on the system favors the existence of the liquid state, thus depressing the temperature of freezing.

3. Triple Points of Water

In contrast to the temperature variability of the ice I/water phase equilibrium line, there are several temperature-invariant phase equilibria in Figure 1. They are the intersection points where three phases can coexist. These are called *triple points;* they hold a special significance for thermometry. According to the Gibbs phase rule, first proposed in 1876, there is a relationship among the numbers of degrees of freedom F, of components C, and of phases P that is given by the equation[2,9]

$$F = C - P + 2 \tag{2}$$

Table 2
TRIPLE POINTS OF WATER

Phases in equilibrium	Pressure (MPa)	Temp (°C)
Ice I, liquid, vapor	00.000611	+00.0099
Ice I, liquid, ice III	207.4	−22.0
Ice I, ice II, ice III	212.8	−34.7
Ice II, ice III, ice V	344.2	−24.3
Ice III, liquid, ice V	346.2	−17.0
Ice V, liquid, ice VI	625.7	+00.16
Ice VI, liquid, ice VII	2199.	+81.6
Ice VI, ice VII, ice VIII	(2077.)	(1)

Gibbs derived Equation 2 by considering a system with C chemical components and P physical phases. In order to completely specify the state of such a heterogeneous system at equilibrium, one must in general know the temperature, the pressure, and the composition of each phase, or (2 + CP) variables. However, since the total number of moles of each component that is distributed among the P phases is fixed and since, at equilibrium, the chemical potential of a given constituent is the same in all phases, there are P + C(P − 1) equations that relate the variables to one another. Gibbs simply noted that the number of free variables, F, is just the difference between the total number of variables and the number of available equations that interrelate them.

Whenever three phases coexist (i.e., at all triple points) in a one-component system, Equation 2 predicts that there are no further degrees of freedom in specifying the state of the system. That is, neither temperature nor pressure can be a variable. Thus, specification of the coexistence of the three phases ice I, ice III, and water defines an invariant value for both temperature and pressure. In contrast, for the coexistence of the two phases ice I and water (F = 1 − 2 + 2 = 1), an additional degree of freedom, the magnitude of either the temperature or the pressure, is allowed. The difference between the two-phase line and the three-phase point is illustrated clearly in Figure 1. The triple points existing in the H_2O system include all of those listed[10] in Table 2.

Each of the triple points listed above conceivably could serve as a temperature reference point or, for that matter, as a pressure reference point. Each one exists only with a unique value of temperature and of pressure in the pure, one-component H_2O system. However, any thermometrist who mentions "the triple point of water" always is referring to the vapor/liquid/ice-I point at 0.0099°C and 611 Pa. The alert reader will have noticed already that Figure 1 contains no information at all on the vapor phase of H_2O. The reason for this omission is a simple one; on the scale of Figure 1, the vapor phase has nearly no extent at all, lying entirely on the left-hand ordinate axis above T = 273 K.

Figure 2 shows the very small portion of the H_2O system that is of primary interest in thermometry. The vapor/liquid/ice-I triple point region is expanded still further in the inset of Figure 2, showing clearly the triple point as the only point that is common to all three phases. The axes of the inset were chosen also to illustrate the relatively larger variation in the vapor-pressure curve, over the liquid as well as over ice I, compared to the ice I/liquid melting line.

The point CP in Figure 2 designates the "critical point," the point at which the liquid and gas densities become equal and the phases become indistinguishable. Although the critical point represents a unique temperature, it is not easily achieved or maintained; therefore, it is not used in thermometry as a reference temperature.

4. The "Triple-Point" Cell

The vapor/liquid/ice-I water triple point is the principal defining temperature for ther-

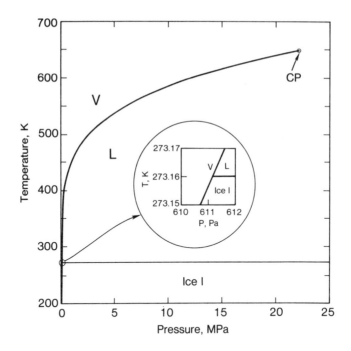

FIGURE 2. The phase diagram of H_2O in the neighborhood of the ice-I/water/water-vapor triple point. The letters CP identify the critical point of water.

mometry, as we noted in Chapter 2. Its temperature is defined as 273.16 K or 0.010°C. It is a remarkably simple and inexpensive temperature reference to realize.[11] Pure, air-free water is introduced into a borosilicate glass tube such as the one pictured in Figure 3, and the filling tube is closed without the introduction of air or other impurities. A thin mantle of ice (ice I, of course) is formed around the central thermometer well by the introduction into it of dry ice or of a similar cooling agent. Once the mantle is formed, the well is warmed slightly, creating a very thin layer of water between the well and the ice mantle; the presence of this water layer can be verified by rotating the cell and noting that the ice mantle is free of the well. The practice of preparing the water triple point in this way is critical to the cell's performance as a temperature fixed point. As we shall see presently, impurities in the water system modify its phase equilibrium temperatures. Moreover, in common with many substances, the liquid state of water accommodates inpurities more readily than does the solid (crystalline) state. Therefore, by first freezing the ice mantle around the cell's thermometer well, the operator selectively forces into the outer liquid layer many of the impurities that inadvertently were included in the cell during its construction; by then melting the layer of the purified ice that surrounds the thermometer well, the operator provides the closest approach possible for that cell to the ideal water triple point.

We note in Figure 3 one feature of the use of the water triple-point cell that may not be obvious at first glance — the temperature-sensing element of the thermometer (its coiled Pt resistor, discussed in Chapter 6, Section IV) actually is located some distance below the position of the triple point itself in the cell. Since the triple-point condition requires that all three phases be present, only the region near D in Figure 3 should actually attain a temperature of 0.01°C. Temperatures obtained further down in the well of the triple-point cell should be reduced by small amounts corresponding to the pressure created by the hydrostatic head of the water column above the measuring position (the hydrostatic head correction is about -7×10^{-6}°C per cm of water). It is routine practice to reduce the measured temperature according to the depth of immersion of the thermometer coil.

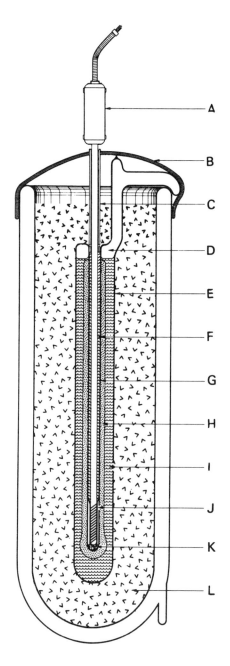

FIGURE 3. Water triple-point cell as maintained in ice in a light-shielded Dewar vessel. B is an opaque cloth. C is a thermometer guide tube of plastic or glass. D, H, and I are, respectively, pure water vapor, pure ice I, and pure water. E is a borosilicate glass envelope with a reentrant thermometer well G. F, water from the ice-water bath L, surrounds the thermometer A in order to promote thermal equilibrium. J is a metal bushing intended to provide good thermal tempering between the tip of the thermometer A and the lower end of the well. K is a soft pad meant to cushion the thermometer as it is seated in the well. (Taken from Riddle, J. L., Furukawa, G. T., and Plumb, H. H., *Platinum Resistance Thermometry*, National Bureau of Standards Monogr. 126, Section 7.1, April 1973.)

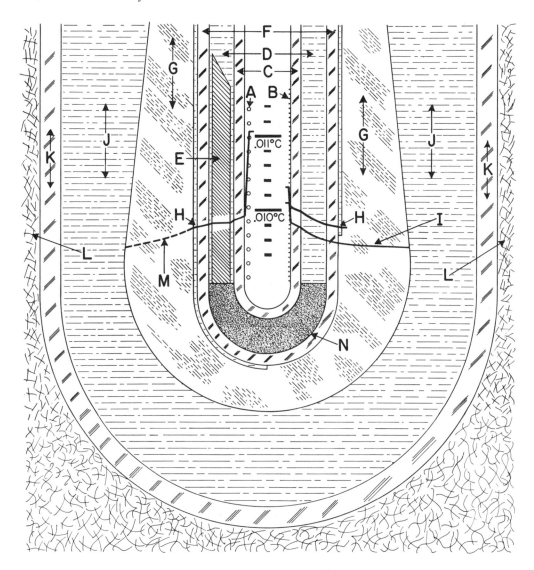

FIGURE 4. The temperature distribution near the bottom of a water triple-point cell. A and B indicate the positions of two different types of winding in the thermometer resistor. C is the thermometer sheath. D is the ice-bath water surrounding the thermometer. E is a metal bushing intended to improve thermal contact between the thermometer sheath and the well F. G is the ice mantle, H the inner melt, and J the pure water of the cell. I is the temperature profile in the event that no inner melt or bushing are present. K is the glass cell envelope. L is the ice bath, and M is the temperature profile in the event that both an inner melt and a metal bushing are present. Note that the temperature gradient through the ice mantle is somewhat steeper if an inner melt is used; note also that the metal bushing reduces the gradient in the water layer D. The self-heating effects of coils A and B reflect 1 mA measuring current in each, but a closer thermal coupling between the thermometer sheath and coil B. (Taken from Riddle, J. L., Furukawa, G. T., and Plumb, H. H., *Platinum Resistance Thermometry*, National Bureau of Standards Monogr. 126, Section 7.1, April 1973.)

The triple point can be maintained nearly indefinitely by enclosing the cell in a protective layer of an ice-water mixture, as indicated in Figure 3, thus minimizing the heat exchange between the cell and its surroundings; in addition one should use care in adjusting to approximately 0°C the temperature of any thermometer before placing it in the well for measurement.

Figure 4 shows in detail the origins of temperature gradients likely to be found within a water triple-point cell during its use.[12] The inner melt H at the position of the thermometer tip is characterized by a temperature very slightly lower than the triple-point temperature 0.010°C (reduced by perhaps 0.2 mK by the hydrostatic head of water). The temperature of the resistor wire is raised markedly, on the other hand, by the ohmic heating of the measuring current. This heat is dissipated more or less well by flow through the gas of the thermometer and the various glass, metal, ice, and water components of the triple-point cell. The *zero-current* resistance value of the thermometer usually is obtained by extrapolation of two measured currents using the relation $R(I) = R(0) + kI^2$.

Experienced operators report using the same mantle for periods of one month or longer and obtaining temperature readings that are constant within 0.1 mK over that period of time. It should be noted that temperature variations of that magnitude often are reported to have taken place during the first day or two of use, presumably as a result of the growth of crystals in the ice mantle. In truth, if there can be a criticism of the ice/water vapor/water triple point as a temperature standard, it is that measurements of the temperature of a particular cell, made with temperature accuracy near the limit of modern thermometry, show a markedly slower recovery from thermal disturbances than do relatively high-thermal-conductivity metal systems such as the triple point of gallium or mercury.[13] At this time, however, it is doubtful that any fundamental measurement standard in science can be made at the $^1/_2$-ppm level of accuracy with less trouble and expense than that obtained using the triple point of water.

Increasing the pressure in a water triple-point cell immediately changes the state of the system away from the triple point, as mentioned above. The pressure can be increased by increasing the height of the cell, thus increasing the hydrostatic head of water, or by increasing the gas pressure on the cell. McAllan,[14] wishing to measure directly the effect of atmospheric pressure on the triple-point cell temperature, introduced an atmosphere of air into a cell initially at the triple point and measured the change in the cell temperature before the relatively slow diffusion of the air into the water had time to take place. His experiments yielded -7.474 ± 0.01 mK for the change in temperature occasioned by increasing the pressure in a triple-point cell from 611 Pa to 101,325 Pa. The magnitude of this result agrees quite well with the calculated value of dT/dP quoted earlier, adjusted for the proportional pressure increase. Careful experiments of this type are necessary not only to increase the understanding of thermometry, but also to provide the data from which accurate phase diagrams can be constructed.

B. Isotopic Effects on the Water Triple Point

Strictly speaking, pure water itself can be treated as a multicomponent system. This situation arises when it is necessary to distinguish between samples of water in which the isotopic abundances of the hydrogen or oxygen vary. Ordinary water contains three isotopes of hydrogen. The mass-1 isotope is known simply as hydrogen (H), the mass-2 isotope is called deuterium (D), and the mass-3 isotope is referred to as tritium (T). According to Dorsey,[10] their relative molar abundances are, respectively, 6500, 1, and vanishingly small. There are also three isotopes of oxygen, with masses 16, 17, and 18.

No complete study of the effects of isotopic composition on the phase transitions of water has been made. There is substantial information showing that the transition temperatures vary with isotopic composition, however. Bridgman[15] has provided data on the temperature of the triple points of D_2O similar to that given in Table 2. We present some of these data in Table 3. We see that in every case the D_2O triple-point temperature is 2.5 to 3.7°C above its normal-water counterpart.

The Supplementary Information section of the IPTS-68 text provides data on the variation of the water triple-point temperature with the source of the water.[16] This information is given in Table 4. We see that natural variations in the isotopic composition of water are quite small. Nevertheless, variations in the resulting triple-point temperatures are detectable at present levels of thermometric precision.

Table 3
TRIPLE POINTS OF ORDINARY WATER AND OF D$_2$O

Phases in equilibrium	Water		D$_2$O		t(D$_2$O) − t(H$_2$O) (°C)
	t (°C)	P (MPa)	t (°C)	P (MPa)	
Vapor, liquid, ice I	+0.0099	0.000611	—	—	—
Liquid, ice I, ice III	−22.0	207.4	−18.75	220.2	+3.25
Ice I, ice II, ice III	−34.7	212.8	−31.0	224.6	+3.7
Ice II, ice III, ice V	−24.3	344.2	−21.5	347.2	+2.8
Liquid, ice III, ice V	−17.0	346.2	−14.5	348.6	+2.5
Liquid, ice V, ice VI	+0.16	625.7	+2.6	628.1	+2.44
Liquid, ice VI, ice VII	+81.6	2200.0	—	—	—
Liquid, ice IV, ice VI	—	—	−6.2	530.5	—

Table 4
COMPOSITION AND TRIPLE-POINT TEMPERATURES OF TERRESTRIAL WATER SAMPLES

Source	Composition	t (°C)
Ocean water[a]	^{17}O; 0.04 mol/100 mol ^{16}O ^{18}O; 0.2 mol/100 mol ^{16}O ^{2}H; 0.016 mol/100 mol ^{1}H	0.0099
Continental water	^{2}H; 0.015 mol/100 mol ^{1}H	0.0099 − 0.00004
Polar water	^{2}H; 0.010 mol/100 mol ^{1}H	0.0099 − 0.00025

[a] Water of this isotopic composition is recommended for use in realizing the IPTS-68 water fixed points.

Questions of isotopic composition will arise again as we continue our discussion of temperature reference points.

C. Impurity Effects in the H$_2$O System

It is trivially easy to transform the one-component H$_2$O system into a multicomponent system, with corresponding changes in all the phase equilibrium temperatures. One need only bubble air through the water, thus dissolving in the water substantial quantities of N$_2$, O$_2$, CO$_2$, and the other components of air.

The changes in the ice melting temperature brought about by the introduction of one atmosphere of various gaseous impurities into the ice-water system were measured by Ancsin.[17] Ancsin also showed the close correlation between the measured freezing-point depressions and the solubilities of the respective impurities. These data are summarized in Table 5.

Clearly, there is a depression of the melting temperature of ice arising from an interaction between the ice-water system and the various gases introduced in these experiments. From the data of Ancsin, the melting point depression in each case is directly proportional to the solubility of each of the gases in water, with the following relation:

$$\Delta T/\Delta P \text{ (mK/atm)} = -7.5 - 0.7 S \tag{3}$$

where S is the solubility of the gas in water in cm^3/100 g.

The theory of the influence of impurities on the melting points of their solvents was developed by Raoult, who showed that the major influence of the impurity is to reduce the

Table 5
FREEZING-POINT DEPRESSIONS AND SOLUBILITIES OF VARIOUS IMPURITIES IN WATER[17]

Nature of gas introduced at 1 atm pressure	Depression of ice melting temperature (m°C)	Solubility in cm³/100 g water
He	−7.5	1
Air	−10.0	3
O_2	−10.5	5
Ar	−11.5	5.5
CH_4	−12.0	6
Kr	−16.5	12
Xe	−25.0	24
CO_2	−130.	170

solvent vapor pressure. We shall discuss the relationship of impurities to the temperatures of phase transitions further as we continue our study of temperature fixed points.

D. Dilute Solutions. Raoult's Law

Throughout this book, we discuss triple points, freezing and boiling temperatures, and other phase equilibria as if the substance under study were a perfectly pure, one-component system. Indeed, it is an objective of the research thermometrist to make it so. Before we further discuss one-component systems, however, it will be useful for us to consider the consequences of man's failure to achieve perfection in the preparation of one-component systems. It is by far the more usual case that the thermometrist must use a dilute solution instead of a pure substance to realize a temperature reference point.

The trouble with solutions, from the point of view of the thermometrist, is that they melt and boil at different temperatures than does the pure solvent. Raoult treated this problem from the point of view of the behavior of ideal solutions composed of a pure solvent to which a small amount of nonvolatile impurity has been added.[18] In 1886, Raoult stated the principle that the partial vapor pressure of any volatile component of an ideal solution is equal to its vapor pressure in the pure state at the same temperature multiplied by its mole fraction in the solution. Stated in mathematical form,

$$P(A) = P(A_0)X(A) \qquad (4)$$

where $P(A)$ is the partial vapor pressure of volatile component A in the (ideal) solution, $P(A_0)$ is the vapor pressure of the same component in the pure state, and $X(A)$ is the mole fraction of A (the number of moles of component A in the solution divided by the total number of moles of all components of the solution). One can readily create the mental image of a liquid surface full of evaporating and condensing (but noninteracting) molecules in which part of the surface is occupied by solute molecules, thus reducing the probability of solvent evaporation in direct proportion to its mole fraction.

How does this change in the vapor pressure of the major component of a solution translate into modified boiling and freezing temperatures? The schematic drawing in Figure 5 illustrates this effect. The solid curves in the drawing indicate the melting, boiling, and sublimation lines of the pure substance; the dashed lines show the effect of a small fraction of a nonvolatile solute. At normal atmospheric pressure, the pure-substance melting temperature T_{f0} is reduced to T_f and the pure-substance boiling temperature is raised from T_{b0} to T_b.

In the very dilute solutions that comprise most "one-component" fixed points, Raoult's

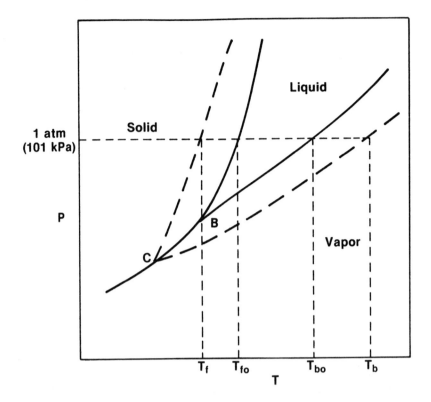

FIGURE 5. Effects of a nonvolatile solute on freezing and boiling points. Solid curves, pure substance; dashed curves, dilute solution; B, triple point of pure substance; C, initial freezing point of solution; T_{fo}, normal (1-atm) freezing point of pure substance; T_f, normal freezing point of solution; $(T_{fo} - T_f)$, normal freezing-point depression; T_{bo}, normal boiling point of pure substance; $(T_b - T_{bo})$, normal boiling-point elevation (cf. Figure 6-5 in Moore[18]).

law is followed rather well. Therefore, for nonvolatile impurities, the pure-substance boiling points are raised and their freezing points are depressed as the curves of the figure show. We note, however, that point C in the figure does *not* indicate a new triple point for the solution. According to the Gibbs phase rule, point C can only mark the initial freezing temperature of the solution when it is in equilibrium with the pure-solvent solid and the pure-solvent vapor. The details of the freezing process depend upon the thermal history of the solution; as the solution freezes, the equilibrium compositions of its phases change according to the solubility of the impurity in the solid.

One can calculate the freezing-point depression and the boiling-point elevation from the approximate equations[19]

$$\Delta T_f = K_f X = \frac{RT_f^2}{\Delta H_f} X \tag{5}$$

$$\Delta T_b = K_b X = \frac{RT_b^2}{\Delta H_v} X \tag{6}$$

where K_f and K_b are called the cryoscopic constant and the ebullioscopic constant, respectively; X is the mole fraction of dissolved impurity in the liquid phase; R is the gas constant; T_f and T_b are, respectively, the freezing or triple-point and boiling-point temperatures, in kelvins, of the pure liquid; and ΔH_f and ΔH_v are the latent heats of fusion and vaporization,

Table 6
CRYOSCOPIC CONSTANTS FOR SEVERAL SUBSTANCES

Substance	T_f (K)	$-K_f$[a]
e-H_2	13.81	13.5
Ne	24.56	15.2
O_2	54.36	55.2
N_2	63.15	46.1
Ar	83.80	49.3
Hg	234.31	199
H_2O	273.16	103
Ga	302.92	137
Sn	505.12	122
Zn	692.73	526
Al	933.61	670
Ag	1235.08	1120
Au	1337.58	1173

[a] Kelvin per unit mole fraction.

Table 7
EBULLIOSCOPIC CONSTANTS FOR SEVERAL SUBSTANCES

Substance	T_b (K)	K_b[a]
e-H_2	20.28	3.82
N_2	77.34	8.91
Ar	87.29	9.74
O_2	90.19	9.94
H_2O	373.15	28.5

[a] Kelvin per unit mole fraction.

respectively. Tables 6 and 7 contain values of the cryoscopic and ebullioscopic constants for representative fixed-point substances.

Solutions comprised of small quantities of nonvolatile solute actually are not typical of boiling-point systems. More common are solutions in which both components are volatile, but with different pressures P_0. In that case, Equation 4 is modified to the form

$$P(tot) = X(A)P_0(A) + X(B)P_0(B) \qquad (7)$$

where $X(B) = [1 - X(A)]$, and the total vapor pressure over the solution is simply a linear combination of the two pure-component vapor pressures. Thus, for most impure systems, depending upon the nature of the impurity, the boiling temperature may either be higher or lower than it is for the pure substance.

In the case of melting and freezing point studies, too, the common situation is that the pure substance and its impurities form solid solutions over some or all of their concentration range, with the result that the melting and freezing temperatures are not the same and furthermore vary with solute concentration. In Figure 6 we illustrate the case in which the presence of the solute depresses the pure-solvent melting/freezing temperature, shown as T_0 (one can equally well find examples in which the solution melting and freezing temperatures are *higher* than T_0).

For a solid solution, the range of freezing temperatures found for solutions of varying impurity concentrations commonly will form a "liquidus" curve such as that shown by the upper (light) curve in the figure; similarly, the range of melting points will provide a "solidus" curve such as the lower one. Choosing a particular impurity concentration "a", we can describe the equilibrium freezing and melting processes; at the same time, we can illustrate the form that the equilibrium cooling and heating curves should take.

Starting in the fully liquid region, the solution with impurity concentration "a" can be allowed to cool to the temperature indicated by T_{la} without the initiation of freezing. (If the solution is subject to the phenomenon of "undercooling", then its behavior at T_{la} will not correspond to the equilibrium process.) At temperature T_{la}, solid of impurity concentration "b" will appear; as cooling continues, the solid gradually will change composition according to the heavy portion of the "solidus" curve. At the same time, the composition of the

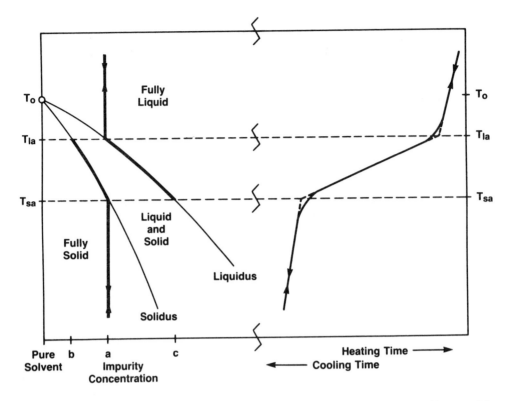

FIGURE 6. Equilibrium solid-solution phase diagram and temperature-time curve. Left-hand side: part of the phase diagram of a two-component solid solution, showing the progress of the phase concentrations during equilibrium heating or cooling. Right-hand side: equilibrium heating and cooling curve that might accompany the system shown at the left.

remaining solution will change according to the heavy part of the "liquidus" curve, terminating with liquid of impurity concentration "c". When the temperature T_{sa} is reached, all of the liquid will have frozen and cooling will continue with the overall solid characterized by impurity concentration "a".

In order for the reverse process to occur on heating the (solid) sample, solid-state diffusion must have equalized its impurity concentration. In that case, the melting process begins at T_{sa} with liquid of initial impurity concentration "c" appearing, and both phases follow the heavy curves in the diagram.

The heating/cooling curve that one might expect in the case just described is shown on the right-hand side of Figure 6. The flatter part of the curve denotes the extent of the two-phase region. If equilibrium is maintained throughout, then the temperature-time data are independent of whether the sample is melted or frozen. Small departures from equilibrium appear as slight rounding of the plateau at one or both ends, as indicated by the dashed lines. Larger departures can result in the formation of entirely different curves on heating and cooling the sample.

The reader should note well that congruence of the heating and cooling curves is *not* evidence for a pure fixed-point sample, *nor* is it evidence that the "true" melting/freezing temperature lies within the limits imposed by T_{la} and T_{sa}. Methods exist by which one can attempt to analyze heating and cooling data to deduce the impurity concentration,[20] but the technique is not a straightforward one.

E. The Steam Point

It is clear from Figure 2 that the boiling temperature of water depends upon the operator's choice of pressure. Whereas the water/water vapor/ice-I triple point provides an invariant temperature, and whereas the temperature of the melting line separating the ice I and water phases shows only a very weak dependence upon pressure (-7.5 mK/atm), the boiling point of water increases by about 30 mK if the pressure is increased by 10^{-3} atm. Water can be made to boil at any temperature from 273.161 K to 648 K (the critical point) by the appropriate choice of pressure. Many cooks use this principle by employing the "pressure cooker" to raise the boiling temperature of meats or vegetables and thus to cook them quicker. The "normal boiling point" of water, 373.15 K (100°C) on the IPTS-68, is simply the temperature at which water boils under a pressure of 101.325 kPa (1 atm). The curve of Figure 2, in the region near 373 K, shows in graphical form the relation between the boiling temperature of water and its pressure that is represented mathematically[21] by Equation 8.

$$t(68) = \{[100 + 28.0216\,(p - p_0)/p_0] - 11.642[(p - p_0)/p_0]^2 + 7.1\,[(p - p_0)/p_0]^3\}\ °C \qquad (8)$$

The noticeable dependence of the boiling temperature of water upon pressure is reflected in the relative difficulty of using the "steam point" as a reference temperature.[22]

Berry[23] has presented a detailed discussion of the temperature reproducibility achievable by use of the steam point. Controlling the "bumping" that is common to boiling systems, carefully shielding the measuring thermometers, and maintaining the purity of the water in the boiling cell, Berry was able to achieve results that were reproducible within about 0.1 mK.

Note that Raoult's law (Figure 5) illustrates the effect of nonvolatile impurities in raising the boiling temperature of water. Once again, one can use the analogy of solute molecules displacing their mole fraction of water molecules from the evaporating surface, so that the boiling temperature of the solution must be increased slightly to reach a given pressure.

V. HIGH-PRESSURE CELLS FOR LOW-TEMPERATURE TRIPLE POINTS

The common gases comprising ordinary air are mutually soluble to an extent that is troublesome for precise thermometry. Because this is true, the melting and boiling temperatures of N_2, O_2, CO_2, H_2, H_2O, and the noble gases tend to be quite sensitive to the presence of other gas species.

In order to obtain reproducible temperature reference points by the use of the common and noble gases, it is necessary to maintain the samples in a highly pure state (impurity levels near 1 ppm). We have seen that water is an exception to this rule; one readily can realize the "ice point" by freezing water in the presence of air. The resulting freezing point, of course, is less precise than the triple point. In addition, it is lower than the triple-point temperature, not only by the 7.5 mK arising from the triple-point/pure-water-freezing-point pressure difference, but, in addition, by some 2.5 mK of temperature that represents the sum of the freezing-point-depression contributions arising from the solubility of the various gases that comprise ordinary air.

Because of the high sensitivity to pressure evidenced by the boiling temperatures of the common and noble gases, special consideration has been given in fixed-point work below 0°C to the study of triple points.

In recent years, considerable attention has been given to preparing sealed, transportable cells for the observation of triple points below 0°C. The widespread use of sealed water triple-point cells has long permitted statistically-based studies of the reproducibility of this fundamental temperature reference point. These studies have shown that, by employing careful techniques, one can expect to achieve a temperature that is reproducible within 0.2

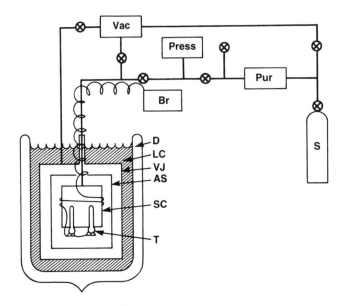

FIGURE 7. Schematic drawing of low-temperature boiling-point calibration apparatus. S, storage tank for pure gas; Pur, chemical purification system; Press, pressure measurement instrument; SC, sample chamber; AS, adiabatic shield; VJ, vacuum jacket; LC, liquid cryogen; D, Dewar vessel; Vac, vacuum system; T, thermometers; Br, thermometer bridge.

mK (less than 1 ppm) by the careful preparation and use of sealed water triple-point cells. Of course, this achievement is made easier by the fact that the cell can be observed visually and manipulated readily while it is in the triple-point condition. On the other hand, the difficulty of realizing low-temperature triple points and boiling points without the use of sealed cells has been for a long time a stumbling block to their effective use as reference temperatures for calibration.

Typically, the apparatus used for many years for this purpose might appear in schematic form like that shown in Figure 7. The objective, to obtain precise thermometer readings while the thermometers T are maintained in thermal equilibrium with the triple-point or boiling-point temperature of the sample chamber SC, is complicated considerably by the necessity of preparing the gas for the measurement. Once assembled, the sample-handling system must be pumped clean by use of a vacuum system Vac. Then the pure-gas sample from the storage tank S can be admitted to the sample chamber through the chemical purification system Pur. When the Dewar vessel D has been cooled to the proper temperature, the operator must monitor the filling of the sample chamber to determine that the correct amount of sample is introduced into the chamber. Depending upon the nature of the phase equilibrium, the calibration process would be accomplished either during a heating cycle or during a cooling cycle, making use of the latent heat of transition to provide an approximately constant-temperature environment for the thermometer despite the slow change in the cryostat temperature. If the reference temperature is a boiling point, the pressure must be monitored carefully because of its sizable influence upon the boiling temperature.

Once a calibration has been performed in the manner described above, the sample gas usually is returned to its storage tank. The only feasible method for comparing precisely the hotness provided by two such calibrations has been to compare the readings of the thermometers during a subsequent calibration, hoping that they will not have drifted to a new temperature dependence by then.

A good example of the effort and care required to obtain precise reference temperatures by the use of open triple-point and boiling-point cells can be found in the work of Compton.[24] He found it necessary to pay careful attention to verifying the purity of the starting gases and to preserving their purity during the measurements, as well as to monitoring carefully the accuracy of his pressure measurements.

The use of sealed high-pressure cells for low-temperature triple points simplifies the sample-preparation work of the calibration operator at the expense of complicating the work of the person who prepares the cells. The essence of this method lies in the suppositions that the cell design and construction can be accomplished so as to avoid long-term contamination of the enclosed gas; that the reliability of the cells can be examined in side-by-side testing; that the cells will permit repeated realizations of their triple points without substantial variation in the measured temperatures; and that eventual standardization of cell design will permit standardization of the measurement procedures to further promote reproducible calibrations.

At this time, the sealed cell technique still is in the experimental stage; there are many different designs in use. Figure 8 shows most of the features that are common to all designs, however. Good discussions on the progress towards perfecting this technique can be found in papers written by Pavese,[25] Bonnier,[26] and Furukawa;[27] measurements in different laboratories have agreed within ±0.2 mK.

There are three major difficulties in the use of sealed cells for low-temperature triple points. First, sufficient amounts of the sample gases must be loaded in order to perform a calibration at the triple point. This requirement implies a high filling pressure in order to increase the total latent heat of the sample, a very light cell construction, or, most probably, a compromise between these conflicting requirements. Second, minimal heating of the cell and essentially adiabatic conditions within the cryostat must be carefully accomplished in order to obtain reproducible results at the triple point. This requirement implies the use of a rather sophisticated cryostat and careful attention to the tempering of all thermometer leads. Third, the use of minimal sample quantities requires unusually careful attention to sample purity levels, to cleanliness of the sample cell, and to sealing it.

Despite these difficulties, the advantages accruing to the use of sealed, transportable triple-point cells — the ready comparison between laboratories using similar cells, and the possibility of long-term stability — have impelled many thermometrists to study this technique.

As is the case with the water triple point, low-temperature triple points, strictly speaking, are reached only where the three phases are contiguous. At points lower in the cell, the temperature deviates from the triple-point value because of the hydrostatic head. In contrast to the accessibility of the thermometer in the water triple-point cell, however, the thermometer placed in the sealed high-pressure triple-point cell generally is beyond the reach of the operator. An effort usually is made in the design of the cell to minimize its height, using an extended well of high-conductivity metal, often copper, to bring the thermometer into equilibrium with the sample.

The triple point itself commonly is achieved by the pulse-heating technique, monitoring the small change in cell temperature as the solid part of the sample liquefies. Figure 9 illustrates the use of this technique as practiced by Ancsin.[28] Notice that the "plateau" achieved by gradually melting the sample does not provide a single temperature, but rather a narrow range of temperatures spanning, in this case, about 1 mK. Notice also that the use of uninterrupted heating at the level chosen by Ancsin for this experiment would lead to considerable distortion of the observed melting curve. Referring again to the idealized curves of Figure 6, we note that Ancsin's results show the difficulty of obtaining perfect thermal equilibrium in practical situations.

The substances that have been examined in sealed cells include all of those listed in Table 8.

A considerable variety of cell designs has been used in cryogenic sealed-cell measurements

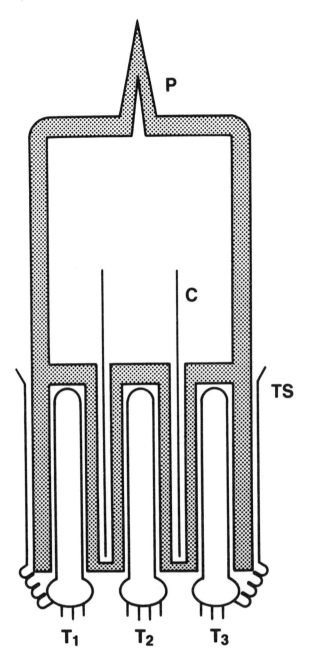

FIGURE 8. Typical design of a high-pressure sealed cell for the realization of low-temperature triple points. The cell is made with a heavy wall to withstand the filling pressure (up to 20 MPa, or 200 atm); often the material is stainless steel. After filling the cell with 0.05 to 0.2 mol of the desired gas, the filling tube P is pinched shut, forming a permanent seal. Often, there is provision for the simultaneous calibration of several thermometers T_i; they are placed in wells in the bottom of the cell and their electrical leads are tempered to the outside of the cell by the use of thermal shunts TS. Upon cooling, the gas liquefies, filling the crevices at the bottom of the cell. A mesh or a tube C of a material that is a good thermal conductor — copper, for example — helps to equalize the internal temperature of the cell.

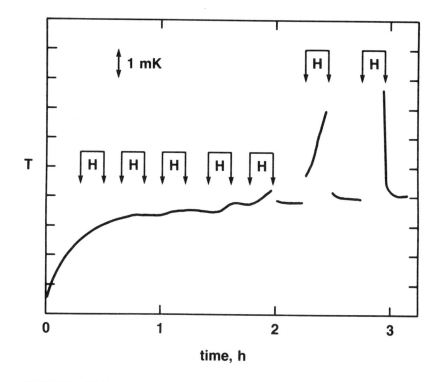

FIGURE 9. Melting curve at the triple point of oxygen obtained by the pulse-heating method. As the melted fraction of the sample exceeded about 0.5, strong temperature disequilibrium appeared upon continued heating. On cessation of the heating pulse, equilibrium quickly was reattained. (Data from Ancsin.[28])

Table 8
SUBSTANCES STUDIED IN SEALED TRIPLE-POINT CELLS

Gas	Triple point T (K)	Ref.
e-Hydrogen[a,c]	13.81 (0.5 mK)[b]	25
e-Deuterium	18.7 (5 mK)	25, 29
Neon[c]	24.6 (0.3 mK)	25, 26
Oxygen[a]	54.361 (0.6 mK)	25, 26
Nitrogen	63.1 (0.2 mK)	25, 26
Argon[a]	83.798 (0.6 mK)	25—27
Propane	85.5 (25 mK)	25
Ethane	90.4 (12 mK)	25
Methane	90.7 (4 mK)	25
Xenon	161.4 (2 mK)	30
Carbon dioxide	216.6 (50 mK)	25
Water[a]	273.16	26

[a] Defining temperature for IPTS-68.
[b] Values in parentheses represent typical plateau ranges reported in Reference 25.
[c] Defining temperature for EPT-76.

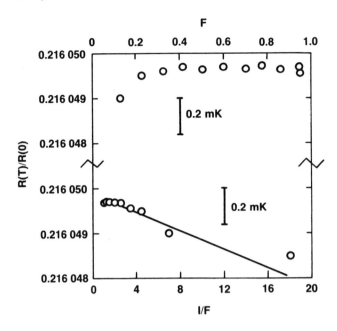

FIGURE 10. Melting data obtained by the pulse-heating method with a sealed argon triple-point cell. (Data from Furukawa.[27])

up to the present time. Surprisingly, however, there has been little discussion of the "inner-melt" technique that has been so successful in producing precise water triple-point results. Instead, melting curves have been generated by pulsed heating applied to the body of the cells. These melting curves usually are converted to curves of temperature vs. the reciprocal of the sample fraction melted, as shown in Figure 10. The best cell-to-cell reproducibility is derived by comparing temperatures extrapolated to the 100% melted condition ($1/F = 1$). As we noted in Section IV.D, however, the best fixed-point curves are characterized by flat plateaus and congruent heating and cooling data. Obtaining such curves requires the use of materials of very high purity and procedures that provide nearly equilibrium conditions, usually by means of essentially adiabatic control of the cell.

Before leaving the topic of low-temperature triple points, we take further note of hydrogen. The hydrogen system is an interesting one, because it is a multicomponent system. Besides the very noticeable effect of the presence of its isotopes ^{3}H (tritium) and ^{2}H (deuterium), which must either be accounted for or be eliminated from the common isotope ^{1}H in preparing a particular sample, the pairing of the nuclear spins in the molecule ^{1}H-^{1}H changes the triple-point temperature of hydrogen.

Ordinary hydrogen contains a small amount of deuterium and of HD (^{1}H-^{2}H) as well as a vanishingly small amount of HT. In order to predict the triple-point temperature of a particular sample, its composition must be known. Whenever the isotopic purity of a sample is not stated, the relative abundances are assumed to be those occurring naturally (99.985% ^{1}H, 0.015% ^{2}H).

The nuclear-spin-pairing property of hydrogen is a separate concern in precise thermometry; errors in the triple-point temperature from this source[31] can exceed 0.1 K. The two possible spin configurations, parallel and antiparallel, have been given the names *ortho* and *para,* respectively. The equilibrium ratio of *ortho* to *para* hydrogen molecules at room temperature — so-called "normal hydrogen" — is approximately three to one, reflecting the fact that the *ortho* configuration is made up of three quantum states, each with the same energy, whereas the *para* configuration is composed of only one quantum state. Although the single *para* state lies slightly lower in energy than the three *ortho* states, the level of thermal excitations above about 220 K is sufficiently high that all four states are about equally likely to be occupied. The "equilibrium" ratio — the *ortho-para* ratio that is consistent with a particular temperature and with the energy separation of the two configurations — decreases at lower temperatures. Below 20 K, equilibrium hydrogen is composed of approximately 99.8% *para* and 0.2% *ortho*.

The influence of the *ortho-para* composition on the hydrogen triple-point temperature is quite noticeable; the normal hydrogen triple-point temperature is 13.96 K, whereas the equilibrium hydrogen triple-point temperature is 13.81 K.

Compounding the problems that the thermometrist encounters in using hydrogen in cryogenic fixed points is the fact that direct transitions between the *ortho* and *para* states are forbidden by quantum selection rules. Thus the change from the "normal" composition to the "equilibrium" composition must involve collisions with a wall of the container or with a catalyzing surface. In practice, inclusion of catalysts such as charcoal or ferric hydroxide guarantees that the hydrogen will attain its equilibrium composition within a few hours after reaching a given temperature.

VI. MELTING, FREEZING, AND TRIPLE POINTS OF METALS

We have seen that, below 0°C, triple points of the noble gases and those comprising ordinary air offer a large selection of reproducible reference temperatures. Triple points are much to be preferred to boiling points whenever feasible because of their freedom from the need for pressure measurements.

Above −40°C, one can use metallic liquid-solid fixed points. Metals offer some worthwhile advantages as fixed-point materials over nonmetals; these advantages include a much-reduced sensitivity to the presence of impurity gases and a considerable increase in thermal diffusivity. The relatively low solubilities of the noble gases in the fixed-point metals allows one to choose between realizing the triple point and the melting/freezing point. The differences between melting/freezing-point temperatures and triple-point temperatures generally are small for metals; for example, for Hg, Ga, and In, the differences t(M/F) − t(Tr) are 0.006°C, −0.002°C, and 0.005°C, respectively.

The thermal diffusivity α is defined by the equation

$$\alpha = \frac{\lambda}{\rho C_p} \tag{9}$$

where λ is the thermal conductivity, ρ is the density, and C_p is the heat capacity at constant pressure. The thermal diffusivity of a substance is significant because, more than the thermal conductivity, it measures the rate at which energy is distributed internally in response to a temperature change at the surface.[32] The relatively high thermal diffusivities of metals help to promote equilibrium within the metal fixed-point cell more quickly and in the presence of larger stray heat fluxes than can be the case for nonmetals. Table 9 contrasts the thermal diffusivities for several high-purity temperature reference point materials. Note that the five metallic substances are superior both in thermal diffusivity and in thermal conductivity to the two nonmetals.

The primary reason for the thermal-conductivity superiority in pure metals is that the conduction electrons contribute strongly to metallic thermal conductivities. In the case of nonmetallic solids, the primary mode of heat transfer is the lattice vibrations, known as "phonons."[33] In the liquid state, convective flow can assist in thermal equilibration.

In construction, metal fixed-point cells resemble the water triple-point cell. The sample encloses a central thermometer well, providing an approximately constant-temperature environment by virtue of its latent heat of fusion. As is the case with water, metallic samples

Table 9
THERMAL DIFFUSIVITIES OF THERMOMETRIC SUBSTANCES

Substance	T (K)	α (T) (m²/sec)	λ (W/m K)	ρ (kg/m³)	C_p (J/kg K)
Hg	200	1.8×10^{-5}	34	1.4×10^4	136
Ga	300	3.9×10^{-5}	85	5.9×10^3	373
Sn	400	3.5×10^{-5}	63	7×10^3	257
Ag	1000	12.2×10^{-5}	376	1.0×10^4	294
Au	1000	9.8×10^{-5}	279	1.9×10^4	147
Ar	77	0.026×10^{-5}	0.31	1.7×10^3	732
H$_2$O	250	0.12×10^{-5}	2	0.92×10^3	1800

generally exhibit a volume change between their liquid and solid states; proper account must be taken of the nature of this volume change in the design of the fixed-point cell or in the technique used in preparing the reference temperature.

A typical tin or zinc metal freezing-point sample is contained in graphite. An apparatus used to fill such a cell at the NBS is shown in Figure 11. The high-purity graphite container D and a high-purity graphite funnel C into which the sample B has been placed are balanced on a borosilicate glass stand G so that a borosilicate glass envelope A can be installed around them. To remove evaporable impurities — especially water — the envelope is pumped while its contents are heated to about 200°C for an hour or so by the induction heater coils E. In view of the high vapor pressure of zinc, argon gas at a pressure of 0.3 atm is introduced to retard evaporation. Then the coil power is raised until the metal melts, filling chamber D.

A method for placing the graphite thermometer well and lid is shown in Figure 12. After purging the tube I with helium gas, it is lowered into a furnace and heated above the melting point of the sample. Then the stainless steel pusher rod is used to place the thermometer well and lid in their proper positions; a slight reduction in the diameter of the tube at the place marked J helps to center the apparatus.

The completed NBS freezing-point cell is shown in Figure 13 with a standard platinum resistance thermometer (SPRT) in place. A layer of Fiberfrax® paper cushions the graphite crucible within its borosilicate tube. Layers of Fiberfrax® insulation separate graphite heat shunts G that are placed along the borosilicate thermometer guide tube F in order to temper the thermometer A. The gas tube B allows the introduction of inert gas at a controlled pressure through the silicone rubber plug D.

A design of a high-temperature furnace used at the NBS for heating the metal freezing-point cells is shown in Figure 14. Precise control of the cell temperature is enhanced by use of three heaters — a main winding N, a separate heater for the top aluminum block G, and another for the bottom block V. An Inconel® sleeve C is designed to fit closely near the heat shunts of the freezing-point cell. Thermocouple thermometers used to control the central heater are contained in Inconel® tubes F. The furnace is purged with inert gas before use.

Using the apparatus described so far, one can prepare the fixed point either in the melting or the freezing condition. If a particular sample is sufficiently pure, then the two techniques will provide identical levels of hotness; for less-pure samples, one must choose the method that will provide the desired calibration temperature. Since most crystalline lattices discriminate against impurities, a questionable sample can be made to maintain a temperature closer to the desired fixed-point value by using the freezing point rather than the melting point.

Stress arising from heating and cooling the freezing-point cells can break them; for this reason, the cells always are heated and cooled very slowly. During the periods when they are to be used routinely in calibrating thermometers, the cells are held within a few degrees of the sample melting temperatures.

The technique used to produce a "freeze" in a particular cell depends upon its type. Tin, for example, may not begin to freeze until the temperature of the liquid is 25°C below the

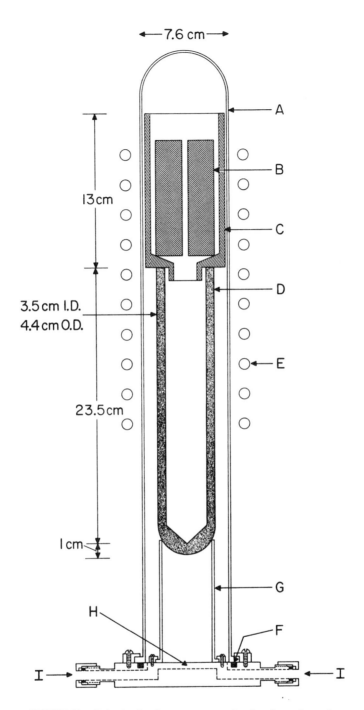

FIGURE 11. Induction-heating apparatus used to install metal samples into graphite crucibles. A, borosilicate glass vacuum envelope; B, high-purity metal sample; C, graphite sample holder; D, high-purity graphite crucible; E, induction heater coils; F, vacuum seal; G, glass stand to position crucible-fill apparatus within coil system; H, pumping port; I, vacuum/filling-gas port. (Figure taken from Furukawa, G. T. et al., National Bureau of Standards Special Publ. 260-77, Section VI, August 1982.)

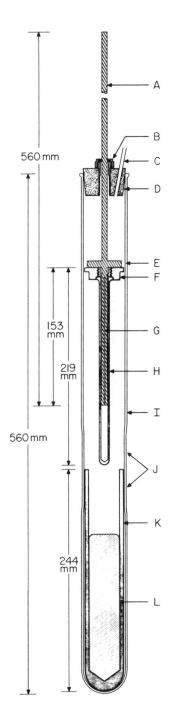

FIGURE 12. One NBS technique for emplacement of the thermometer well. A is a stainless steel pusher rod; B is a gas seal of silicone rubber. Tube C permits the introduction of purging gas through the silicone plug D. Rod A terminates in a flange E and a split rod G that can be slightly sprung to hold the graphite well H and the graphite lid F. The graphite crucible K with its enclosed sample L is held in the borosilicate glass tube I that has a section J of reduced diameter for centering the well. (Figure taken from Furukawa, G. T. et al., National Bureau of Standards Special Publ. 260-77, Section VI, August 1982.)

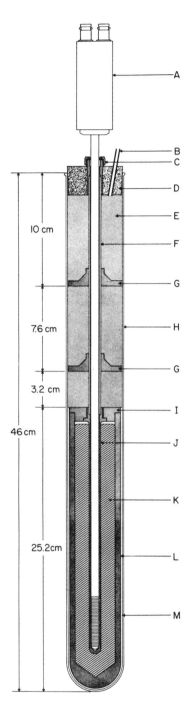

FIGURE 13. Completed NBS-type freezing-point cell with long-stem standard platinum resistance thermometer (A) installed. B, gas filling port; C, thermometer seal; D, silicone seal for borosilicate glass cell (H); E,M, Fiberfrax® insulation; F, glass guide tube; G, graphite heat shunts; I,J,L, graphite sample container; K, high purity fixed-point metal sample. (Figure taken from Furukawa, G. T. et al., National Bureau of Standards Special Publ. 260-77, Section VI, August 1982.)

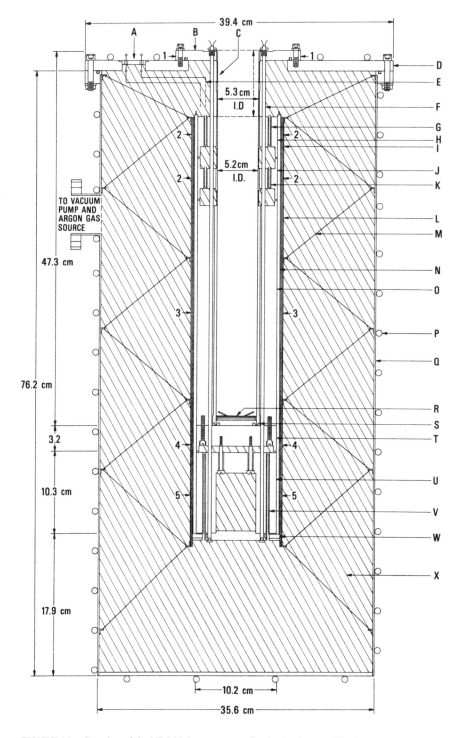

FIGURE 14. Drawing of the NBS high-temperature fixed-point furnace. The furnace can be used to about 800°C. A, insulated heater posts; B,C,R,S, Inconel® flanges, tube, and cell positioner; D,Q, brass cover plate and stainless steel shell; E, leads for heaters G,K,N,V; F, control thermocouples in Inconel® tubes; H,I,J,L,O,T,U, core block assembly; M, core suspension wire; P, cooling water coils; W, main heater support; X, Fiberfrax® insulation. (Figure taken from Furukawa, G. T. et al., National Bureau of Standards Special Publ. 260-77, Section VI, August 1982.)

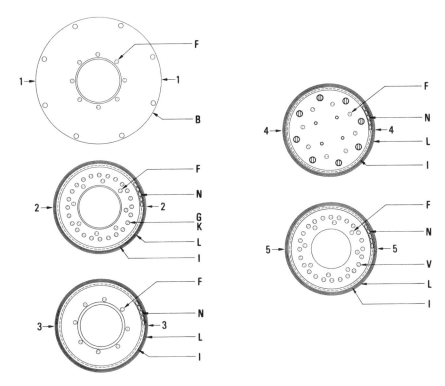

FIGURE 14 continued.

melting temperature; on the other hand, zinc rarely "undercools" more than 0.1°C. McLaren and Murdock[34] have described a technique for initiating freezes in tin cells by temporarily withdrawing them from their furnaces; this technique is not necessary in the case of zinc. A good general discussion of the use of high-purity metals to provide calibration reference points was presented by McLaren.[35]

The methods used at the NBS to prepare tin and zinc freezes are as follows:[36]

1. The sample is completely melted, usually by setting the furnace about 5°C above the melting point late in the day before the freezing point is to be used and leaving the setting there overnight;
2. In the morning, a stable SPRT maintained for the purpose is inserted into the thermometer well of the cell. The furnace is reset to control about a temperature 1 to 5°C below the freezing point;
3. When the thermometer indicates that the cell sample has begun to freeze, the control temperature is set about 1°C below the freezing point. In the tin case, the cell is withdrawn from the furnace for a few moments to initiate the freeze;
4. Two room-temperature borosilicate rods are successively inserted into the thermometer well to induce an inner freeze in the sample immediately surrounding the well;
5. Any thermometers to be calibrated are preheated to about 20°C above the cell melting temperature, inserted into the cell well, and monitored until their resistances are steady. This process may take 15 min or more;
6. Usually six thermometers are calibrated at the NBS during any one freeze. After the last calibration, the monitoring SPRT is preheated and reinserted into the thermometer well. If the cell temperature measured at this time agrees within 0.5 mK with that measured before the calibrations began, the operator assumes that the test thermometers all received valid calibrations. In most cases the difference is about 0.1 to 0.2 mK. Otherwise the calibrations are repeated the next day.

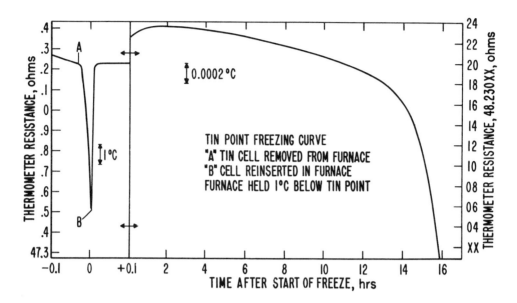

FIGURE 15. Temperature-time curve for a tin freezing-point sample obtained by Furukawa et al.[13a] Left-hand side: record of the cell temperature during removal from the furnace to initiate the freeze; undercooling of about 8°C was observed. Right-hand side: record of the cell temperature over a 16-hr period; during some 12 hr of that time, the cell temperature varied by less than 0.8 mK.

Figure 15 shows a typical temperature-time curve obtained by G. T. Furukawa and W. R. Bigge at the NBS, using the tin fixed-point cell of Figure 13 and a furnace similar to the one shown in Figure 14. Because of the undercooling that is characteristic of tin, the tin cell was withdrawn from the furnace to initiate the freeze. The SPRT, contained within the reentrant well of the tin-point cell, responded to the withdrawal as shown on the left-hand side of the figure. Note that one scale unit there is about 1°C, so that the total cooling was approximately 8°C. Once the freeze was begun, the cell was returned to the furnace, which was maintained at about 231°C. During the next 12 hr, the temperature registered by the SPRT varied by no more than 0.0006°C, as indicated by the much finer temperature scale on the right-hand side of the figure. This long-lasting, stable fixed-point temperature allows ample time for calibration of other thermometers.

Figure 16 shows a temperature-time curve obtained at the NBS using a zinc sample. Zinc is not so prone to undercooling as is tin; therefore, the cell was not withdrawn from the furnace. The level of undercooling in this case amounted to about 0.06°C, as shown by the expanded scale on the left-hand side of the figure. The curve on the right-hand side of the figure shows that the sample temperature was constant within about 0.4 mK over a 12-hr period while the furnace temperature was maintained within about 0.9°C of the zinc melting point.

Aluminum fixed points can provide an excellent reference temperature despite the fact that molten aluminum is chemically reactive — considerably more so than tin or zinc. McAllan and Ammar[37] prepared freezing-point cells using aluminum with as little as 1 ppm impurity. They examined the quality of the cooling curves using various procedures to initiate the freezing condition. These techniques included: a slow cooling in which the aluminum cell was not disturbed in any way; inducing a shell of frozen aluminum on the graphite thermometer well by cooling it; inducing a shell on the graphite crucible by temporary withdrawal of the cell from the furnace; and a combination of the two latter techniques to induce a double shell of frozen aluminum. They concluded that the method of cooling the thermometer well to initiate the freezing of the aluminum sample was the best one.

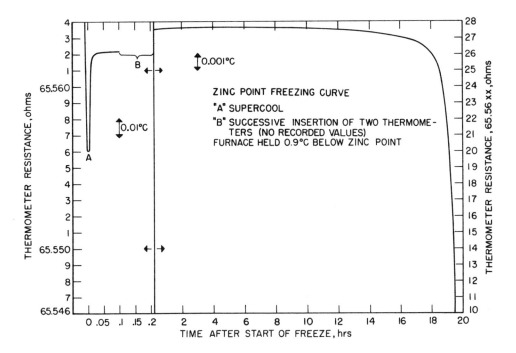

FIGURE 16. Temperature-time curve for a zinc freezing-point sample obtained by Furukawa et al.[13a] After an approximately 0.06°C undercool, the cell temperature remained constant within about 0.4 mK over a 12-hr period.

In studying the techniques for preparing stable aluminum freezing-points at the NBS, Furukawa[38] encased the sample, aluminum with about 1 ppm total impurity, within a graphite crucible as he had done with tin and zinc samples described earlier. Three different types of outer enclosure were prepared, however. We show in Figure 17 an early sealed fused-quartz envelope used to contain the graphite crucible. Before sealing the quartz pumping tube, Furukawa pumped the envelope at a pressure of about 10^{-3} Pa for three days with the apparatus maintained above the Al melting temperature; then he filled the cell with sufficient Ar gas to maintain approximately an atmosphere of pressure at the melting point. The cell then was inserted into its Inconel® protective case and completed by the addition of the thermometer guide tube, shunts, and insulation.

Furukawa induced the freezing of the aluminum sample by cooling the thermometer well. Figure 18 shows typical freezing curves obtained by this technique. Note that the curves have been displaced on the figure to show their details clearly; with only one exception, every curve contains a plateau that is constant within 1 mK over an 8-hr period.

Several other metals have received systematic study as temperature fixed points. These include Ga, Hg, In, Cd, Ag, Au, and the Cu-(71.9% Ag) eutectic. Results obtained in these studies are summarized in Table 10.

VII. FIXED POINTS FOR THE LIFE SCIENCES

The temperature needs of the life-science community (medicine, biology, biochemistry) are focused in the 0°C to 100°C range. Using liquid-in-glass thermometers, pure-metal resistance thermometers, or tiny thermistor thermometers, thermometrists working in this area have indicated various needs for thermometry that reflects the International Practical Temperature Scale of 1968 (see Chapter 4) accurately within about 0.01°C. Berger et al.[50] point out that because of the great sensitivity and convenience of thermistor thermometers, it is useful to create fixed-point devices especially for them.

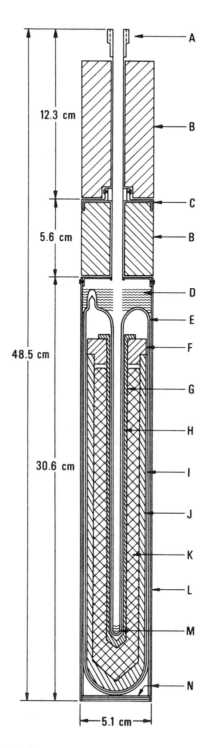

FIGURE 17. Early version of aluminum fixed-point cell used by Furukawa.[38] A,C,L, Inconel® guide tube, shunts, and protective case; B, MgO insulation; D,N, Fiberfrax® insulation; E,H, fused-quartz envelope and thermometer well, sealed in this version but later opened to an Ar gas-handling system; F,G,J, graphite sample container; I,M, fused-quartz cushioning material; K, aluminum sample.

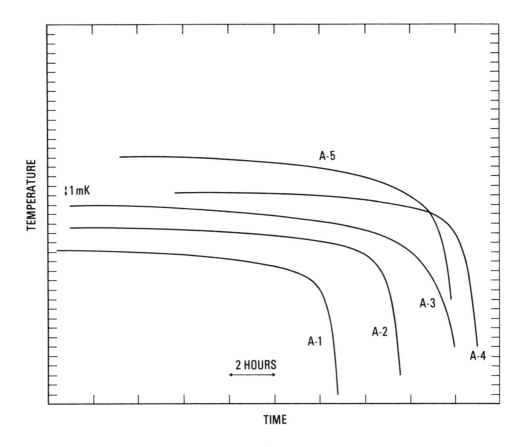

FIGURE 18. Five cooling curves obtained by Furukawa[38] on aluminum samples contained in the fixed-point cell of Figure 17. Note that the curves have been displaced both in time and in temperature for the sake of clarity.

Several such devices, with melting or triple points in the range 0 to 100°C, have been developed. In some cases, the fixed-point cells are particularly adapted for use with small thermometers. Table 11 contains a summary of the fixed points thus examined.

In Chapter 6 we discuss the use of fixed points of the types listed above to calibrate thermometers for laboratory scales.

VIII. SOLID-SOLID EQUILIBRIA AS TEMPERATURE REFERENCE POINTS

It is at once an advantage and a disadvantage that triple points and melting points involve the liquid state of the reference substance. Inasmuch as a crystallizing substance commonly prefers its own kind of atom, the freezing process tends to concentrate impurities in the liquid that remains behind. This fact implies the happy situation that the early stage of a carefully-prepared "freeze" offers a rather purer solid than the original composition warranted, with a beneficial effect on the uniformity of the freezing or triple-point temperature. Furthermore, the presence of liquid — particularly an "inner melt" formed from the purified solid immediately surrounding a thermometer well — helps to foster thermal equilibrium with a test thermometer. Still another advantage for triple and freezing points is that they possess a latent heat of fusion; this fact permits their use as constant-temperature chambers with which a test thermometer can exchange heat until (one hopes) thermal equilibrium occurs. Not all transitions are endowed with latent heats.

There are disadvantages accruing to the presence of the liquid state, too. For one, the triple point or the freezing point prepared at one time is not in fact the same as that prepared

Table 10
METAL FIXED POINTS

Metal	t(IPTS-68) (°C)	Experimental reproducibility (mK)	(Range, mK)	Ref.
Cu	1084.9	6	(<5)	39
Au	1064.43	2.0	(5.5)	40
Au	1064.43	—	(13)	41
Ag	961.93	1.5	(8.2)	40
Ag	961.93	—	(16)	41
Cu-Ag eu	779.57	50	(<100)	42
Cu-Ag eu	779.898	10	(30)	43
Al	660.462	1.0	(5)	37
Al	—	0.2	(2)	38
Zn	419.58	0.2	(0.2)	40
Zn	419.58	0.2	(2)	35
Cd	321.108	0.8	—	44
Cd	321.108	0.7	(<1)	45
Cd	321.108	0.1	(<0.5)	46
Sn	231.968	0.1	(0.4)	40
Sn	231.968	<0.1	(0.1)	36
In (tp)	156.630	0.3	(<0.3)	47
Ga (tp)	29.774	0.2	—	48
Ga (tp)	29.774	<0.1	(<0.1)	49
Ga	29.772	0.2	—	48
Hg (tp)	−38.841	<0.1	(0.1)	13

Note: All data for freezing points except where (tp) denotes triple-point studies. Where applicable, IPTS-68 assigned temperatures are quoted. Plateau reproducibility and range values are given in mK.

Table 11
FIXED POINTS FOR THE LIFE SCIENCES

Substance	Phase equilibrium temperature (°C)	Range (mK)	Ref.
H_2O (tp)	0.010	0.1	51
Phenoxybenzene (tp)	26.871	1	51
Gallium[a] (mp)	29.772	<0.1	52
$Na_2SO_4 \cdot 10\ H_2O$; Na_2SO_4[b]	32.374	1	53
n-Icosane (tp)	34.494	0.5	51
1,3-Dioxolan-2-one (tp)	36.324	0.5	51
Rubidium[a] (tp)	39.265	14	54, 56
$KF \cdot 2H_2O$; KF[b]	41.422	3	53
$Na_2HPO_4 \cdot 7H_2O$; $Na_2HPO_4 \cdot 2H_2O$[b]	48.222	2	53
Succinonitrile[a] (tp)	58.080	0.5	55

[a] Samples available from the NBS OSRM.
[b] Equilibrium involving two phases with different hydration levels.

Table 12
LATENT HEATS, TRANSITION TEMPERATURES, AND UNCERTAINTIES OF OXYGEN TRANSITIONS[59]

Point	ΔH (J mol^{-1})	T (K)	ΔT (K)[a]
Triple point	439	54.361	±0.000 3
Beta-gamma	811	43.801	0.010
Alpha-beta	103	23.867	0.005

[a] Uncertainty in transition temperature arising from variation in sample cooling rate.

the next. Differences in the technique for initiating a transition can result in measurable changes in the equilibrium temperature. Indeed, if the cell is emptied after each use, as has commonly been the case with cryogenic triple and boiling points, there may arise the possibility that the overall impurity level of the system does not remain the same from time to time. Finally, use of a liquid-state temperature reference device can be awkward and bulky if more than one calibration is to be performed or if more than one reference point is to be used.

Solid-solid phase equilibria include the coexistence of two crystalline phases such as the alpha-beta and beta-gamma phases in oxygen, of superconductive and normal phases in metals, of ferromagnetic-paramagnetic and antiferromagnetic-paramagnetic phases and of ferroelectric-dielectric phases. These and other entirely solid-state phase equilibria offer their own advantages and disadvantages.

A. Solid-Solid Equilibria in Oxygen

There are three stable crystalline configurations in solid oxygen. From the freezing point at 54.4 to 43.9 K, the stable form is cubic, known as the gamma phase. A rhombohedral phase, the "beta" phase, is stable from 43.9 to 23.9 K, below which temperature oxygen transforms into a monoclinic (alpha) phase. The alpha-beta and beta-gamma phase transitions have been studied by several thermometrists.[57-59]

Both of the oxygen solid-state transitions occur at temperatures where a reference temperature is needed for the calibration of the standard scale-defining thermometers (see Chapter 4). Note, however, that — barring storage in a permanently refrigerated cell — the solid oxygen sample to be used for this purpose would necessarily be refrozen each time that a calibration apparatus would be cooled from room temperature. Thus the technique for realizing either equilibrium state must include a reproducible method for reforming the sample *in situ*.

The apparatus to be used in calibrating thermometers with solid-solid equilibria in oxygen might resemble that in Figure 8; however, we know of no such use up to the present time.

Both transitions are believed to be first order, and thus to be accompanied by a latent heat of formation of the new phase. The latent heats and transition temperatures, and their uncertainties as determined by Ancsin[59] are shown in Table 12. Ancsin, in particular, found that the quality of his results depended strongly upon the rate of change of temperature during the phase transition; the apparent temperature of the beta-gamma equilibrium varied by more than 10 mK depending upon the cooling rate of the sample. Ancsin observed the most nearly constant beta-gamma plateau when he cooled the sample at about 5°C/min. The strong dependence of Ancsin's results upon the cooling rate may reflect the fact that the crystal configurations are substantially different in the two phases. Achievement of the equilibrium condition may be even slower in a solid-solid transition than it is in a liquid-solid transition because of the former's relatively slower rate of diffusion.

Ancsin measured the heat capacity of the alpha-beta transition and found a sharp peak as expected. The temperature at which the peak of the heat capacity curve occurred varied by about 5 mK, again depending upon the rate of sample cooling.

Whether variations from the techniques used in oxygen studies thus far might improve the temperature stabilities of the solid-solid equilibria sufficiently to enhance their use as practical fixed points one cannot say; at this time, the solid oxygen phase equilibria appear suited for use as temperature reference points for calibration only at the 5 to 10 mK level of precision.

B. Superconductive-Normal Phase Equilibria in Metals

The transition from normal electrical resistivity to superconductivity occurs during the cooling of numerous metallic elements, alloys, and compounds.[60] The temperature range covered by superconductive transitions is 0 to 23 K; thus they are of direct use to thermometrists only in the field of cryogenics.

Although in theory a given superconductive transition can be expected to extend over no more than about 1 µK in temperature, the transitions in most materials are broader than that. In some materials the transitions extend over 1 K or more. In general, purer samples exhibit narrower transitions, as do more carefully annealed samples.

The superconductive transitions in five high-purity elements, Pb, In, Al, Zn, and Cd, contained in a fixed-point device developed at the National Bureau of Standards[61] constitute five of the eleven defining temperatures of the 1976 Provisional 0.5 to 30 K Temperature Scale[62] (see Table 13, Chapter 4). It is instructive to examine the extent to which superconductive transitions can provide useful reference temperatures, since never before have fully solid-state transitions been specified as defining points in an internationally recognized temperature scale.

When it occurs in the absence of a magnetic field, the superconductive transition is a second-order one and thus is not accompanied by a latent heat. It is important to understand the consequences of this fact. With no latent heat of transition to utilize, the experimenter cannot provide the test thermometer with a constant-temperature environment consisting solely of the fixed-point substance; instead, a high-conductivity metal block commonly is used in conjunction with a heater and temperature regulator to provide this environment.

Since the "temperature plateau" method used to reach the characteristic temperature of first-order phase transitions is not available in the case of superconductors, some other parameter must be monitored to mark the transition temperature. Among the properties that change abruptly when a metal crosses the boundary between the superconductive and normal phases, three have received more attention than the others. These three properties are the electrical conductivity, the magnetic susceptibility, and the heat capacity.

One of the hallmarks of the superconductive state is its perfect conductivity. It was this property that Onnes first observed when studying the behavior of pure mercury at low temperatures.[63]

An independent phenomenon discovered later by Meissner and Ochsenfeld is that magnetic flux is excluded by superconductive materials.[64]

A half-century passed between the discovery of superconductivity and the publication of a successful microscopic theory by Bardeen, Cooper, and Schrieffer.[65] According to the "BCS" theory, the superconductive state is one in which the conduction electrons form a macroscopic quantum state, extending over many crystal lattice sites. Electrons of equal and opposite momentum become paired so that they act in some ways as spinless, single particles. A gap appears in the spectrum of quantum states available near the Fermi energy, so that normal-state scattering mechanisms are ineffective. So long as a limiting "critical current" is not exceeded, this mechanism leads to the phenomenon of perfect conductivity with respect to steady electrical currents.

The BCS equation describing the magnitude of the superconducting transition temperature T_c illustrates the fact that crystalline interactions are important for our purposes here:

$$kT_c = 1.14 \ (h/2\pi)\omega \ \exp[N(o)V]^{-1} \qquad (10)$$

Here, k is Boltzmann's constant, h is Planck's constant, ω is a characteristic lattice phonon frequency, N(o) is the density of electron states in the metal at the Fermi surface, and V characterizes the strength of the electron-lattice interaction.

We first point out that $(h/2\pi)\omega$ is approximately equal to $k\theta_D$, the Debye energy of the material; it is proportional to the square root of the average molecular weight of the material. From these equivalences we can deduce that T_c varies with the lattice Debye temperature and with the isotopic abundances of the elements that comprise the superconductor.

Much more influential on T_c is its exponential variation with the product of the electronic density of states and the electron-lattice interaction. Realizing the sensitive relationship between impurities and lattice defects in a crystal and the electrical resistivity of the crystal (which also reflects the electron-lattice interaction), one must expect very strong effects on T_c from the presence of impurities of various kinds[66] and from strains or other imperfections in the lattice.

In a multiply connected superconductor (such as a ring or a solid block), attempts to impose small magnetic fields result in the formation of surface currents of superconducting electrons in such a fashion as to counteract exactly the magnitude of the impressed field. Thus, in magnetic fields smaller than a temperature-dependent "critical field" $H_c(T)$, the superconductor constitutes a perfectly diamagnetic material. The superconductive state is destroyed by steady electric currents larger than the critical current, by steady magnetic fields larger than the critical magnetic field, and by high-frequency electromagnetic radiation. The temperature dependence of the critical magnetic field is given approximately by the relation

$$H_c(T) = H_c(o)[1 - (T/T_c)^2] \qquad (11)$$

Thermodynamic arguments can be used to show that a discontinuous drop in heat capacity should accompany the transition of a sample from the superconductive to the normal state. This drop is easily measured using standard heat-capacity techniques.

It is reasonable to expect that each of the three measurements mentioned above — electrical conductivity, magnetic susceptibility, and heat capacity — should yield the same temperature for a given superconductive transition. To what ultimate level of temperature sensitivity is this agreement likely to persist? For one substance — the element indium — a careful set of experiments was performed at the NBS with the results[67] shown in Figure 19. This figure provides evidence that measurements of the three properties can agree within 1 mK on the definition of the transition temperature of a superconductor.

Experiments of the type shown in Figure 19 are especially important in the case of superconductive fixed points because, with no latent heat, any of these parameters or any other property that undergoes a change during the transition to the superconductive state might be selected to define the transition temperature.

The practical technique for realizing a fixed temperature with a superconductive sample depends upon regulating the temperature of a high-conductivity mounting block that holds both the fixed-point device and any thermometer to be calibrated. The operator can vary the block temperature slowly while observing the chosen parameter; having displayed the temperature-dependent distribution of the parameter, the operator then can adjust the block

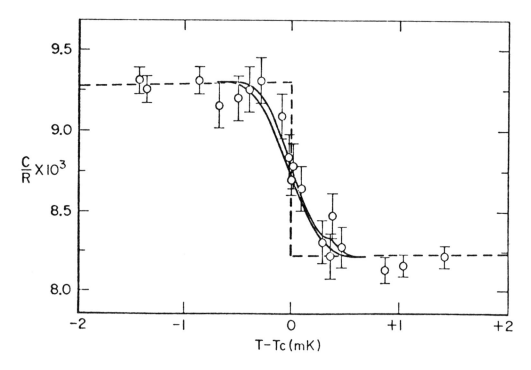

FIGURE 19. The heat capacity (open circles), magnetic susceptibility (solid curves), and electrical resistance (dashed curve) of an indium sample, showing the level of agreement among the three independent measurements of the superconductive transition. The data have been normalized vertically. Data from Soulen and Colwell.[67]

temperature until "T_c" — which the operator may define as he wishes — of the chosen parameter is reached. Any thermometer calibration can be accomplished while the temperature thus defined is held constant.

In selecting the property to use in defining the transition midpoint, the operator may find certain properties to be easier to use than others. In any case, the transition temperature will be found to depend upon the magnetic field impinging upon the sample, as shown by Equation 11. Even the magnitude of the earth's magnetic field is sufficiently large to depress the transition temperature according to the coefficient dT_c/dH; values of this coefficient are given for several materials in Table 13. If the electrical resistance of the sample is observed as a function of temperature, one often finds that the drop to zero resistance occurs at temperatures above those defined by other properties. This result arises because a superconductive path between the voltage leads placed on the sample may develop at temperatures for which most of the sample still exhibits normal properties. Many people agree that the heat capacity is the most reliable guide to the state of a superconductive sample, since it involves the thermal response of the whole sample; unfortunately, the heat capacity can be observed only in consequence of heat flow, a situation that runs counter to the desire for thermal equilibrium throughout the sample/block/thermometer system.

Thermometrists at the NBS and elsewhere have studied for several years the problem of providing temperature reference points based on superconductive transitions. Because of the simplicity of the method, the magnetic susceptibility technique generally has been used to study superconductive fixed points. In this type of measurement, the sample can be mounted so as to protrude from a hole in a copper tempering block, or else it can be enclosed within a bundle of fine copper wires. In either case, a pair of mutual inductance coils surrounding the sample can provide a strong signal indicating the change in state of the sample. The influence of magnetic field on the sample is prevented by enclosing the cryostat with Helmholtz coils or shielding it with carefully demagnetized ferromagnetic foil.

Table 13
TRANSITION TEMPERATURES, CRITICAL MAGNETIC FIELDS, AND MAGNETIC-FIELD SENSITIVITIES OF SEVERAL SUPERCONDUCTORS

Metal	T_c (K)[a]	$H_c(o)$ (mT)	dT_c/dH (mK/μT)	Ref.
Nb	9.288	206.0	0.01	69—72
Pb	7.200	80.3	0.05	69, 72, 73
In	3.4145	28.2	0.06	69, 72, 73
Al	1.181	10.5	0.06	69, 72
Zn	0.850	5.4	0.09	69, 72
Cd	0.520	2.8	0.11	69
AuIn$_2$	0.205—0.208	1.6	0.08	74
AuAl$_2$	0.160—0.161	1.2	0.10	74
Ir	0.098—0.100	1.6	0.08	74
Be	0.021—0.024	0.11	0.14	74
W	0.015—0.017	0.12	0.08	74

[a] T_c values above 0.5 K are expressed on the EPT-76. T_c values below 0.5 K were derived from noise thermometry (see Chapter 5).

The output of a simple mutual inductance bridge[68] is shown against temperature in Figure 20. The data of Figure 20 were obtained on two different samples of Cd by Schooley of the NBS. The two curves, obtained during the same experiment, show several interesting features of fixed points based upon superconductive transitions. In the first place, one can see that not all samples show the same transition width, as one might expect from Equation 10. In fact, the presence of broad transitions and of hysteresis commonly afflicts solid-state properties even apart from superconductivity. To counteract this tendency, the samples shown in Figure 20 were prepared from cadmium stock for which the impurity level is about 1 ppm; the samples were vacuum-cast to the shape of a cylinder of about 1 mm diameter and 3 cm length and then carefully annealed at a temperature near their melting point for as long as 100 hr. Despite this care, occasional samples exhibit widths of several millikelvins.

A second notable feature of the curves of Figure 20 is that the samples display nearly identical T_c values despite the disparity in their transition widths. For most of the materials listed in Table 13, this level of agreement in T_c for samples of different widths was found by Schooley so long as the width did not exceed a characteristic magnitude (however, see also the discussion in References 70 and 71).

Many elements, compounds, and alloys have been tested for suitability as superconductive fixed points. The metals listed in Table 13 all have shown promise for use as calibration standards.

Although superconductive fixed points have been employed for the calibration of thermometers in cryogenics for more than a decade, information still is lacking on the long-term reliability with which this type of fixed point can provide universal reference temperatures. The work of Schooley[69] has included measurements performed upon samples provided by Astrov of the U.S.S.R. and by Borovicka of Czechoslovakia; in several cases, the sample-to-sample disparity is no larger than 0.5 mK. Similar studies have been performed by Fellmuth et al.[70] and by Inaba.[71] For the lower-temperature materials (below 0.5 K), Soulen[74] has found sample-to-sample variations as given in Table 13. Certainly, the relative ease and convenience of calibration that is available in the use of superconductive fixed-point devices ensure their use for second-level laboratory thermometry at a minimum.

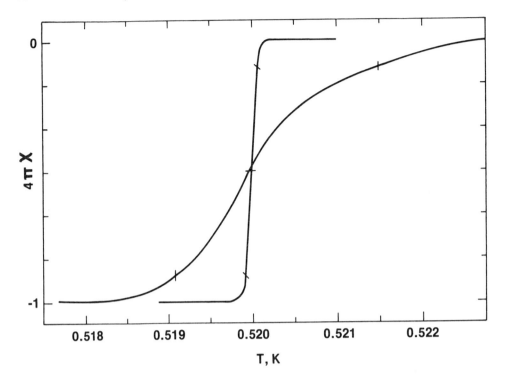

FIGURE 20. Magnetic susceptibility of two cadmium samples, derived from mutual inductance measurements. The midpoint of the magnetic susceptibility transition is defined as the superconductive transition temperature T_c; the temperature range spanned by the central 80% of the transition is defined as the transition width W. Note that, although the sample "W" values differ by a factor ten, the T_c values are identical within 0.1 mK.

GENERAL REFERENCES

Riddle, J. L., Furukawa, G. T., and Plumb, H. H., *Platinum Resistance Thermometry,* U.S. National Bureau of Standards Monogr. 126, April 1973, section 7.

Preston-Thomas, H., Bloembergen, P., and Quinn, T., *Supplementary Information for the IPTS-68 and the EPT-76,* International Bureau of Weights and Measures, Sevres, France, July 1983, section 2.

Furukawa, G. T., Riddle, J. L., Bigge, W. R., and Pfeiffer, E. R., *Application of Some Metal SRM's as Thermometric Fixed Points,* National Bureau of Standards Special Publication 260—77, Aug. 1982.

Zemansky, M. W., *Heat and Thermodynamics,* 5th ed., McGraw-Hill, New York, 1968, chap. 15.

Schooley, J. F., Ed.-in-Chief, *Temperature, Its Measurement and Control in Science and Industry,* Vol. 5, American Institute of Physics, New York, 1982, section 3.

Bedford, R. E., Bonnier, G., Maas, H., and Pavese, F., Recommended values of temperature for a selected set of secondary reference points, *Metrologia,* 20, 145, 1984.

Crovini, L., Bedford, R. E., and Moser, A., Extended list of secondary reference points, *Metrologia,* 13, 197, 1977.

Reisman, A., *Phase Equilibria,* Academic Press, New York, 1970.

Adkins, C. J., *Equilibrium Thermodynamics,* McGraw-Hill, New York, 1968.

REFERENCES

1. See, for example, **Moore, Walter J.,** *Physical Chemistry,* 4th ed., Prentice-Hall, New York, 1972, chap. 5; also see **Glasstone, S.,** *The Elements of Physical Chemistry,* Van Nostrand, New York, 1946, sect. 22.
2. **Gibbs, J. W.,** *Collected Works,* Vol. 1, Longmans, Green, & Co., New York, 1928, 144.
3. See, for example, **Zemansky, M. W.,** *Heat and Thermodynamics,* 5th ed., McGraw-Hill, New York, 1968, chap. 15.

4. **London, F.**, *Superfluids,* Vol. 1, John Wiley & Sons, New York, 1950.
5. **Schrieffer, J. R.**, *Theory of Superconductivity,* W. A. Benjamin, New York, 1964, chap. 1.
6. **Wilks, J.**, *The Properties of Liquid and Solid Helium,* Clarendon Press, Oxford, 1967, chap. 2.
7. Adapted from **Whalley, E.**, Structure problems of ice, in *Physics of Ice,* Proc. Int. Symp. Physics of Ice, Munich, Germany, Sept. 9—14, 1968, Riehl, N., Bullemer, B., and Engelhardt, H., Eds., Plenum Press, New York, 1969.
8. **Strong, H. M.**, *Am. Sci.,* 48, 58, 1960.
9. See, for example, **Zemansky, M. W.**, *Heat and Thermodynamics,* 5th ed., McGraw-Hill, New York, 1968, chap. 18.
10. **Dorsey, R.**, Properties of the ordinary water-substance, *American Chemical Society Monograph 81,* Reinhold, New York, 1940, table 2653. Compare **Zemansky, M. W.**, *Heat and Thermodynamics,* 5th ed., McGraw-Hill, New York, 1968, Table 2.2, p. 32.
11. See, for example, **Riddle, J. L., Furukawa, G. T., and Plumb, H. H.**, *Platinum Resistance Thermometry,* National Bureau of Standards Monogr. 126, Section 7.1, April 1973.
12. Taken from Reference 11. Compare with **Stimson, H. F.**, Precision resistance thermometry and fixed points, in *Temperature, Its Measurement and Control in Science and Industry,* Vol. 2, Wolfe, H. D., Ed., Reinhold, New York, 1955, Fig. 2, p. 144.
13. **Mangum, B. W. and Thornton, D. D.**, Determination of the triple-point temperature of gallium, *Metrologia,* 15, 201, 1979.
13a. **Furukawa, G. T., Riddle, J. L., Bigge, W. R., and Pfeiffer, E. R.**, Application of Some Metal SRM's as Thermometric Fixed Points, National Bureau of Standards Special Publ. 260—77, Section VI, August 1982.
14. **McAllan, J. V.**, The effect of pressure on the water triple-point temperature, in *Temperature, Its Measurement and Control in Science and Industry,* Vol. 5, Schooley, J. F., Ed.-in-Chief, American Institute of Physics, New York, 1982, 285.
15. **Bridgman, P. W.**, *Proc. Am. Acad.,* 47, 441, 1912; and *J. Chem. Phys.,* 3, 597, 1935, and 5, 964, 1937; see also **Ricci, J. E.**, *The Phase Rule and Heterogeneous Equilibrium,* Dover Publications, New York, 1966, 43.
16. The International Practical Temperature Scale of 1968 Amended Edition of 1975, sect. III. 4, *Metrologia,* 12, 7, 1976.
17. **Ancsin, J.**, Melting curves of H_2O, in *Temperature, Its Measurement and Control in Science and Industry,* Vol. 5, Schooley, J. F., Ed.-in-Chief, American Institute of Physics, New York, 1982, 281.
18. See, for example, **Moore, W. J.**, *Physical Chemistry,* Prentice-Hall, New York, 1950, 118; also see **Prutton, C. F. and Maron, S. H.**, *Fundamental Principles of Physical Chemistry,* Macmillan, New York, 1951, 153.
19. **Prutton, C. F. and Maron, S. H.**, *Fundamental Principles of Physical Chemistry,* Macmillan, New York, 1951, chap. 6.
20. **Smit, W. M.**, *Rec. Trav. Chim. Pay-Bas,* 75, 1309, 1956.
21. **Preston-Thomas, H.**, The International Practical Temperature Scale of 1968 Amended Edition of 1975, *Metrologia,* 12, 7, 1976.
22. **Preston-Thomas, H., Bloembergen, P., and Quinn, T.**, *Supplementary Information for the IPTS-68 and the EPT-76,* International Bureau of Weights and Measures, Sevres, France, July 1983, sect. 2.2.2.
23. **Berry, R. J.**, The reproducibility of the steam point, *Can. J. Phys.,* 36, 740, 1958.
24. **Compton, J. P.**, The realization of low temperature fixed points, in *Temperature, Its Measurement and Control in Science and Industry,* Vol. 4, Plumb, H. H., Ed.-in-Chief, Instrument Society of America, Pittsburgh, 1972, 195.
25. **Pavese, F. and Ferri, D.**, Ten years of research on sealed cells for phase transition studies of gases at IMGC, in *Temperature, Its Measurement and Control in Science and Industry,* Vol. 5, Schooley, J. F., Ed.-in-Chief, American Institute of Physics, New York, 1982, 217; see also **Pavese, F.**, On the use of first-generation sealed cells in an intercomparison of triple-point temperatures of gases, in *Temperature, Its Measurement and Control in Science and Industry,* Vol. 5, Schooley, J. F., Ed.-in-Chief, American Institute of Physics, New York, 1982, 209. A complete summary of this work has been published by the International Bureau of Weights and Measures in Paris. See also **Pavese, F. et al.**, An international intercomparison of fixed points by means of sealed cells in the range 13.81 K—90.686 K, *Metrologia,* 20, 127, 1984.
26. **Bonnier, G. and Hermier, Y.**, Thermal behavior of thermometric sealed cells and of a multi-compartment cell, in *Temperature, Its Measurement and Control in Science and Industry,* Vol. 5, Schooley, J. F., Ed.-in-Chief, American Institute of Physics, New York, 1982, 231.
27. **Furukawa, G. T.**, Reproducibility of the triple point of argon in sealed transportable cells, in *Temperature, Its Measurement and Control in Science and Industry,* Vol. 5, Schooley, J. F., Ed.-in-Chief, American Institute of Physics, New York, 1982, 239.

28. **Ancsin, J.**, The triple point of oxygen and its change by noble gas impurities, *Metrologia*, 6, 53, 1970. Note that these measurements were not performed on a sealed cell; however, the conclusions still apply.
29. **Kemp, R. C.**, The triple points of equilibrium and normal deuterium, in *Temperature, Its Measurement and Control in Science and Industry*, Vol. 5, Schooley, J. F., Ed.-in-Chief, American Institute of Physics, New York, 1982, 249.
30. **Kemp, R. C., Kemp, W. R. G., and Smart, P. W.**, The triple point of natural xenon, in *Temperature, Its Measurement and Control in Science and Industry*, Vol. 5, Schooley, J. F., Ed.-in-Chief, American Institute of Physics, New York, 1982, 229.
31. **Hoge, H. J. and Brickwedde, F. G.**, Establishment of a temperature scale for the calibration of thermometers between 14°K and 83°K, *J. Res. Natl. Bur. Stand.*, 22, 351, 1939. For a discussion of the quantum states of ortho and para hydrogen, see **Kennard, E. G.**, *Kinetic Theory of Gases*, McGraw-Hill, New York, 1938, sect. 151. For discussions of the properties of ortho and para hydrogen, see **Farkas, A.**, *Orthohydrogen, Parahydrogen, and Heavy Hydrogen*, Cambridge University Press, 1935; and **McCarty, R. D.**, *Hydrogen: Its Technology and Implications*, CRC Press, Boca Raton, Fla., 1975.
32. See, for example, **Carslaw, H. S. and Jaeger, J. C.**, *Conduction of Heat in Solids*, 2nd ed., Oxford University Press, London, 1959.
33. **Kittel, C.**, *Introduction to Solid State Physics*, 4th ed., John Wiley & Sons, New York, 1971, chap. 5.
34. **McLaren, E. H. and Murdock, E. G.**, The freezing points of high purity metals as precision temperature standards. V. Thermal analysis on 10 samples of tin with purities greater than 99.99%, *Can. J. Phys.*, 38, 100, 1960.
35. **McLaren, E. H.**, The freezing points of high purity metals as precision temperature standards, in *Temperature, Its Measurement and Control in Science and Industry*, Vol. 3, Herzfeld, C. M., Ed.-in-Chief, Reinhold, New York, 1962, part 1, 185.
36. **Furukawa, G. T., Riddle, J. L., and Bigge, W. R.**, Investigation of freezing temperatures of National Bureau of Standards tin standards, in *Temperature, Its Measurement and Control in Science and Industry*, Vol. 4, Plumb, H. H., Ed.-in-Chief, Instrument Society of America, Pittsburgh, 1972, 247.
37. **McAllan, J. V. and Ammar, M. M.**, Comparison of the freezing points of aluminum and antimony, in *Temperature, Its Measurement and Control in Science and Industry*, Vol. 4, Plumb, H. H., Ed.-in-Chief, Instrument Society of America, Pittsburgh, 1972, 275.
38. **Furukawa, G. T.**, Investigation of freezing temperatures of National Bureau of Standards aluminum standards, *J. Res. Natl. Bur. Stand.*, 78A, 477, 1974; see also Reference 13a.
39. **McLachlan, A. D., Uchiyama, H., Saino, T., and Nakaya, S.**, The stability of the freezing point of copper as a temperature standard, in *Temperature, Its Measurement and Control in Science and Industry*, Vol. 4, Plumb, H. H., Ed.-in-Chief, Instrument Society of America, Pittsburgh, 1972, 287.
40. **Evans, J. P. and Wood, S. D.**, An intercomparison of high temperature platinum resistance thermometers and standard thermocouples, *Metrologia*, 7, 108, 1971.
41. **Jung, H.-J.**, Determination of the difference between the thermodynamic fixed-point temperatures of gold and silver by radiation thermometry, in *Temperature Measurement, 1975*, Billing, B. F. and Quinn, T. J., Eds., Conference Series Number 26, The Institute of Physics, London, 1975, 278.
42. **Bedford, R. E. and Ma, C. K.**, Measurement of the melting temperature of the copper 71.9% silver eutectic alloy with a monochromatic optical pyrometer, in *Temperature, Its Measurement and Control in Science and Industry*, Vol. 5, Schooley, J. F., Ed.-in-Chief, American Institute of Physics, New York, 1982, 361.
43. **McAllan, J. V.**, Reference temperatures near 800°C, in *Temperature, Its Measurement and Control in Science and Industry*, Vol. 5, Schooley, J. F., Ed.-in-Chief, American Institute of Physics, New York, 1982, 371.
44. **McLaren, E. H.**, *Can. J. Phys.*, 37, 422, 1959.
45. **McAllan, J. V. and Connolly, J. J.**, The use of the cadmium point to check calibrations on the IPTS, in *Temperature, Its Measurement and Control in Science and Industry*, Vol. 5, Schooley, J. F., Ed.-in-Chief, American Institute of Physics, New York, 1982, 351.
46. **Furukawa, G. T. and Pfeiffer, E. R.**, Investigation of the freezing temperature of cadmium, in *Temperature, Its Measurement and Control in Science and Industry*, Vol. 5, Schooley, J. F., Ed.-in-Chief, American Institute of Physics, New York, 1982, 355.
47. **Sawada, S.**, Realization of the triple point of indium in a sealed glass cell, in *Temperature, Its Measurement and Control in Science and Industry*, Vol. 5, Schooley, J. F., Ed.-in-Chief, American Institute of Physics, New York, 1982, 343.
48. **Chattle, M. V., Rusby, R. L., Bonnier, G., Moser, A., Renaot, E., Marcarino, P., Bongiovanni, G., and Frassineti, G.**, An intercomparison of gallium fixed point cells, in *Temperature, Its Measurement and Control in Science and Industry*, Vol. 5, Schooley, J. F., Ed.-in-Chief, American Institute of Physics, New York, 1982, 311.

49. **Mangum, B. W.**, Triple point of gallium as a temperature fixed point, in *Temperature, Its Measurement and Control in Science and Industry*, Vol. 5, Schooley, J. F., Ed.-in-Chief, American Institute of Physics, New York, 1982, 299.
50. **Berger, R. L., Clem, T. R., Harden, V. A., and Mangum, B. W.**, Historical development and newer means of temperature measurement in biochemistry, chapter in *Methods of Biochemical Analysis*, Vol. 30, Glick, D., Ed., John Wiley & Sons, New York, 1984, 269.
51. **Cox, J. D. and Vaughan, M. F.**, Temperature fixed points: Evaluation of four types of triple-point cell, in *Temperature, Its Measurement and Control in Science and Industry*, Vol. 5, Schooley, J. F., Ed.-in-Chief, American Institute of Physics, New York, 1982, 267.
52. **Mangum, B. W. and Thornton, D. D., Eds.**, The Gallium Melting Point Standard, National Bureau of Standards Spec. Publ. 481, June 1977.
53. **Magin, R. L., Mangum, B. W., Statler, J. A., and Thornton, D. D.**, Triple points of the hydrates of Na_2SO_4, Na_2HPO_4, and KF as fixed points in biomedical thermometry, *J. Res. Natl. Bur. Stand.*, 86, 181, 1981.
54. **Mangum, B. W.**, SRM 1969: Rubidium triple-point standard — a temperature reference standard near 39.30°C, National Bureau of Standards Spec. Publ. 260—87, December 1983.
55. **Mangum, B. W.**, The succinonitrile triple-point standard: a fixed point to improve the accuracy of temperature measurements in the clinical laboratory, *Clin. Chem.*, 29, 1380, 1983.
56. **Figueroa, J. M. and Mangum, B. W.**, The triple point of rubidium: a triple point for biomedical applications, in *Temperature, Its Measurement and Control in Science and Industry*, Vol. 5, Schooley, J. F., Ed.-in-Chief, American Institute of Physics, New York, 1982, 327.
57. **Orlova, M. P.**, Temperatures of phase transitions in solid oxygen, in *Temperature Its Measurement and Control in Science and Industry*, Vol. 3, Part 1, Herzfeld, C. M., , Ed.-in-Chief, Reinhold, New York, 1962, 179.
58. **Kemp, W. R. G. and Pickup, C. P.**, The transition temperatures of solid oxygen, in *Temperature, Its Measurement and Control in Science and Industry*, Vol. 4, Plumb, H. H., Ed.-in-Chief, Instrument Society of America, Pittsburgh, 1972, 217.
59. **Ancsin, J.**, Crystalline transformations of oxygen, in *Temperature Measurement, 1975*, Billing, B. F. and Quinn, T. J., Eds., The Institute of Physics, London, 57.
60. **Roberts, B. W.**, Survey of superconductive materials and critical evaluation of selected properties, *J. Phys. Chem. Ref. Data*, Vol 5(3), 581, 1976; see also **Roberts, B. W.**, *Properties of Selected Superconductive Materials 1978 Supplement*, U.S. National Bureau of Standards Tech. Note 983, October 1978.
61. **Schooley, J. F., Soulen, R. J., Jr., and Evans, G. A., Jr.**, *Preparation and Use of Superconductive Fixed Point Devices SRM 767*, U.S. National Bureau of Standards Spec. Publ. 260—44, December 1972.
62. The 1976 Provisional 0.5 K to 30 K Temperature Scale, *Metrologia*, 15, 65, 1979, table 1.
63. **Onnes, H. K.**, *Commun. Phys. Lab. Univ. Leiden*, Suppl. no. 34, 1913.
64. **Meissner, W. and Ochsenfeld, R.**, *Naturwissenschaften*, 21, 787, 1933.
65. **Bardeen, J., Cooper, L. N., and Schrieffer, J. R.**, *Phys. Rev.*, 108, 1175, 1957.
66. See, for example, **Maple, M. B.**, Superconductivity, a probe of the magnetic state of local moments in metals, *Appl. Phys.*, 9, 179, 1976.
67. **Soulen, R. J., Jr. and Colwell, J. H.**, The equivalence of the superconducting transition temperature of pure indium as determined by electrical resistance, magnetic susceptibility, and heat-capacity measurements, *J. Low Temp. Phys.*, 5, 325, 1971.
68. **Soulen, R. J., Jr., Schooley, J. F., and Evans, G. A., Jr.**, Simple instrumentation for the inductive detection of superconductivity, *Rev. Sci. Inst.*, 44, 1537, 1973.
69. **Schooley, J. F.**, BIPM Document (CCT/84-2), "NBS SRM 767a Superconductive Fixed Point Devices"; see also **Schooley, J. F. and Colwell, J. H.**, Superconductive temperature reference points above 0.5 K, Paper BP8, 17th Conference on Low Temperature Physics, Karlsruhe, August 1984.
70. **Fellmuth, B., Maas, H., Elefant, D.**, BIPM Document (CCT/84-23), Progress in Research Concerning the Superconducting Transition Temperature of Niobium as a Temperature Reference; See also **Fellmuth, B., Maas, H., and Elefant, D.**, Investigations of the superconducting transition points of Nb, Pb, and In as reference temperatures, *Metrologia*, to be published.
71. **Inaba, A.**, *Jap. J. Appl. Phys.*, 19, 1553, 1980.
72. **Pogorelova, O. F., Kytin, G. A., Astrov, D. N.**, BIPM Document (CCT/84-35), On the Superconductive Fixed Point Temperatures.
73. **Fellmuth, B.**, BIPM Document (CCT/84-24), Investigations Concerning the Superconductive Transition Temperatures of Pb and In.
74. **Schooley, J. F. and Soulen, R. J., Jr.**, Superconductive thermometric fixed points, in *Temperature, Its Measurement and Control in Science and Industry*, Vol. 5, Schooley, J. F., Ed.-in-Chief, American Institute of Physics, New York, 1982, 251.

Chapter 4

INTERNATIONAL SCALES OF TEMPERATURE

I. INTRODUCTION

In this chapter, we portray certain properties of temperature scales, including not only thermodynamic scales, but also those chosen to serve more limited purposes. Then we discuss the origins and features of the temperature scales that have been given international sanction.

II. ELEMENTS OF TEMPERATURE SCALES AND THERMOMETRY

We began Chapter 2 with four simple definitions — those of "temperature," "fixed point," "temperature scale," and "thermometer." In succeeding sections we probed more deeply into the meaning of "temperature." We have seen that temperature is an intensive property of all matter, that there is a natural zero of temperature, and that temperature enters thermodynamic and statistical-mechanical discussions in a basic way. In Chapter 3, the natural role of "fixed points" in helping to anchor measurements in temperature appeared. In this chapter, we discuss some of the elements that make a "temperature scale" and "thermometers" useful.

A. Characteristics of a Temperature Scale

Clearly, the most meaningful form of temperature scale for scientific use would be one that matches well the temperature used in thermodynamic and statistical-mechanical calculations. Both thermodynamic and statistical-mechanical models exist for such an "absolute" scale.

In fact, there are many possible choices for thermodynamic temperature scales. To emphasize this point, we present two such scales:

The "iron scale" — This hypothetical thermodynamic scale is defined by the existence of an absolute zero of temperature and by the assignment of 2000 A to the temperature provided by the liquid-solid phase equilibrium of iron at a pressure of 101,325 Pa. The unit of temperature is called the "amonton," abbreviated "A"; it is equal to 1/2000 of the hotness of the "iron point." All values of temperature except the defining point are to be determined from Carnot-cycle efficiencies or from other thermodynamic or statistical-mechanical relations.

The "decigrade scale" — This hypothetical thermodynamic scale is defined by the existence of exactly 10 A difference in temperature between the triple-point temperature of mercury and its normal boiling-point temperature. The unit of temperature is called the "amonton," abbreviated "A"; it is equal to 1/10 the difference in hotness between the two defining fixed-point temperatures. All values of temperature except that of the normal boiling point of mercury are to be determined from Carnot-cycle efficiencies or from other thermodynamic or statistical-mechanical relations.

It should be obvious to the readers of Chapters 2 and 3 that these hypothetical temperature scales have been defined so as to mimic respectively the Kelvin and the centigrade thermodynamic scales. So much the better to illustrate their essential features:

- Each scale depends upon thermodynamic thermometry for its evaluation at all temperatures except one.

- The fact that a temperature interval, rather than a particular fixed-point temperature, lies at the heart of the "decigrade" scale implies that thermometrists will be forever searching for the "correct" values of its two pivotal reference temperatures. Any effort to stabilize the scale by assigning a fixed value to the mercury triple-point temperature must be avoided; such a step would overspecify the scale.

Of the many possible choices, the thermodynamic temperature scale proposed by Kelvin (the KTTS, first discussed in Chapter 2) has been chosen as the basis for present-day internationally recognized temperature scales. Beginning with the awareness of an absolute zero of temperature and designating as 273.16 K the equilibrium temperature that pure water reaches in the presence of ice and its own vapor, the scientific community has adopted the principle of defining temperatures on the basis of a satisfactory understanding of the science of thermodynamics. At the present time, however, the most reproducible methods for obtaining values of temperature experimentally are not the thermodynamic ones.

In order to compare in the most precise way those temperatures measured by different individuals in different laboratories, thermometrists have found it necessary to prescribe the use of specially prepared, calibrated thermometers. "Laboratory temperature scales" have been constructed by assigning temperature values to selected fixed points and prescribing interpolation techniques for use with the standard thermometers. The relations between these laboratory scales and the Kelvin thermodynamic scale have not always been clear.

Let us examine for a moment the characteristics of temperature scales without regard to their origin. Later on, we can contrast the features of the Kelvin thermodynamic scale with one that is chosen in a more arbitrary way.

Certain characteristic elements in the definition of any temperature scale are particularly important. Unfortunately, the names given to these elements are not used in the same way by all thermometrists. The major elements of a temperature scale can be described as follows:

- The extent of the scale in temperature (its *range*)
- The provision in the scale of procedures to obtain a temperature value for a particular level of hotness (its method of *realization*)
- The range of temperature values that use of the scale permits at a particular level of hotness (its *non-uniqueness*)
- The faithfulness of the scale temperature values to the basic Kelvin thermodynamic scale (its *uncertainty with respect to the KTTS*)

In addition to these properties of the scale itself, there are uncertainties that necessarily are encountered in the measurement process. These include the following elements:

- Differences in hotness of fixed points (*uncertainty of fixed-point realization*)
- Differences among repeated measurements with a particular thermometer (*experimental imprecision*)[1]

Perhaps the most elementary quality of a temperature scale is addressed by the question, "What constitutes a temperature scale?" The idea has been suggested[2] that actual thermodynamic temperatures (see Chapter 5) cannot be considered to provide a temperature *scale* in the same way that is possible with, say, the electrical resistance of a platinum wire.* The argument, in brief, is that any scale of measurement should provide a continuous standard for reference over its whole range. This is a goal that thermometrists never have achieved

---

* This suggestion runs counter to many discussions, including the language of the International Practical Temperature Scale of 1948 (to be discussed later in this chapter), but it is nonetheless a valid point.

in thermodynamic temperature measurements; instead, they have evaluated the thermodynamic values of selected reference temperatures. In contrast, a practical temperature scale prescribes methods for the measurement of temperature as a continuous quantity. Recourse to a dictionary provides the information that the definition of the word "scale" involves the notion of a series of measuring marks (presumably drawn from a reference continuum). One can argue also that the "thermodynamic thermometrist" could, in principle, evaluate the KTTS temperature of any arbitrary level of hotness (within the range of the experiment, of course), thus fulfilling the spirit of the strictest definition.

Without further discussion at this moment about the difficulty of determining thermodynamic temperatures accurately and completely, we herein use the term *Kelvin thermodynamic temperature scale* (abbreviated KTTS) to refer to a unique set of temperature values. They are the ones that, in principle, can be assigned, one by one, to all levels of hotness by reference to the laws of thermodynamics and by reference to one fixed point above the absolute zero of temperature, that one point being the triple point of water defined as providing the temperature 273.16 K.

We construe a *practical temperature scale* to be composed of a set of fixed points of temperature, each with a definite assigned value, and one or more definite methods by means of which a temperature evaluation may be made for any level of hotness encompassed by the scale.

The *range* of a particular practical temperature scale generally depends upon the working limits of the apparatus used to realize it rather than upon the limits of applicability of the physical law that underlies the measurements. In Chapter 5 we discuss several experimental thermodynamic temperature measurements; each of these is limited to a certain range by some feature of the experimental apparatus or of the procedures used in measurement. The *range* of the KTTS, of course, is 0 to $+\infty$. On the other hand, temperature scales that are constructed artificially — one might call them "consensus scales" or "laboratory scales" rather than "practical scales" — are designed specifically to cover a definite *range*.

Thermodynamic temperatures can be realized in a variety of ways. For example, the range around room temperature can be realized in terms of the ideal gas law, the Planck law, the Nyquist theory of Johnson noise, and the Stefan-Boltzmann law. Thus we can say that, in general, thermodynamic temperature scales are *multiply realizable*. On the other hand, consensus or practical scales can be constructed so that only one procedure is specified for evaluating any particular level of hotness, thus providing any number of *singly realizable* scales; or they can allow two or more methods to be used in some or all of their ranges, providing *multiply realizable* scales.

Kelvin wanted to create a thermodynamic scale of temperature that would be independent of material properties. The principles of thermodynamics and statistical mechanics are rooted in the idea that the temperature parameter used in all physical laws should be one and the same. Therefore we think of the Kelvin or any other specific thermodynamic temperature scale as a perfectly *unique* entity — all definitions must lead to equal values of temperature at equal levels of hotness. This property of perfect overall uniqueness cannot apply, however, to consensus scales. Although they may be considered to be perfectly unique at their defining fixed-point temperatures, they must be characterized elsewhere by levels of *nonuniqueness* that vary with temperature. The reason for this theoretical difference between the Kelvin thermodynamic scale and consensus scales is that the latter offer to the thermometrist "prescriptions" for obtaining temperature values at particular levels of hotness. In following a particular prescription, an operator may use any approved thermometer; the group of approved thermometers, identical with respect to satisfying the scale requirements, in fact possess small differences that inevitably will result in detectable differences in the numbers obtained using the same prescription to evaluate the same hotness. The nonuniqueness of a consensus scale describes the levels of disagreement that appear in determinations of temperatures between the fixed points of the scale, even allowing for perfectly accomplished calibrations of the standard thermometers and for perfectly accomplished measurements at the test temperature.

Despite its theoretical uniqueness, Kelvin's thermodynamic temperature scale never has been realized perfectly. We discuss in Chapter 5 the problems that accompany the determination of thermodynamic temperatures — suffice it here to say that typical levels of *uncertainty with respect to the KTTS* that accompany Kelvin temperature determinations are far larger than typical levels of *nonuniqueness* that accompany the evaluation of temperatures on the best practical scales. For many measurements, the uncertainty with respect to the KTTS makes little difference; the potato farmer suffers no loss if his community's current air temperature is reported as 296 K (''73°F'') instead of 297 K (''75°F''). The same latitude does not exist for a great many technical and scientific temperature measurements, however. In assessing the worth of careful temperature measurements performed according to the dictates of a particular scale, one frequently must estimate the level of disagreement between one's practical-scale temperature measurements and the KTTS. Such an estimate can be credible only if a conscientious effort has been made to judge the uncertainty with respect to the KTTS of the fundamental measurements on which the practical scale was based and to evaluate the level at which the scale allows one to approach Kelvin thermodynamic temperatures in an actual measurement.

We note above that consensus or practical scales can be said to be unique at the temperatures defined by the scale's fixed points, since these are assigned values and thus are not subject to variation. The experimental fact that determinations of fixed-point temperatures in general are found not to agree with one another if sufficiently careful measurements are performed can be described as *uncertainty of realization of fixed-point temperatures*. By this phrase, we imply that there is in fact only one temperature defined by a given phase equilibrium, but that the limitations of chemical purity, measurement uncertainty, and laboratory technique cause one to realize that temperature imperfectly.

Underlying all of our efforts to make use of thermometers in accordance with the demands of a physical law or the prescriptions of a consensus temperature scale is the problem of *experimental imprecision*.[3] In the previous paragraph we alluded to the devils that harass measurements of all types:

- Instrument resolution — The level at which an instrument can no longer allow one to discriminate between neighboring values of the measured parameter
- Measurement imprecision — The variability of a particular repeated measurement. For most measurements, the results are distributed in approximately symmetrical fashion about a central value; furthermore, small deviations from the central value are found more frequently than large deviations. In this case, a ''Gaussian,'' or ''normal,'' measurement frequency curve would be generated by the process of repeating the measurement a great many times. One can estimate the imprecision of a single measurement by computing the so-called ''standard deviation about the mean'' of one of a relatively small number of measurements. The equation for this standard deviation (SD or σ) is

$$SD = \left[\sum_{i=1}^{N} \frac{(x_i - \bar{x})^2}{N - 1} \right]^{1/2} \tag{1}$$

where $\bar{x}$ is the average of N determinations, $(x_i - \bar{x})$ are the differences between the individual determinations and the average, and the sum extends over all N measurements. Use of Equation 1 presumes that the measurement results would exhibit a

Table 1
HYPOTHETICAL RESISTANCE THERMOMETER CALIBRATION DATA

Measurement no.	Ag (1235 K)	Au (1338 K)	Cu (1358 K)
1	10.698 086	11.408 315	11.544 205
2	10.698 243	11.408 685	11.543 850
3	10.698 185	11.408 499	11.543 937
4	10.698 113	11.408 278	11.544 341
5	10.698 005	11.408 654	11.544 218
6	10.698 315	11.408 561	11.544 113
7	10.698 116	11.408 748	11.544 024
8	10.698 272	11.408 367	11.544 400
9	10.698 208	11.408 540	11.544 175
10	10.698 057	11.408 483	11.543 987
$\bar{R}$ (Ω)	10.698 160	11.408 513	11.544 125
SD ($\mu\Omega$)	100.4	158	176

normal distribution if repeated a great many times. The significance of the calculated standard deviation is that there is a 68% probability that one more measurement will produce a result that agrees within one standard deviation with the average of the N results accumulated so far. If one increases the "imprecision band" to [$\bar{x} \pm 3$ (SD)], then the probability that a subsequent measurement will yield a result within the calculated limits rises to more than 99%.

In an analogous way, the quantity (SD)/$\sqrt{N}$, called the "standard deviation of the mean," provides a measure of the imprecision of the mean of a set of measurement results.

Assuming that a measurement process will yield a normal distribution of results, one can describe its level of imprecision in a manner that is readily understandable by others simply by stating the calculated standard deviation of a particular set of N results or the standard deviation of the mean of that set.

B. Temperature-Scale Thermometry

We now illustrate the meanings of many of the terms listed above by an example which, though hypothetical, is not unrealistic.

Let us suppose that the Kelvin thermodynamic temperature of the melting point of gold is to be determined. Let us suppose, too, that this is a team effort involving four scientists whom we shall designate as A, B, C, and D.

Scientists A and B must prepare gold cells in which the melting point can be observed; then they must determine the suitability of the cells for use in thermodynamic determinations of the melting temperature by scientists C and D.

Scientist A prepares a gold cell using well-understood materials and construction techniques and fills the cell with gold that has been reclaimed from an earlier experiment. In order to evaluate the quality of the cell, A needs a thermometer. He selects a resistance thermometer (which we discuss at length in Chapter 6; for the moment we shall describe it only as a thermometer which provides a continuous resistance-temperature relation). He measures the resistance of his thermometer ten times in a silver melting-point cell, ten times in his gold melting-point cell, and ten times in a copper melting-point cell. The approximate thermodynamic temperatures of the melting points and A's hypothetical calibration data are listed in Table 1.

82 *Thermometry*

Using a simple quadratic relation

$$R(T) = a + bT + cT^2 \qquad (2)$$

and the data of Table 1, A obtains the coefficients $a = 0.622\,053\ \Omega$ (ohms), $b = 0.932\,38 \times 10^{-2}\ \Omega/K$, and $c = -0.943\,326 \times 10^{-6}\ \Omega/(K)^2$. By differentiating the quadratic equation, he obtains the equation

$$(dR/dT) = b + 2cT \qquad (3)$$

At $T = 1338$ K, $(dR/dT) = 6.80 \times 10^{-3}\ \Omega/K$; this is the approximate sensitivity of his thermometer at the temperature of melting gold.

Now A is ready to evaluate his gold melting-point cell. Heating the cell to a temperature slightly lower than the gold melting point in a furnace, he waits until his thermometer shows a resistance of 11.388 Ω, some three kelvins below the gold melting point. Then, increasing the furnace power very slowly, he records the thermometer resistance as a function of time. His data are shown in Figure 1 as the upper (b) curve.

In the meantime, scientist B also has prepared a gold cell, using materials and techniques similar to those used by A; his gold sample, however, is the purest available. Using a thermometer similar to A's for which the values of $\bar{R}$ and (dR/dT) at the gold melting point are respectively 11.506 625 Ω and $7.03 \times 10^{-3}\ \Omega/K$, B prepares a melting curve for his gold cell. His data are shown in Figure 1 as the lower (a) curve.

The four scientists discuss the data of Figure 1 and agree that B's cell probably is the better one to use for the thermodynamic temperature determination, since the noticeable change in temperature of A's cell during its melting period is characteristic of an impure sample. As a final check on the relative quality of the two cells, A and B observe both cells sequentially in the same furnace, using first one thermometer and then the other to monitor the melting process. After forty "melts", the thermometer resistances measured 3 hr after the melting began comprise the data shown in Table 2.

Now it is time for scientists C and D to begin *their* work of determining the thermodynamic temperature of melting gold. They decide to work in sequence, using the gold cell prepared by B. Scientist C chooses to perform an experiment using the ideal gas law. He will compare the pressure exerted by a constant volume of gas when it is maintained at the triple point of water to the pressure when the same volume is exposed to the temperature of melting gold. The gold melting temperature is to be approximated in his experiment by the inclusion of thermometer B, using the resistance found during the experiments on cell B. For our purpose here, we need not discuss the difficulties surmounted by C in this work; some of these will be revealed in Chapter 5. Let us assume for now that C performs the experiment as well as the state of the art permits and that his ten pressure ratios, adjusted for deviations from the ideal gas law and for other experimental imperfections, are listed in Table 3. Here, T_0 is 273.16 K, the thermodynamic temperature of the triple point of water.

Finally, scientist D performs his experiment, using the Stefan-Boltzmann law to derive the gold melting temperature from the observed ratios of the power radiated from a black body at that temperature to the power radiated from a black body maintained at the triple point of water (cf. Chapter 5). His data, hypothetical but plausible like all the data of this example, are shown in Table 4.

Some of the details of the experiments mentioned here will be presented in Chapter 5. The hypothetical results shown in Tables 1 to 4 complete the basic data for our illustrative example on the properties of temperature scales.

Perhaps the most important feature demonstrated by our example is that no scale of temperature derived from thermodynamic experiments is singly realizable. We have men-

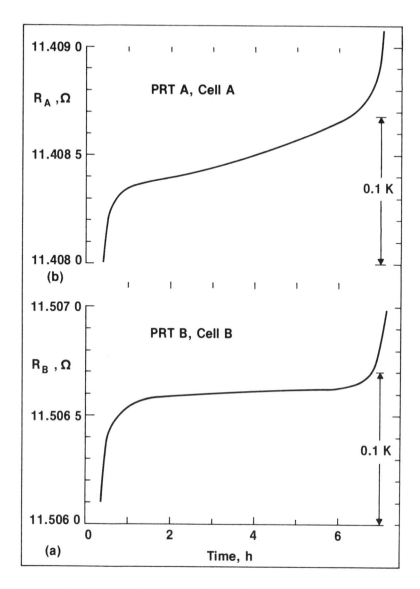

FIGURE 1. Hypothetical melting curves for two gold samples of unequal impurity levels, as observed using two different thermometers.

tioned two thermodynamic methods for determining the melting temperature of gold on the KTTS. We could have mentioned others that stand on an equal footing; it is not possible to name a level of hotness that can be measured by only one thermodynamic experiment.

Furthermore, the value of temperature derived from a single thermodynamic determination on the KTTS is not perfectly certain, even though it is unique in theory. Repeated determinations of the same temperature by the same thermodynamic procedure will yield values that deviate from the first one — still less likely is agreement between the results of two different types of thermodynamic temperature determinations.

These facts of multiple realization and relatively large uncertainty mean that, in general, thermodynamic temperatures are ambiguous; we cannot know *a priori* which measurement is correct. Let us compare the results of Tables 3 and 4. The two values of $\overline{T}$(Au) are the averages of two sets of experimental determinations using different thermodynamic laws. The standard deviation is a mathematical method for approximating the distribution of the

Table 2
HYPOTHETICAL RESISTANCE THERMOMETER
DATA FOR TWO GOLD FIXED-POINT CELLS[a]

Measurement no.	A,A	A,B	B,A	B,B
1	11.408 456	11.408 845	11.506 087	11.506 683
2	11.408 397	11.408 946	11.506 355	11.506 546
3	11.408 584	11.408 953	11.506 241	11.506 624
4	11.408 655	11.408 874	11.506 274	11.506 598
5	11.408 413	11.408 942	11.506 298	11.506 576
6	11.408 521	11.408 844	11.506 136	11.506 615
7	11.408 627	11.408 846	11.506 294	11.506 642
8	11.408 564	11.408 878	11.506 103	11.506 573
9	11.408 385	11.408 908	11.506 327	11.506 611
10	11.408 598	11.408 924	11.506 085	11.506 609
$\bar{R}$ (Ω)	11.408 520	11.408 896	11.506 220	11.506 608
SD ($\mu\Omega$)	100.4	44.0	106.1	38.4

[a] The thermometer is identified first, then the cell.

Table 3
HYPOTHETICAL GOLD-POINT GAS
THERMOMETER RESULTS

Determination	$P(Au)/P(T_0)$	$T(Au) = T_0 P(Au)/P(T_0)$
1	4.895 885	1337.36
2	4.896 25	1337.46
3	4.896 07	1337.41
4	4.896 36	1337.49
5	4.896 18	1337.44
6	4.895 85	1337.35
7	4.896 14	1337.43
8	4.895 96	1337.38
9	4.896 32	1337.48
10	4.896 03	1337.40
$\bar{T}$ (K)		1337.42
SD (K)		.05

Note: P(T); Corrected value of the gas-bulb pressure at temperature T.

results of a given type of measurement. Note that the gas thermometry SD is smaller than the Stefan-Boltzmann SD; from this comparison, we can say that the "precision" of the gas thermometry measurements is better than the precision of the radiation thermometry measurements. (Note well, though, that this statement may not hold true for measurements at other temperatures or for measurements made by other scientists, but only for the two hypothetical sets of data under consideration here.)

Note also that the difference between the two values is larger than the sum of their standard deviations. This inequity is to be expected. The uncertainty of either gold-point-temperature determination is composed only partly of its experimental imprecision — the quantity described by the standard deviation of a series of measurements. Besides the experimental

Table 4
HYPOTHETICAL GOLD-POINT TOTAL-RADIATION THERMOMETER RESULTS

Determination	$P(Au)/P(T_0)$	$T(Au) = T_0[P(Au)/P(T_0)]^{1/4}$
1	574.992	1337.62
2	575.353	1337.83
3	575.199	1337.74
4	574.890	1337.56
5	575.147	1337.71
6	575.233	1337.76
7	574.925	1337.58
8	575.405	1337.86
9	575.045	1337.65
10	575.303	1337.80
$\overline{T}$ (K)		1337.71
SD (K)		0.10

Note: P(T); Corrected value of the power radiated by the black body at temperature T.

imprecision, the uncertainty of a temperature determination must reflect systematic errors that are peculiar to the particular experiment. These errors might include improperly calibrated resistance bridges, pressure gages, or radiation detectors; such instruments provide precise but inaccurate measurements that spoil the accuracy of the temperature determinations with respect to the KTTS. As a matter of usual practice in measurement science, the effect of "experimental imprecision," "random error," or "thermometric imprecision" can be reduced nearly to any level one might wish by repeating a given type of measurement over and over again. "Systematic error" or "experimental bias," on the other hand, usually is unaffected by such repeated efforts. It is for this reason that thermometrists prefer strongly to base thermodynamic temperature assignments on the results of experiments that involve different systematic errors.

The uncertainty of the thermodynamic temperature of melting gold as determined in our example here could be assigned on the basis of the gas thermometry results, since they show the smaller SD. A more realistic approach, however, is to assess the contribution to the measured $\overline{T}$ of each known systematic error. Such systematic uncertainties generally are substantially larger than the standard deviation of the mean of individual sets of thermodynamic temperature determinations.

Turning our attention to Table 2, we see that resistance thermometers A and B provided measurements of the gold melting point for which the SD values are substantially smaller than it was for either of the two thermodynamic experiments. We can say that the resistance thermometers appear to be more precise than either the gas thermometer or the radiation thermometer. By repeating such measurements as those in Table 2 over a long period of time, one can examine the "stability," "long-term reproducibility," or "long-term precision" of the thermometers. Clearly if scientists A, B, C, and D were to decide to assign a temperature value to the gold melting temperature, resistance thermometers could represent it more precisely than could the thermodynamic thermometers.

The sensitivities $(dR/dT)_{Au}$ determined from Equation 3 for thermometers A and B measure the response of the thermometer to a small change in temperature. The reader can verify the fact that the sensitivities of the two thermometers used in our example vary with temperature. The resolutions of the thermometers are indicated by the least significant figure in Tables 1 and 2. From the data given, it is apparent that the system composed of each resistor and the resistance-measuring instrument can resolve ± 1 $\mu\Omega$, which corresponds to $\pm 10^{-6}$ $\Omega/6.89 \times 10^{-3}$ $\Omega/K = \pm 0.145$ mK and $\pm 10^{-6}$ $\Omega/7.03 \times 10^{-3}$ $\Omega/K = \pm 0.142$ mK, respectively.

Returning to the discussion at the beginning of this section, we suggest that scientists A, B, C, and D might assign the temperature of melting gold on the KTTS as follows: with no *a priori* knowledge of the systematic errors incorporated into the two gold-point determinations, the assigned temperature can be derived from a weighted average of the experimental values, giving proportionately greater weight to the more precise measurement. This procedure gives 1337.52 K as the (hypothetical) assigned temperature of the melting point of gold. In assigning an uncertainty to this value, one simply could use the differences between the assigned fixed-point temperature and the experimental values ($+0.19$ K and -0.10 K) or, more conservatively, one could expand the uncertainty range to include numbers above the higher value and below the lower value (say, $+0.38$ K and -0.20 K).

In trying to construct a practical temperature scale, one well might choose to define it in such a way as to take advantage of the improved precision and ease of measurement that can be achieved by use of resistance thermometers. One way to accomplish this result would be to assign an approximately correct (in thermodynamic terms) but arbitrary temperature to the melting point of gold and to all other useful fixed points, and to prescribe a relation such as Equation 2 to determine temperatures between the fixed points. This procedure would have two worthwhile advantages over a strictly thermodynamic scale; it would produce a more precise scale by employing more precise thermometers, and it would produce a singly realizable scale by specifying only one technique for evaluating any temperature.

In fact, four of the five international temperature scales that will be described later have been constructed in just this way. Temperatures obtained from the results of thermodynamic determinations have been assigned to a number of "defining fixed points" scattered over the range to be covered by the scale, and definite methods have been specified for determining values of temperature between the fixed points. The accuracy of the various portions of each of the scales with respect to temperatures on the KTTS has been a matter for continued reassessment; on the other hand, the nonuniqueness levels of temperatures provided by the scales have depended upon the quality of the fixed points and of the specified thermometer systems.

III. CONFERENCE ON THE METER AND ITS CONSEQUENCES

The Diplomatic Conference on the Meter, which resulted in the signing of the Convention of the Meter, was convened in Paris in March 1875 at the invitation of the French government. Its purpose was to regularize international standards of weights and measures. Twenty nations attended the three-month-long session.[4] (For a well-written English history of the conference and its consequences, see Reference 5.)

The conference was a striking success; the framework conceived then for establishing international agreement on standards of measurement not only has survived until the present time, but it operates yet today as the final arbiter in matters of scientific metrology.

The Convention of the Meter, signed on May 20, 1875, by representatives of 17 (including the U.S.) of the 20 nations attending the conference, established an International Bureau of Weights and Measures (BIPM, the Bureau International des Poids et Mesures), a continuing General Conference on Weights and Measures, and a subsidiary International Committee for Weights and Measures. The financial support for building the bureau facilities and maintaining it and the administrative expenses connected with the treaty apparatus were to be borne by the treaty nations in proportion to their populations.

The initial responsibility of the BIPM staff was to verify the accuracy of new prototypes of the meter and the kilogram before their distribution to the member nations.[6]

The BIPM mission today includes establishing basic standards and scales for the principal physical quantities and fundamental constants; to maintain prototypes in all the principal measurement areas; to carry out comparisons of national and international standards of measurement; and to propagate standard techniques of measurement. The staff of the BIPM is not nearly numerous enough to accomplish all of its mission — most metrological research is done in the national laboratories such as the National Bureau of Standards in the U.S. and its counterparts in other countries. However, the BIPM staff is highly competent, so that, in certain areas, very creative metrological work is done. Furthermore, the BIPM serves the metrological world by providing administrative support to consultative committees, by printing and circulating agendas and working papers, and by providing meeting space for treaty organizations. Since 1960, the journal *Metrologia,* a publication entirely devoted to measurement science, has been sponsored by the BIPM; currently the editorial offices for this journal are located there.

The General Conference on Weights and Measures, (CGPM, the Conférence Générale des Poids et Mesures) is a diplomatic conference rather than a technical one. Its mission is to propagate the metric system and its modern successor, the international system of units; to ratify results of new metrological determinations; to adopt scientific resolutions of international importance; and to monitor the organization and development of the BIPM. All of the member nations of the Treaty of the Meter may send delegates to the General Conferences. Since 1889, the General Conference has met 17 times.[7]

The 18-member International Committee for Weights and Measures (CIPM, the Comité International des Poids et Mesures) directly supervises the work of the BIPM and prepares proposals for the ratification of the General Conference. It meets approximately biennially. (See Reference 8 for an example of the proceedings.) The members of the CIPM are metrological experts elected individually by the CGPM. Although initially it was intended that the CIPM should prepare the details of all metrological proposals, including temperature scales, by 1927 it was clear that the increasing complexity of metrological research could be mastered only with specialized assistance. Therefore, consultative committees in various specialized areas were created; in the meetings of the consultative committees, the full significance of particular experiments and their results could properly be assessed. At this time, there are seven consultative committees, operating in the areas of electricity, photometry and radiometry, thermometry, length, time, ionizing radiation, and units of measurement. Each of the consultative committees is made up of specialists working in the particular area covered; occasionally, further subdivisions (working groups) undertake detailed projects for the committees.

The Consultative Committee for Thermometry (CCT, the Comité Consultatif de Thermométrie) was created in 1937. It met during 1984 for the 15th time.[9] The delegates attending any particular session represent laboratories from treaty nations. Usually they are working scientists who know from personal experience the limits of measurements in thermometry. The committee's responsibility is to recommend to the CIPM actions that will improve the usefulness of the international standards for temperature measurement.

It is a testament to the soundness of the treaty structure that the work of advancing international agreement on measurement science has persisted despite the recurrence of wars and other political and economic turbulence throughout the world. The preparation of new standards of measurement requires scientific cooperation of a very detailed nature, often over a period of many years. Yet the work of the treaty member nations has continued and it has been fruitful.

IV. INTERNATIONAL TEMPERATURE SCALES

In principle, the General Conference on Weights and Measures could provide a scale of temperature simply by specifying the numerical value of the gas constant, R; in turn, the value of R would fix the size of the "degree" of temperature. Since Boltzmann's constant, k, is related to the gas constant by the equation $kN_o = R$, where N_o is Avogadro's constant, any convenient thermodynamic or statistical-mechanical law involving either constant could then be used to evaluate the temperature associated with a particular level of hotness. An equivalent procedure would be the specification of the exact temperature of a particular fixed point, such as the triple point of water.

Some thermometrists are confident that the international scale of temperature eventually can be defined much as the KTTS is defined today:[10] "Given that the water triple-point temperature is 273.16 K, any other temperature may be found by the employment of gas thermometry, noise thermometry, total radiation thermometry, or any other convenient method that is based upon an explicit physical law." However, as the 20th century began, one could obtain temperatures with substantially better precision by means of platinum resistance thermometers or thermocouple thermometers than one could accomplish by using any thermodynamic thermometer. That situation persists today, although the "precision gap" has narrowed somewhat.

Because precision in thermometry often is more valuable to the user than is thermodynamic accuracy, the General Conference on Weights and Measures consistently has approved the most precise temperature scale available. We now discuss the nature of this type of scale.

Four international scales of temperature have been adopted by the CGPM. They were promulgated in 1889, 1927, 1948, and 1968. After the first, each has been designated by the year of its adoption. The first, the normal hydrogen scale, was based upon gas thermometry; the following three scales were constructed by assigning values to a number of fixed-point temperatures, by specifying equations by means of which the platinum resistance thermometer and the platinum vs. platinum-rhodium alloy thermocouple thermometer could be used to derive values of temperature between the fixed points, and by recommending a radiation thermometry law for use above the highest assigned temperature. In 1976, a low-temperature scale was given provisional status to provide methods for obtaining temperature values between 0.5 and 30 K. We note only briefly the features of the older scales, describing the present ones in more detail.

A. The Normal Hydrogen Scale

In the first years of the BIPM, its primary interest was to promulgate accurately the meter and the kilogram, as we already have noted. It was discovered early in this project that accurate length measurement required better temperature measurement and control than was available at that time.

Chappius, a member of the BIPM staff, responded to this need by developing a gas thermometer scale based upon temperature fixed points provided by melting ice and boiling water. These points were given the usual assignments of 0 and 100°C, respectively. The procedure of Chappius was not different in essence from the earlier work described in Chapter 2 (cf. Figure 5, Chapter 2). Using hydrogen gas of normal room-temperature composition (see Section III.D) as the working fluid, Chappius built a gas-thermometer calibration facility that operated in the range -25 to $+100$°C. In this facility, mercury-in-glass thermometers were calibrated for the use of various standards laboratories.

The scale developed by Chappius was proposed by the CIPM in 1887 for adoption as an international temperature scale. During the first General Conference on Weights and Measures, the scale was given official approval.

B. The International Temperature Scale of 1927 (ITS-27)

Work on the 1927 temperature scale was begun around 1913. The completion of the scale was delayed by the outbreak of World War I, however.

By the time that the ITS-27 was proposed, the platinum resistance thermometer was acknowledged as the most precise thermometer in the range −190 to +660°C. Following the introduction of platinum resistance thermometers by Siemens, Callendar[11] and others[11a] had demonstrated that wire resistors made of high-purity platinum could be mounted in a nearly strain-free manner by wrapping them on light, insulating supports or formers. In this condition, their electrical resistances provided sensitive and reproducible indications of their temperatures. Callendar himself had painstakingly compared platinum resistance thermometers with a gas thermometer built along the lines proposed by Regnault. Callendar's efforts were contemporary with those of Chappius, described above. His formulation of an equation to express the differences between the temperature obtained with the gas thermometer and with platinum resistors was fundamental to the use of platinum resistance thermometers in a practical temperature scale.

The introductory paragraphs of the ITS-27 contain the statement that the new scale was not intended to displace the thermodynamic temperature scale as the ultimate reference for science, but only to approximate it with a scale that could be used more easily for temperature calibrations.[12,12a]

The ITS-27 was a true centigrade scale, being based upon an interval of exactly 100 degrees between its two major fixed points. The six assigned ("defining") fixed points were

°C

- Oxygen boiling point   −182.97
- Ice point   0.00
- Steam point   100.00
- Sulfur boiling point   444.60
- Silver melting point   960.5
- Gold melting point   1063.

1. The Range −190 to +660°C

Temperatures from −190 to +660°C were to be determined by platinum resistance thermometers (PRTs). The recommended diameter of the wires in the thermometers was 0.05 mm to 0.2 mm. Limiting ratios of the thermometer resistances were specified at three temperatures: $[R(-183°C)/R(0°C)] < 0.250$, $[R(100°C)/R(0°C)] \geq 1.390$, and $[R(444.6°C)/R(0°C)] \geq 2.645$. By specifying these limits for the temperature dependence of the platinum resistance thermometers, those who formulated the ITS-27 expected that the approved thermometers all would provide nearly equal values of temperature throughout the PRT range.

From −190 to 0°C, a variation of the Callendar-Van Dusen equation[13] was chosen as the defining formula for temperature:

$$R(t) = R(0)[1 + A_1 t + B_1 t^2 + C_1(t - 100)t^3] \tag{4}$$

The coefficients $R(0)$, A_1, B_1, and C_1 were to be evaluated after calibrations were performed at the oxygen point, the ice point, the steam point, and the sulfur point. Simultaneous solution of the four resulting equations provided values of the coefficients.

From 0 to 660°C, Callendar's simple quadratic equation was modified to an equivalent form for use in deriving scale temperatures:

$$R(t) = R(0)(1 + A_2 t + B_2 t^2) \tag{5}$$

Calibrations of a standard platinum resistance thermometer at the ice, steam, and sulfur points would yield three equations; these then could be solved readily to evaluate the coefficients of Equation 5.

The careful reader will have noticed that the platinum resistance thermometer range of the ITS-27 extended beyond the calibration temperatures in both directions. As might be supposed, the use of this procedure caused large temperature uncertainties to exist at each end of the platinum resistance thermometer range.

2. The Range 660 to 1063°C

The relatively large uncertainty in scale temperatures above 500°C was particularly troublesome for the ITS-27 in the range 660 to 1063°C. Of the three calibration points recommended for use in that range, only two, the silver and gold points, actually were defining points of the scale; the third, the antimony point (assigned the value 630.5°C as a secondary fixed point in the ITS-27 text), was in principle to have been evaluated by platinum resistance thermometry. Thus the scale definition in this range depended upon temperatures extrapolated from 444.6°C.

The standard interpolating instrument in this range was the platinum vs. platinum-rhodium thermocouple thermometer. The requirements for a standard thermocouple were the following:

- $[R(100°C)/R(0°C)] \geq 1.390$ for the platinum wire
- Diameter of each wire to be 0.35 to 0.65 mm
- Composition of the alloy wire to be 90% Pt, 10 wt% Rh
- The thermocouple electromotive force (emf) at 1063°C to lie between 10.200 and 10.400 mV

The interpolation equation in the range 660 to 1063°C was a quadratic:

$$e = a + bt + ct^2 \tag{6}$$

where e is the thermocouple emf at temperature "t" and the coefficients a, b, and c were to be determined for the standard thermocouple thermometer by calibration at the three fixed points mentioned above.

3. The Range Above 1063°C

Above 1063°C, temperatures were defined by the Wien law of radiation

$$\ln\left(\frac{L_2}{L_1}\right) = \frac{c_2}{\lambda}\left[\frac{1}{1336} - \frac{1}{(t + 273)}\right] \tag{7}$$

that describes the ratio of the radiant intensity (L_2/L_1) emitted by a black body at the wavelength λ cm. The constant c_2 was taken equal to 1.432 cm · °C. The wavelength to be used was to be governed by the restriction

$$[\lambda (t + 273)] < 0.3 \text{ cm} \cdot °C \tag{8}$$

By the late 1930s, it was recognized that ITS-27 could be improved substantially in its precision and in its accuracy with respect to the centigrade thermodynamic scale. However, Wensel, a thermometrist working at the NBS, noted that the ITS-27 already had been a great step forward for thermometry: "Its greatest usefulness lies in the fact that its universal use has practically eliminated ambiguities in the specification of temperature. The first requirement of any scale is that it provide means by which any thermal state may be unambiguously specified."[14]

C. The International Temperature Scale of 1948 (ITS-48) and the International Practical Temperature Scale of 1948 (IPTS-48)

As the faults of the ITS-27 became clearer, interest in a replacement temperature scale grew stronger. In 1937, the International Committee for Weights and Measures approved the formation of a Consultative Committee for Thermometry (CCT). This committee met for the first time in 1939. During this first session, a scale revision was proposed, but the eruption of World War II delayed consideration of this proposal for nearly a decade.

The CCT met again in 1948, once more proposing a replacement for the ITS-27. In the fall of the same year, both the CIPM and Ninth General Conference met and approved the new scale, to be known as the International Temperature Scale of 1948 (ITS-48).[15,15a]

The ITS-48 was based explicitly upon an overspecified thermodynamic scale. The recommended thermodynamic temperature of the ice point was 273.15 °K, and the thermodynamic interval between the ice point and the steam point was given as 100 degrees. No discussion was given regarding this inconsistency in the definition of the thermodynamic scale.

Although the triple-point temperature of water was recognized by 1948 as being realizable with better precision than was the ice point, the latter was retained in the ITS-48 as its fundamental defining point.

Scale temperatures were expressed in °C. In the text, the name "centigrade" was attached to the degree symbol, but after approving the scale text, the Ninth General Conference decided that the name "Celsius" was preferable.

The ITS-48 was defined by the same set of six fixed points as its predecessor; the only change in fixed point assignments was that the freezing point of silver was assigned a new value, 960.8°C.

Once again, the standard PRT — now with a more restrictive limit of 1.391 on the resistance ratio R(100)/R(0) — was specified as the means to define temperatures over the bulk of the scale.

1. The Range −183 to +630°C

The lower limit of the ITS-27, −190°C, was raised in the ITS-48 to coincide with the oxygen boiling point, −182.970°C, thus eliminating the temperature uncertainty caused by extrapolation in the earlier scale.

Between −182.97 and 0°C, Equation 4 was preserved:

$$R(t) = R(0)[1 + A_1 t + B_1 t^2 + C_1(t - 100)t^3] \quad (4)$$

Its equivalent, the form actually proposed by Van Dusen,[13] also was recommended for use:

$$t = \frac{1}{\alpha}\left(\frac{R(t)}{R(0)} - 1\right) + \delta\left(\frac{t}{100} - 1\right)\frac{t}{100} + \beta\left(\frac{t}{100} - 1\right)\left(\frac{t}{100}\right)^3 \quad (9)$$

In this equation,

$$\alpha = \frac{1}{100}\left[\frac{R(100)}{R(0)} - 1\right]$$

For either of the above equations, the coefficients were to be evaluated by simultaneous solution of the four equations resulting from calibrations of the standard thermometer at the ice point, the steam point, and the sulfur and oxygen boiling points.

The relations between the coefficients of Equations 4 and 9 are as follows:

$$A = \alpha\left(1 + \frac{\delta}{100}\right) \qquad B = -\alpha\delta(10^{-4}) \qquad C = -\alpha\beta(10^{-8})$$

$$\alpha = A + 100B \qquad \delta = -\frac{B(10^4)}{A + 100B} \qquad \beta = \frac{-C(10^8)}{A + 100B} \qquad (10)$$

The uncertainty at the upper end of the platinum resistance thermometer range in the ITS-27 was diminished by reducing the upper junction temperature to "the freezing point of antimony . . . not lower than 630.3°C." Unfortunately, the antimony point still was not assigned a scale temperature.

From 0°C to the antimony point, the ITS-48 was defined by the quadratic equation

$$R(t) = R(0)(1 + A_2 t + B_2 t^2) \qquad (5)$$

or its equivalent (the original Callendar equation)

$$t = \frac{1}{\alpha}\left(\frac{R(t)}{R(0)} - 1\right) + \delta\left(\frac{t}{100} - 1\right)\left(\frac{t}{100}\right) \qquad (11)$$

The equivalency relations given in Equation 10 also apply to Equations 5 and 11.

The more precise triple point of water (t = 0.01°C) was suggested as a calibration reference in place of the ice point (t = 0°C), although all resistances were still to be divided by the t = 0°C resistance value for that thermometer.

2. The Range 630 to 1063°C

From the antimony freezing point to 1063°C, the platinum vs. platinum-10 wt% rhodium thermocouple thermometer again was specified to define temperatures between the fixed points. New constraints were placed upon the emf of the standard thermocouple thermometer. These can be summarized as follows:

$$E(Au) = 10.30 \pm 0.05 \text{ mV}$$

$$E(Au) - E(Ag) = 1.185 + 1.58 \times 10^{-4}[E(Au) - 10.31] \pm 0.003 \text{ mV}$$

$$E(Au) - E(Sb) = 4.776 + 6.31 \times 10^{-4}[E(Au) - 10.31] \pm 0.005 \text{ mV} \qquad (12)$$

All of these limits were to be satisfied using an ice-point thermocouple reference junction.

The relevant interpolating equation was

$$E(t) = a + bt + ct^2 \qquad (6)$$

with the coefficients a, b, and c to be determined by measurements at the freezing points of Sb, Ag, and Au. Once again, the antimony freezing point temperature was to be determined by extrapolation of the platinum resistance equation from 444.6°C.

3. The Range Above 1063°C

Above 1063°C, the Wien equation Equation 7 (/) of the ITS-27 was replaced by the Planck law. This change improved the thermodynamic consistency of temperatures in this range and allowed use of the ITS-48 to higher temperatures than its predecessor could provide. In addition, the numerical value of the second radiation constant (c_2) was changed from 1.432 cm · °C to 1.438 cm · °C.

The Planck law was written as follows:

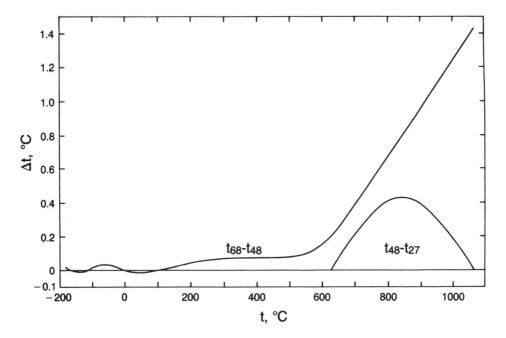

FIGURE 2. Differences in temperature up to 1063°C resulting from the replacement of the International Temperature Scale of 1927 by the International Temperature Scale of 1948 ($t_{48} - t_{27}$), and by the replacement of the latter scale by the International Practical Temperature Scale of 1968 ($t_{68} - t_{48}$). Cf. Figure 3, p. 123 of Reference 16.

$$\frac{L(T)}{L(Au)} = \frac{\exp[c_2/(\lambda\ 1336.15)] - 1}{\exp[c_2/(\lambda T)] - 1} \qquad (13)$$

Here L(T) is the radiant energy per unit wavelength interval (at wavelength λ) emitted per unit time by a unit area of a black body at Kelvin temperature T. The wavelength λ, in cm, was specified as part of the visible spectrum. The two scales ITS-27 and ITS-48 produced different values for the same thermal state only in the range above 630°C. Figure 2 shows the divergence of the temperatures $t_{48} - t_{27}$ between 630 and 1063°C. Scale differences above 1063°C became quite large. They can be found in several places. (See, for example, Reference 16.)

4. A "Practical" ITS-48

In 1960, a text revision of the ITS-48 was approved by the 11th General Conference on Weights and Measures.[17] The changes reflected in the revised text resulted from discussions that had taken place in several meetings of the CCT, CIPM, and the General Conferences.

Perhaps the most far-reaching change was the specification of the water triple-point temperature as 273.16 K to create the present Kelvin thermodynamic scale and the inclusion of the water triple point as a defining point (t = +0.01°C) of the revised scale. (The ice point, although no longer a defining calibration point, still was to be the basic point for the interpolating equations of the scale.) No longer was the thermodynamic scale a true "centigrade" (exactly 100 degrees separating the ice-point and steam-point temperatures) scale, although there was no evidence in 1960 that a thermodynamic inconsistency exists between the assignment of the water triple-point temperature as 273.16 K and the assignment of the ice and steam points as 0 and 100°C, respectively.

In order to avoid confusion between temperatures expressed on the ITS and within the framework of the then-new "International System of Units" (the "Systéme International"

Table 5
STANDARD INSTRUMENTS FOR THE IPTS-48(60)

Standard platinum resistance thermometer
  $[R(100)/R(0)] \geq 1.3920$
Standard thermocouple thermometer (reference junction temperature = 0°C)
  $E(Au) = 10.300 \pm 0.050$ mV
  $[E(Au) - E(Ag)] = 1.183$ mV $+ 1.58 \times 10^{-4}[E(Au) - 10.300]$ mV ± 0.004 mV
  $[E(Au) - E(630.5)] = 4.766$ mV $+ 6.31 \times 10^{-4}[E(Au) - 10.300]$ mV ± 0.008 mV

or SI), the name of the scale was changed to the "International Practical Temperature Scale of 1948". As might have been expected from the foregoing discussion, the name of the temperature unit was changed; it became "Celsius" in place of "centigrade".

To take account of the fact that the temperature of the equilibrium between the solid and liquid phases of pure zinc could be realized with better precision than could the sulfur boiling-point temperature, the zinc point was recommended as a replacement calibration temperature. Its assigned value was 419.505°C.

New restrictions were placed upon the standard platinum resistance thermometer and the standard thermocouple thermometer. They are given in Table 5.

It was expected that no change in scale temperature would result from any of the modifications to the ITS-48 that were accomplished in 1960.

In his discussion of the 1960 revision of the ITS-48,[17] Stimson drew attention to the difference between fixed point temperatures on the Kelvin thermodynamic scale and fixed point temperatures on the IPTS-48. Two paragraphs particularly seem significant for our purposes:

Taking the triple point of water as one of the defining fixed points of the international scale thus makes it the one defining fixed point which is common to both the international scale and the Kelvin thermodynamic scale. Its value is now adopted as 0.01°C on both the international and the thermodynamic Celsius scales, and the ice point is still 0°C within the present experimental error.

The international scale is a practical scale wherein the values assigned to the fixed points (except for the triple point) only approximate the true thermodynamic values, and the formulas specified for interpolation are empirical. It is extremely unlikely, therefore, that any arbitrarily chosen temperature has exactly the same value on the international scale that it has on the thermodynamic scale. The temperature of 0°C on the international scale, for example, is not exactly the same temperature as 0°C on the thermodynamic Celsius scale. The difference between these temperatures is too small to be detected experimentally but it may be estimated by interpolation sometime in the future when the relations of the two scales are better known.

We shall return to Stimson's ideas again in later sections.

D. The International Practical Temperature Scale of 1968 (IPTS-68) and the IPTS-68(75)

The development, approval, and actual realization of the IPTS-68 provide a sequence of events and personal interactions that is worthy of attention by the most accomplished writer of mysteries. An unrealistic timetable for the extension to lower temperatures or outright replacement of the IPTS-48 was coupled with strong desires to correct thermodynamic and other deficiencies in the older scale, and a dash of political intrigue was added to make an almost wholly new temperature scale. Given the circumstances surrounding its construction, it is surprising that the IPTS-68 was completed at that time. It is a testament to the expertise of the people who produced it that the IPTS-68, all things considered, is a fairly good temperature scale. (A fascinating, yet factual account of the genesis of the IPTS-68 is given in Reference 18.) Like its predecessor, the IPTS-68 text was amended later without changing the valuation of any defined temperatures; this amendment took effect in 1975.

The period of time from the end of World War II to the mid-1960s was characterized in thermometry by substantial improvements in measurement methods, by a corresponding increase in the levels of precision and accuracy required by science and industry, and by rapidly expanding interest in measurements at very low and very high temperatures. Many more people became involved in precision measurements of all kinds, and they found at their disposal more money and better equipment.

As might be supposed, this general expansion of scientific and technical effort exposed many of the weaknesses in the IPTS-48. By 1965 it was clear that the international temperature scale should reach deep into the regime of cryogenics, that its deviations from KTTS temperatures should be reduced, and that its fixed-point temperature assignments and interpolation formulations should be improved.

Many research results demonstrating these problems with the IPTS-48 appeared in the international symposia on temperature measurement and control[19,19a] held in 1954 and 1961 and in the journal *Metrologia* that began publication of articles on measurement science in 1965 under the auspices of the International Committee for Weights and Measures.[20]

It may be worthwhile to mention some of the scale-thermometry research projects of this period to portray more vividly the marked increase in the amount of thermometry work and to facilitate further study of the subject.

In the area of thermodynamic measurements there were studies of gas thermometry, still the principal method for realizing accurate thermodynamic temperatures.[21-29] There were suggestions for corrections to and extensions of the temperature scale,[21-23,25-44] and modifications to the standard thermometers and the interpolation formulas.[40,41,43,45-54]

As Preston-Thomas aptly indicated,[18] the difficulty in modifying the IPTS-48 during 1968 lay not with a lack of worthwhile information, but rather with its careful digestion and conversion into a reliable temperature scale.

The principal impetus in 1967 for the replacement of the IPTS-48 was to extend the international temperature scale to lower temperatures. The low-temperature limit was reduced by extending the PRT nearly to its minimum useful temperature. The electrical resistivity of pure platinum follows an approximately linear dependence upon temperature above 40 to 50 K, as shown in Figure 3, but at low temperatures, the resistivity approaches a constant value dictated by residual impurities and other crystalline imperfections. This property of platinum resistance thermometers required replacement of the Callendar-Van Dusen equation for precise thermometry below ~100 K. By providing several calibration points over a relatively short temperature range, the scale designers were able to reduce scale-temperature imprecision below the 1 mK level down to 13.8 K.

The number of fixed points defining the IPTS-48 was nearly doubled in the definition of the IPTS-68. Table 6 contains a list of the IPTS-68 defining fixed points with their temperature assignments both in kelvins and in degrees Celsius.

Figure 4 shows in schematic form the fixed points and the various temperature ranges of the IPTS-68.[55,56]

The IPTS-68 requires that standard PRTs be constructed as four-lead resistors mounted as nearly strain-free as possible and that they exhibit $W(100) = R(100)/R(0)$ values not smaller than 1.39250.

Two types of standard PRT have come into general use (see, for example, Reference 57) for defining temperatures on the IPTS-68. These two types are compared in Figure 5. The long-stem thermometer is used to realize temperatures over the range 90 to 900 K and the capsule type is used from 13.8 to 400 K. Both are composed of four-lead resistors of $R(0)$ ~ 25 Ω; if calibrated correctly, both should yield nearly identical temperature values in the overlapping range.

The long-stem PRT is well-suited for insertion into furnaces, baths, and other high-temperature apparatus. The stem provides a long path for heat flow between the region

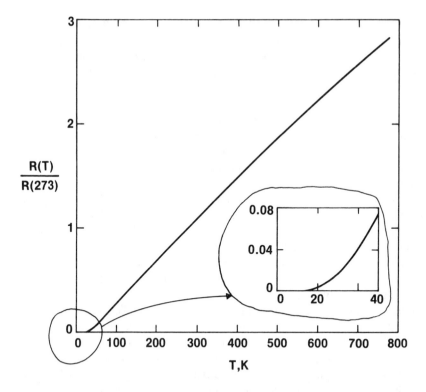

FIGURE 3. The resistance of a standard platinum resistance thermometer divided by its ice-point resistance vs. temperature. At very low temperatures, the resistance approaches a constant value (the "residual resistance") that is about 0.1% of its room-temperature resistance. Throughout most of its range, the platinum exhibits a nearly linear resistance with a small negative contribution of higher order.

around the sensing coil and room temperature. This heat flow can be retarded further by the "tempering effect" (i.e., the encouragement of thermal equilibrium between two objects by placing them in physical contact with each other) of baffles in the insertion hole.

The capsule PRT, on the other hand, is designed for use in cryogenic apparatus. Often, the thermometer is mounted in a metal tempering block that contains other thermometers, fixed-point devices, or other objects under study, as shown in Figure 6. The thermometer leads are connected to tempered cryostat signal wires chosen to minimize heat leaks from room temperature. Given the variety found in cryostat construction, the shortened form of the capsule PRT provides it with needed versatility.

In the calibration of capsule platinum resistance thermometers above 273 K, it is helpful to use special holders like the one shown in Figure 7. These holders allow the operator to calibrate the capsule thermometers in the water triple-point cell and in other higher-temperature cells using the same techniques employed for long-stem thermometers.

1. The Range 13.81 to 273.15 K

Temperatures on the IPTS-68, denoted T_{68} and expressed in kelvins, from 13.81 to 273.15 K are defined by the equation*

$$W(T_{68}) = W_{CCT\text{-}68}(T_{68}) + \Delta W_i(T_{68}) \qquad (14)$$

---

* In expressing the interpolation relations of the IPTS-68, we have modified the textual notation in several places in the interest of clarity. In general, the most useful formulations are to be found in the amended text of 1975.

Table 6
IPTS-68 DEFINING FIXED POINTS, INCLUDING CHANGES INDICATED BY THE 1975 AMENDMENT

Equilibrium state	T_{68} (K)	t_{68} (°C)
Triple point of equilibrium hydrogen	13.81	−259.34
Liquid and vapor phases of equilibrium hydrogen at 33,330.6 Pa (25/76 atm)	17.042	−256.108
Liquid and vapor phases of equilibrium hydrogen[a]	20.28	−252.87
Liquid and vapor phases of neon[a]	27.102	−246.048
Triple point of oxygen	54.361	−218.789
Triple point of argon[b]	83.798	−189.352
Liquid and vapor phases of oxygen[a]	90.188	−182.962
Triple point of water	273.1600	0.0100
Liquid and vapor phases of water	373.150	100.000
Solid and liquid phases of tin[b]	505.1181	231.9681
Solid and liquid phases of zinc	692.73	419.58
Solid and liquid phases of silver	1235.08	961.93
Solid and liquid phases of gold	1337.58	1064.43

Note: Except as noted, all points are to be realized at 101,325 Pa pressure (1 atm). The IPTS-68 text specifies isotopic abundance data where necessary.

[a] The possible existence of small temperature differences caused by fractionation of isotopes or impurities suggests the use of boiling points (disappearance of the vapor phase) for hydrogen and neon, and the use of the condensation point (disappearance of the liquid phase) for oxygen.

[b] The triple point of argon and the freezing point of tin are offered as alternative fixed points to the condensation point of oxygen and the boiling point of water, respectively.

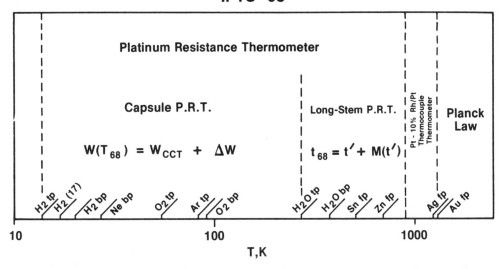

FIGURE 4. The International Practical Temperature Scale of 1968 in schematic form. The defining fixed points of the scale are listed along the bottom of the figure at their appropriate temperatures. The limits of the capsule PRT (13.8 K to 273.15 K), the long-stem PRT (273.15 K or 0°C to 903.89 K or 630.74°C), the (Pt + 10 wt % Rh) vs. Pt thermocouple thermometer (903.89 K or 630.74°C to 1337.58 K or 1064.43°C), and the Planck law (all higher temperatures, by ratio to the temperature of freezing gold) are indicated, as well as the character of the interpolation relations for the two types of PRT.

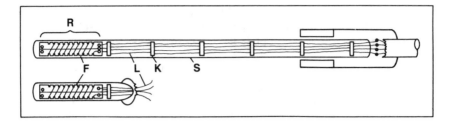

FIGURE 5. Line drawing to illustrate the features found in most standard platinum resistance thermometers. In each type, a resistor R of high-purity platinum wire of ~25 Ω resistance is wound on an insulating former F. The wire may be wound in a simple bifilar helix so as to minimize both the electrical inductance and the thermal equilibration time, or it may be wound in a bifilar compound helix (a "door spring" configuration) to minimize thermal stress; other configurations also are used. Four platinum leads L are employed so as to restrict the thermometric response to the coil region. In the long-stem thermometer, the tempering of the leads is accomplished by a large immersion length, with the wires retained in position by spacers K; in the capsule type, the leads commonly are tempered by cementing extension wires to the thermometer block that holds the thermometer. The sheath S may be made either of glass or metal; the fill gas usually is dry air for the long-stem thermometer, but may be helium gas in the capsule type.

For a given thermometer, $W(T_{68})$ is the ratio of its resistance at temperature T_{68} divided by its resistance at 273.15 K. The term W_{CCT} is a reference function that approximates the temperature dependence of the resistance ratio of an "average" capsule platinum resistance thermometer. The reference function defines a scale of temperature that is different from the IPTS-48 scale, as is shown by the low-temperature portion of the upper curve in Figure 2. The ΔW_i are deviation functions for four different temperature ranges, obtained for the particular PRT being calibrated.

The reference function W_{CCT} was derived from measurements of several actual platinum resistance thermometers in comparison with low-temperature gas thermometers. This technique allowed the interpolation equations to closely approximate KTTS temperatures below room temperature, but caused the reference function to be more difficult to fit than would have been the case had only one thermometer been represented in its formulation.

In the original version of the IPTS-68, Equation 15 was used to generate values of the reference function W_{CCT-68}:

$$T_{68} = A_0 + \sum_{i=1}^{20} A_i [\ln W_{CCT-68}(T_{68})]^i \qquad (15)$$

This formulation was simplified somewhat in the 1975 amendment to the text; the new generating function is

$$T_{68} = \sum_{j=0}^{20} a_j \left[\frac{\ln W_{CCT-68}(T_{68}) + 3.28}{3.28} \right]^j \qquad (16)$$

Values of the reference function at any T_{68} temperature can be calculated using Equation 16 and the coefficients listed in Table 7. A tabulation of the reference function W_{CCT-68} can be obtained from the BIPM; a table of values at 1-K intervals is printed in the IPTS-68 text. 68 text.

Table 8 shows the values of $W_{CCT-68(75)}$ (from the amended text, which differs slightly from the original table) at the lower defining fixed points of the IPTS-68.

The capsule PRT must be calibrated from the steam point or the tin point downwards in

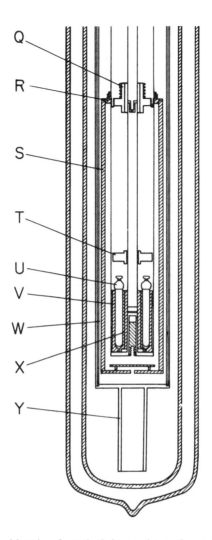

FIGURE 6. Mounting of capsule platinum resistance thermometer in low-temperature apparatus. Q, R, T, V, X, Thermal tie-down points for thermometer lead wires, for control thermometers, and for heaters; V, copper block designed to hold one or more capsule thermometers, miniature fixed-point cells, or secondary thermometers; S, W, thermal shields; U, capsule thermometer; Y, thermal link to cryogenic bath. (Figure taken from Riddle, J. L., Furukawa, G. T., and Plumb, H. H., *Platinum Resistance Thermometry*, U.S. National Bureau of Standards Monogr. 126, April 1973.)

order to match the derivatives of the scale at its junction points. From 273.15 to 90.188 K, the deviation function is

$$\Delta W_4(t_{68}) = A_4 t_{68} + C_4 t_{68}^3 (t_{68} - 100) \tag{17}$$

where t_{68} (the IPTS-68 temperature expressed in °C) equals (T_{68} − 273.15) (T_{68} is the IPTS-68 temperature expressed in K).

The coefficients A_4 and C_4 are determined for a particular thermometer by measuring ΔW_4, the deviations from the reference-function resistance ratio (cf. Table 8) at either the oxygen condensation point or the argon triple point and at either the steam point or the tin freezing point and then solving the two resulting Equations 17 simultaneously.

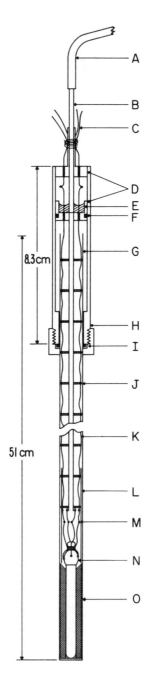

FIGURE 7. NBS-designed adapter for capsule platinum resistance thermometers to facilitate calibration in deep-well fixed-point cells. A, B, Helium gas supply; C, K, electrical extension wires connected to thermometer leads at M; D, E, J, L, tie down points, insulators, and spacers for thermometer leads; F, G, H, I, stainless steel protection tube and handle assembly; N, capsule thermometer; O, aluminum sleeve to promote thermal equilibrium between the thermometer and the fixed-point temperature. (Figure taken from Riddle, J. L., Furukawa, G. T., and Plumb, H. H., *Platinum Resistance Thermometry*, U.S. National Bureau of Standards Monogr. 126, April 1973.)

Table 7
COEFFICIENTS a_j OF THE REFERENCE FUNCTION FOR PLATINUM RESISTANCE THERMOMETERS IN THE RANGE 13.81 TO 273.15 K

j	a_j	j	a_j
0	38.592 76	10	239.502 85
1	43.448 37	11	524.649 44
2	39.108 87	12	−319.799 81
3	38.693 52	13	−787.606 86
4	32.568 83	14	179.547 82
5	24.701 58	15	700.428 32
6	53.038 28	16	29.486 66
7	77.357 67	17	−335.243 78
8	−95.751 03	18	−77.256 60
9	−223.528 92	19	66.762 92
		20	24.449 11

Table 8
VALUES OF $W_{CCT\text{-}68(75)}$ AT IPTS-68 FIXED-POINT TEMPERATURES

Point	T_{68} (K)	$W_{CCT\text{-}68(75)}$
e-H$_2$ tp	13.81	0.001 412 08
e-H$_2$ 17	17.042	0.002 534 45
e-H$_2$ bp	20.28	0.004 485 17
Ne bp	27.102	0.012 212 72
O$_2$ tp	54.361	0.091 972 53
Ar tp	83.798	0.216 057 05
O$_2$ cp	90.188	0.243 799 12
ice	273.15	1.000 000 00
H$_2$O bp	373.15	1.392 596 68
Sn fp	505.1181	1.892 570 86

$W(T_{68})$, the resistance ratio in the range 90.188 to 273.15 K, then can be obtained for the test thermometer by use of Equations 14, 16, and 17. Note that these equations are written in a form that is best suited for constructing a table of reference values of W as functions of T_{68}, rather than the converse.

In the range 90.188 to 54.361 K, the deviation function in Equation 14 is

$$\Delta W_3(T_{68}) = A_3 + B_3 T_{68} + C_3 T_{68}^2 \tag{18}$$

For a particular thermometer, the coefficients A_3, B_3, and C_3 are evaluated by measuring ΔW_3, the deviations from the reference function, at the oxygen triple point and at either the oxygen condensation point or the argon triple point, and by equating the derivative of Equation 17, evaluated at the oxygen condensation point, with the derivative of Equation 18 at the same point.

In the range 54.361 to 20.28 K, the deviation function is

$$\Delta W_2(T_{68}) = A_2 + B_2 T_{68} + C_2 T_{68}^2 + D_2 T_{68}^3 \tag{19}$$

Table 9
**CALIBRATION DATA FOR CAPSULE PLATINUM
RESISTANCE THERMOMETER NO. −1207**

T_{68} (K)	W_{CCT-68}	R_x (Ω)	W_x	ΔW_x
273.15	1.000 000 00	25.546 733	1.000 000 00	0
373.15	1.392 596 68	35.577 309	1.392 636 30	0.000 039 62
90.188	0.243 799 12	6.226 3658	0.243 724 50	−0.000 074 62
54.361	0.091 972 53	2.346 1945	0.091 839 30	−0.000 133 23
27.102	0.012 212 72	0.307 6076	0.012 040 90	−0.000 171 82
20.28	0.004 485 17	0.110 2100	0.004 314 00	−0.000 171 17
17.042	0.002 534 45	0.060 4915	0.002 367 80	−0.000 166 65
13.81	0.001 412 08	0.032 0624	0.001 255 00	−0.000 157 08

with the coefficients to be evaluated by calibration at the boiling point of equilibrium hydrogen, at the boiling point of neon, and at the triple point of oxygen; and by equating the derivatives of Equations 18 and 19 at the triple point of oxygen.

Finally, in the range 20.28 to 13.81 K, the deviation function is

$$\Delta W_1(T_{68}) = A_1 + B_1 T_{68} + C_1 T_{68}^2 + D_1 T_{68}^3 \qquad (20)$$

with the coefficient to be evaluated from calibration data obtained at the triple point of equilibrium hydrogen, at the temperature (17.042 K) obtained by maintaining equilibrium hydrogen at a pressure of 33,330.6 Pa (25/76 standard atmosphere), and at the boiling point of equilibrium hydrogen; and by equating the derivatives of Equations 19 and 20 at the boiling point of equilibrium hydrogen.

To illustrate the manner by which one constructs a reference table of resistance ratio values vs. temperature for a test thermometer in the range 13.81 to 273.15 K, we now present an example using calibration data for an actual capsule platinum resistance thermometer, identified here as No. −1207 and with the subscript x.

Table 9 shows the relevant IPTS-68 fixed-point temperatures, the values of $W_{CCT-68(75)}$ obtained either from Equation 16 or from Table 4 of the IPTS-68(75) text, the values of R_x for the test thermometer obtained by calibration at the fixed points, the values of W_x ($= R/R_o$) at the fixed points, and the resulting values of ΔW_x ($= W_x - W_{CCT}$) at the fixed points.

Using the data in the last column of Table 9, we now can calculate the coefficients of Equations 17 to 20 for the test thermometer. In Table 10 are listed the sets of equations thus obtained for each range of the IPTS-68 below 273.15 K, along with approximate values of the coefficients obtained by simultaneous solution of the equations. In addition, each of the needed first derivatives is evaluated at the proper temperature.

Note that the calibration calculation for each temperature range involves the simultaneous solution of several equations. In order to guarantee that the four sets of solutions join smoothly, the first derivatives of the equations are matched at the junction temperatures. It is necessary to perform the calculations beginning with the range 90 to 273 K and ending with the range 13 to 20 K.

To complete our illustration of the preparation of $W(T_{68})$ vs. T_{68} values for a test thermometer, we show in Table 11 the generation of $W(T_{68})$ values for capsule PRT No. −1207 at four temperatures, one in each of the four low-temperature ranges. In column 1 we list the selected T_{68} temperatures; column 2 contains the values of W_{CCT-68} calculated from Equation 16 for the respective temperatures; the entries in column 3 are the differences $\Delta W(T_{68})$ for thermometer No. −1207 as calculated from Equations 17 to 20 using the

Table 10
CALIBRATION EQUATIONS $\Delta W_i(T_{68})$ AND THEIR SOLUTIONS FOR CAPSULE PRT NO. −1207; TEMPERATURE RANGE 13.81 K TO 273.15 K

Range 90.188 to 273.15 K
　Deviation function

$$\Delta W_4(t_{68}) = A_4 t_{68} + C_4 t_{68}^3 (t_{68} - 100) \tag{17}$$

　Simultaneous equations

$$-7.462 \times 10^{-5} = A_4(-182.962) + C_4(-182.962)^3(-282.962)$$

$$+3.962 \times 10^{-5} = A_4(100) \quad\quad + C_4(100)^3(0)$$

　Approximate coefficients

$$A_4 \cong 3.962 \times 10^{-7}/°C; \quad C_4 \cong -1.229 \times 10^{-15}/(°C)^4$$

　Deviation function derivative

$$[d < \Delta W_4(T_{68}) >/dT] \, (t = -182.962°C)$$
$$= A_4 + 4C_4(-182.962)^3 - 3(100)C_4(-182.962)^2$$
$$\cong 4.3865 \times 10^{-7}/°C$$

Range 54.361 to 90.188 K
　Deviation function

$$\Delta W_3(T_{68}) = A_3 + B_3 T_{68} + C_3 T_{68}^2 \tag{18}$$

　Simultaneous equations

$$-1.3323 \times 10^{-4} = A_3 + B_3(54.361) + C_3(54.361)^2$$
$$-7.462 \times 10^{-5} = A_3 + B_3(90.188) + C_3(90.188)^2$$
$$+4.3865 \times 10^{-7} = B_3 + 2C_3(90.188)$$

　Approximate coefficients

$$A_3 \cong -3.860 \times 10^{-4}; \quad B_3 \cong +6.4665 \times 10^{-6}/K;$$
$$C_3 \cong -3.3418 \times 10^{-8}/(K)^2$$

　Deviation function derivative

$$[d<\Delta W_3(T_{68})>/dT] \, (T = 54.361 \text{ K}) = B_3 + 2C_3(54.361)$$
$$\cong +2.8332 \times 10^{-6}/K$$

Range 20.28 to 54.361 K
　Deviation function

$$\Delta W_2(T_{68}) = A_2 + B_2 T_{68} + C_2 T_{68}^2 + D_2 T_{68}^3 \tag{19}$$

　Simultaneous equations

$$-1.7117 \times 10^{-4} = A_2 + B_2(20.28) + C_2(20.28)^2 + D_2(20.28)^3$$
$$-1.7182 \times 10^{-4} = A_2 + B_2(27.102) + C_2(27.102)^2 + D_2(27.102)^3$$
$$-1.3323 \times 10^{-4} = A_2 + B_2(54.361) + C_2(54.361)^2 + D_2(54.361)^3$$
$$+2.8332 \times 10^{-6} = B_2 + 2C_2(54.361) + 3D_2(54.361)^2$$

Table 10 (continued)
CALIBRATION EQUATIONS $\Delta W_i(T_{68})$ AND THEIR SOLUTIONS FOR CAPSULE PRT NO. -1207; TEMPERATURE RANGE 13.81 K TO 273.15 K

Approximate coefficients

$$A_2 \cong -1.515 \times 10^{-4} \; ; \quad B_2 \cong -1.500 \times 10^{-6}/K \; ;$$
$$C_2 \cong +2.160 \times 10^{-8}/(K)^2 \; ; \quad D_2 \cong +2.239 \times 10^{-10}/(K)^3$$

Deviation function derivative

$$[d<\Delta W_2(T_{68})>/dT] \; (T = 20.28 \text{ K}) = B_2 + 2C_2(20.28) + 3D_2(20.28)^2$$
$$\cong -3.476 \times 10^{-7}/K$$

Range 13.81 to 20.28 K
  Deviation function

$$\Delta W_1(T_{68}) = A_1 + B_1 T_{68} + C_1 T_{68}^2 + D_1 T_{68}^3 \tag{20}$$

Simultaneous equations

$$-1.5708 \times 10^{-4} = A_1 + B_1(13.81) + C_1(13.81)^2 + D_1(13.81)^3$$
$$-1.6665 \times 10^{-4} = A_1 + B_1(17.042) + C_1(17.042)^2 + D_1(17.042)^3$$
$$-1.7117 \times 10^{-4} = A_1 + B_1(20.28) + C_1(20.28)^2 + D_1(20.28)^3$$
$$-3.476 \times 10^{-7} \;\;\; = B_1 + 2C_1(20.28) + 3D_1(20.28)^2$$

Approximate coefficients

$$A_1 \cong -8.729 \times 10^{-5} \; ; \quad B_1 \cong -5.2 \times 10^{-6}/K \; ;$$
$$C_1 \cong -0.8 \times 10^{-7}/(K)^2 \; ; \quad D_1 \cong +6.563 \times 10^{-9}/(K)^3$$

Table 11
$W(T_{68})$ VS. T_{68} FOR CAPSULE PLATINUM RESISTANCE THERMOMETER NO. -1207

T_{68} (K)	W_{CCT-68}	$\Delta W(T_{68})$	$W(T_{68})$
15.0	0.001 745 42	-0.000 161 14	0.001 584 28
40.0	0.041 719 69	-0.000 162 61	0.041 557 08
70.0	0.156 495 43	-0.000 097 09	0.156 398 34
180.0	0.622 969 74	-0.000 037 10	0.622 932 64

appropriate coefficients listed in Table 10; and column 4 contains the resulting tabular values $W(T_{68})$ that we set out to obtain.

In Figure 8 we show graphically the differences $\Delta W(T_{68})$ for thermometer No. -1207 as calculated for the temperatures listed in Table 11. For comparison, we show also the $\Delta W(T_{68})$ differences obtained by calibration at the fixed points (cf. Table 9, column 5). One can see readily from Table 11 that the resistance ratio of the actual platinum resistance thermometer illustrated differs from the reference function W_{CCT-68} by varying amounts — as much as 10% at the lowest temperatures, where the residual-resistance behavior makes such thermometers most variable, and as little as 0.01% near room temperature.

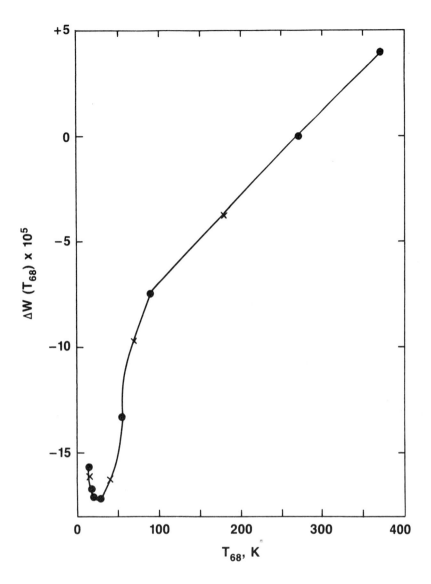

FIGURE 8. Differences $\Delta W(T_{68})$ between the resistance ratio of capsule platinum resistance thermometer No. -1207 and the IPTS-68 low-temperature reference function W_{CCT-68}. Solid circles; differences evaluated by calibration at the IPTS-68 defining fixed points. Crosses; differences calculated from Equations (17 to 20).

2. The Range 273 to 903 K

The Callendar equation used in the IPTS-48 to define the scale in platinum resistance thermometer range above 273 K was modified in the IPTS-68 to accommodate new thermodynamic measurements mentioned earlier.[23,30,41] The difference $(t_{68} - t_{48})$ is shown by the upper curve in Figure 2.

The IPTS-68 is defined in the range 0 to 630.74°C by the equation

$$t_{68}(°C) = t' + M(t') \tag{21}$$

where t' is the Callendar formulation

$$t' = \frac{1}{\alpha}[W(t') - 1] + \delta\left(\frac{t'}{100}\right)\left(\frac{t'}{100} - 1\right) \quad (22)$$

with $W(t') = R(t'°C)/R(0°C)$, and

$$M(t') = 0.45 \left(\frac{t'}{100}\right)\left(\frac{t'}{100} - 1\right)\left(\frac{t'}{419.58} - 1\right)\left(\frac{t'}{630.74} - 1\right) \quad (23)$$

Equation 23 provides for deviations from a quadratic temperature dependence for the resistance ratio of the test thermometer. One can see by inspection that this deviation function vanishes at the temperatures 0, 100, 419.58, and 630.74°C.

Equation 22 is equivalent to the standard quadratic form

$$W(t') = 1 + At' + Bt'^2 \quad (24)$$

The coefficients of Equations 22 and 24 are related by the following equations

$$A = \alpha(1 + 0.01\delta) \qquad B = -\alpha\delta \times 10^{-4} \quad (25)$$

To prepare a table of values of $W [=R(t)/R(0)]$ vs. t_{68} for a test thermometer, one first must evaluate R(0°C) and the coefficients of Equation 22 or 24. By calibration of the test thermometer at the triple point of water, at the steam point, and at the zinc point, one can obtain the values R(0.01°C), R(100°C), and R(419.58°C). The value R(0.01°C) is adjusted to R(0°C) by the relation

$$R(0°C) = R(0.01°C)/1.000039 \quad (26)$$

and the resistance ratios $W(t' = t_{68} = 100°C)$ and $W(t' = t_{68} = 419.58°C)$ are calculated. Then one can determine the coefficients α and δ from Equation 22 or A and B from Equation 24. In either case t' values then can be determined for any $R(t')$ in the range 0 to 630.74°C and Equations 21 and 23 can be used to calculate t_{68}.

If the tin freezing point is used as a calibration point instead of the steam point, then one must use the relation t' (tin freezing point) = 231.9292°C in evaluating the coefficients of Equation 22 or 24. The reason for the difference between t_{68} (231.9681°C) and t' at the tin point is that the $M(t')$ term does not vanish there.

The level of nonuniqueness of the IPTS-68 achieved by the calibration personnel of the National Bureau of Standards in the range 0 to 419.58°C is estimated as ±0.001°C. Because of the fact that t_{68} in the range 419.58 to 630.74°C is determined by extrapolation, the nonuniqueness level may reach ±0.005°C at the upper limit of the platinum resistance thermometer range in NBS calibrations.[57] The levels of *imprecision* achievable with a single PRT in the same laboratory are estimated as only about one tenth of the nonuniqueness levels.

3. The Range 903 to 1337 K

As in the IPTS-48, the platinum-10 wt% rhodium vs. platinum thermocouple thermometer was specified in the IPTS-68 as the defining instrument from 630.74 to 1064.43°C. Once again, the requirements for a standard thermocouple thermometer were changed in the amended text. The present [IPTS-68(75)] specifications are given in Table 12.

The defining equation again is a quadratic

$$E(t_{68}) = a + bt_{68} + ct_{68}^2 \quad (27)$$

Table 12
THE SPECIFICATIONS FOR A STANDARD PT-10 WT% RH VS. PT THERMOCOUPLE THERMOMETER [IPTS-68(75)]

Pt wire $W(100°C) \geq 1.3920$
Pt-Rh alloy composition nominally 90 wt% Pt, 10 wt% Rh
With a 0°C reference junction,

$E(Au) = 10.334 \pm 0.03$ mV

$E(Au) - E(Ag) = 1.186$ mV $+ 1.7 \times 10^{-4}[E(Au) - 10.334$ mV$]$
$\pm 3 \times 10^{-3}$ mV

$E(Au) - E(630.74°C) = 4.782$ mV $+ 6.3 \times 10^{-4}[E(Au) - 10.334$ mV$]$
$\pm 5 \times 10^{-3}$ mV

The coefficients of Equation 27 are to be determined by calibration at 630.74°C as defined by measurements with a platinum resistance thermometer, at the silver freezing point, and at the gold freezing point. The emf, E(t), is to be measured with one junction at 0°C and the other at the calibration temperature.

The relatively large experimental imprecision of the standard thermocouple thermometer (estimated as ±0.2°C in the range 630 to 1064°C) implies that the temperature uncertainty introduced by defining the lowest thermocouple calibration temperature by extrapolation of a platinum resistance thermometer from the zinc point does not significantly degrade the IPTS-68 in the thermocouple range. Nevertheless, the fact that a major scale junction temperature was not defined by a fixed point in the IPTS-68 is a curious one. Perhaps more than any other characteristic of the IPTS-68, this omission illustrates the lack of planning and discussion within the Consultative Committee for Thermometry prior to preparation of the new scale.[18]

4. The Range Above 1337 K
The ratio obtained from the Planck law of radiation

$$\frac{L_\lambda(T_{68}, K)}{L_\lambda(1337.58 \text{ K})} = \frac{\exp[c_2/(\lambda\ 1337.58)] - 1}{\exp[c_2/(\lambda T_{68})] - 1} \quad (28)$$

again was utilized to define the IPTS-68 above the gold point. In Equation 28, $L_\lambda(T)$ is the spectral concentration of the radiance of a black body at the wavelength λ (in vacuum) and at the temperature T expressed in kelvins. The second radiation constant is given the value $c_2 = 0.014\ 388$ meter · kelvin.

5. Supplementary Information
The text of the IPTS-68 offers only a small amount of supplementary information to help the user of the scale. Topics discussed include the following:

- Standard platinum resistance thermometers
- Standard thermocouple thermometers
- Pressure measurements
- Water triple point and boiling point
- Hydrogen triple and boiling points
- Neon boiling point
- Oxygen triple point and condensation point

- Tin, zinc, silver, and gold freezing points
- Secondary reference temperatures
- History of the international temperature scales
- Temperature differences between T_{68} and T_{48}

The calibration of standard thermometers at the most precise levels is a difficult process. The IPTS-68 text provides only a minimum amount of information on calibration techniques. The reader who wishes to learn more about calibration methods and the use of standard thermometers is referred to the appropriate sections in Chapter 6 as well as the References 57 and 58 at the end of this chapter.

E. 1976 Provisional 0.5 to 30 K Temperature Scale (EPT-76)

Capitalizing upon the continued improvements in low-temperature thermometry following the introduction of the IPTS-68, the Consultative Committee for Thermometry recommended in 1976 that a provisional temperature scale be adopted for the range 0.5 to 30 K. Such a scale could relate the older T_{58} (^{4}He) and T_{62} (^{3}He) vapor-pressure scales[59,60] and the lower end of the IPTS-68 to more accurate thermodynamic measurements of temperature[61] (cf. Chapter 5). In addition, it could provide guidance to thermometrists in cryogenics pending the extension to lower temperatures or replacement of the IPTS-68.

Despite the date given the EPT-76 in its title, the scale was not available until 1979.[62,63]

1. Development of EPT-76

The advances that made the creation of the EPT-76 feasible included the gas thermometry studies of Plumb in the range 2 to 20 K,[64] and those of Berry from 2.6 to 27 K;[61] the paramagnetic salt thermometry of Cetas and Swenson,[65] Cetas,[66] and van Rijn and Durieux;[67] the development of iron-doped rhodium (Rh-Fe) resistance thermometers;[68] and the development of fixed points based upon superconductive transitions in pure metals.[69]

A working group of the Consultative Committee for Thermometry had as its task the development of the EPT-76. Its primary objective in creating the scale was to derive a thermodynamically smooth scale that would be continuous with the IPTS-68 at 27.1 K; a secondary objective was that the new scale should agree closely with thermodynamic temperature throughout its range. These two requirements were only slightly incompatible; whereas the differences between the EPT-76 and the older scales (the 1958 and 1962 helium vapor pressure scales, the NBS Acoustic Scale of 1965, and the IPTS-68) were as large as 8 mK, as shown in Figure 9, the EPT-76 itself appears to differ from Kelvin thermodynamic temperatures by no more than 4 mK (see Chapter 5).

Comparisons of national temperature scales in the cryogenic range by Compton and Ward[70] and by Besley and Kemp[71] permitted explicit expressions to be derived for virtually all existing low-temperature scales.

2. Features of the EPT-76

The EPT-76 is composed essentially of a set of 11 fixed points, shown in Table 13, and 4 approved interpolating methods for some or all of its range, shown in Table 14.

It is clear from a glance at Table 14 that the EPT-76 was effective in reconciling a large number of low-temperature scales. Because it is multiply realizable, the EPT-76 scale cannot provide unambiguous temperatures. The level of nonuniqueness thus incurred is not large, however; Pfeiffer and Kaeser[72] have reported agreement among several methods of realization within about ± 1 mK.

The most precise realization of the EPT-76 currently available is provided by Rh-Fe thermometers. The National Physical Laboratory of England has furnished calibrations of these thermometers against the NPL-75 gas thermometry scale; this service provides the user with direct access to the temperature data upon which the EPT-76 is based.

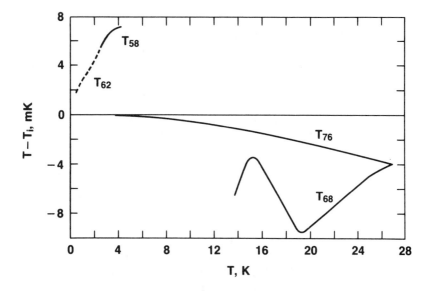

FIGURE 9. Difference between thermodynamic temperatures, represented within about 1 mK by the NPL-75 Gas Thermometer Scale[61] as extended by paramagnetic salt thermometry,[65] and four current practical temperature scales.[56,59,60,63]

Table 13
FIXED POINTS OF THE EPT-76

Reference point	Assigned temperature, T_{76} (K)
Superconducting transition point of cadmium[a]	0.519
Superconducting transition point of zinc	0.851
Superconducting transition point of aluminum	1.1796
Superconducting transition point of indium	3.4145
Boiling point of ^{4}He[b]	4.2221
Superconducting transition point of lead	7.1999
Triple point of equilibrium hydrogen[c]	13.8044
Boiling point of equilibrium hydrogen at 33,330.6 Pa pressure[c]	17.0373
Boiling point of equilibrium hydrogen[b,c]	20.2734
Triple point of neon[d]	24.5591
Boiling point of neon[b,c,d]	27.102

[a] The superconducting transition point is the temperature midway between the superconductive and normal states of the metal in negligible magnetic field as provided by the NBS SRM 767 device.[69]
[b] Unless otherwise stated, the boiling point is to be measured under a pressure of 101,325 Pa (1 standard atmosphere).
[c] Also a fixed point of the IPTS-68, but not necessarily with the same assigned temperature.
[d] The two neon points are to be realized with neon of natural composition (2.7 mmol of ^{21}Ne and 92 mmol of ^{22}Ne per 0.905 mol of ^{20}Ne).

V. INTERNATIONAL TEMPERATURE SCALES OF THE FUTURE

As our ability to determine values of temperature on the KTTS has improved, we have modified and extended our international temperature scales repeatedly. Table 15 shows the progression in temperature assignments given to fixed points in the various scales to date; in Chapter 5, Figures 8, 10, 16, 18, and 20, we portray the apparent deviations of temperatures defined by the IPTS-68 and EPT-76 from those given by the KTTS.

Table 14
APPROVED METHODS FOR REALIZING THE EPT-76

1. Use of a thermodynamic interpolating instrument, such as a gas thermometer or a magnetic thermometer, calibrated at one or more reference points of Table 13
2. Use, above 13.81 K, of the IPTS-68 together with the table of differences between the two scales provided in the EPT-76 text
3. Use, below 5 K, of the 1958 ^{4}He vapor pressure scale or the 1962 ^{3}He vapor pressure scale together with the table of differences between those scales and the EPT-76 provided in the EPT-76 text
4. Use of any of the laboratory scales for which the differences from the EPT-76 are provided in its text (these include the NBS 2 to 20 K scale of 1965; magentic scales originating at Iowa State University, Kamerlingh Onnes Laboratory, National Measurement Laboratory of Australia, Physico-Technical and Radio-Technical Measurements Institute of Russia, and the National Physical Laboratory in England; the NPL-75 Gas Thermometry Scale of the National Physical Laboratory in England; and the NBS version of the IPTS-68)

Table 15
COMPARISON OF TEMPERATURE SCALE FIXED-POINT ASSIGNMENTS (IN KELVINS OR THE EQUIVALENT)

Point	NHS[a]	ITS-27	ITS-48(60)	IPTS-68(75)	EPT-76
Au fp[b]	—	1336.15	1336.15	1337.58	—
Ag fp	—	1233.65	1233.95	1235.08	—
S bp[c]	—	717.75	717.75	—	—
Zn fp	—	—	(692.655)	692.73	—
Sn fp	—	—	—	(505.1181)	—
H$_2$O bp	373	373.15	373.15	373.15	—
H$_2$O tp[d]	—	—	(273.16)	273.16	—
H$_2$O fp	273	273.15	273.15	—	—
O$_2$ bp[e]	—	90.18	90.18	90.188	—
Ar tp	—	—	—	(83.798)	—
O$_2$ tp	—	—	—	54.361	—
Ne bp	—	—	—	27.102	27.102
Ne tp	—	—	—	—	24.5591
H$_2$ bp	—	—	—	20.28	20.2734
H$_2$ bp[f]	—	—	—	17.042	17.0373
H$_2$ tp	—	—	—	13.81	13.8044
Pb st[g]	—	—	—	—	7.1999
^{4}He bp	—	—	—	—	4.2221
In st	—	—	—	—	3.4145
Al st	—	—	—	—	1.1796
Zn st	—	—	—	—	0.851
Cd st	—	—	—	—	0.519

[a] NHS = normal hydrogen scale.
[b] fp = Freezing point.
[c] bp = Boiling point at 101,325 Pa.
[d] tp = Triple point.
[e] Changed in 1975 to the condensation point.
[f] Reduced-pressure boiling point, at P = 33 330.6 Pa.
[g] st = Superconducting transition point.

Early in this chapter we mentioned the desirability of defining temperatures on the international practical temperature scale simply by invoking thermodynamic laws. Such a scale would be attractive, since it could embody exactly the unique Kelvin thermodynamic temperatures. Only one fixed point besides the absolute zero of temperature would require an assigned value. In the next chapter, however, we discuss the level reached in thermodynamic thermometry by current practitioners of this science. It will become clear from a study of these thermodynamic experiments that the present "practical" scales still offer by far the more precise realization of temperature in every range. The convenience enjoyed by users of the practical scales in measuring temperature also is relatively much greater than that provided by the involved and cumbersome apparatus generally used in the more fundamental studies.

Despite their practical superiority to the thermodynamic scale, distinct improvements can be made in the IPTS-68 and the EPT-76. These include the following features:

1. *Revisions in numerical assignments of fixed-point temperatures and interpolation equations to bring the practical scale into closer agreement with thermodynamic temperatures* — The differences between thermodynamic and practical temperatures as currently understood are discussed in Chapter 5; they range from 0.01 K at low temperatures to 0.3 K or more near the gold point.
2. *Extension of the IPTS to 0.5 K or lower* — Thermodynamic temperatures have been measured to temperatures as low as 0.01 K, with estimated uncertainty levels as small as 0.5 mK. Interest in the use of the EPT-76 for commerical and scientific calibrations indicates the need for an IPTS below 14 K.
3. *Replacement of the platinum-10% rhodium vs. platinum thermocouple thermometer as a defining instrument of the IPTS* — The imprecision of the thermocouple thermometer, about 0.2 K, was known even in 1968 to be 5 to 10 times larger than the imprecision of the platinum resistance thermometer above 630°C. Extension of the platinum resistance thermometer range to a fixed-point junction temperature with the radiation-thermometer definition of the IPTS also will eliminate the existence of a range of temperature defined by extrapolation of the platinum resistance thermometer indication beyond its highest calibration fixed-point temperature.
4. *Inclusion, in the IPTS or in its supplementary information, of explicit and realistic scale uncertainty levels* — Statements should be included concerning its uncertainty with respect to the KTTS, its nonuniqueness, and its experimental imprecision in conjunction with specific calibration procedures for the guidance of scale users.

As this book is being written, there still are gaps in our knowledge of the detail needed to formulate a replacement IPTS based upon the features listed above. However, it is safe to say that a very superior IPTS soon will lie within our grasp.

GENERAL REFERENCES

Beattie, J. A. and Oppenheim, I., *Principles of Thermodynamics*, Elsevier Scientific, New York, 1979, chap. 2.

Zemansky, M. W., *Heat and Thermodynamics*, 5th ed., McGraw-Hill, New York, 1968, chap. 8.

1975 Amendment to the IPTS-68, official version: Comptes Rendus de Séances de la Quinziéme Conférence Général des Poids et Mesures, A1-A121; English version: *Metrologia*, 12, 7, 1976.

Preston-Thomas, H., Bloembergen, P., and Quinn, T., *Supplementary Information for the IPTS-68 and EPT-76*, International Bureau of Weights and Measures, Sèvres, France, July 1983.

Durieux, M., Astrov, D. N., Kemp, W. R. G., and Swenson, C. A., EPT-76, the 1976 provisional 0.5 K to 30 K temperature scale, *Metrologia*, 15, 65, 1979.

Riddle, J. L., Furukawa, G. T., and Plumb, H. H., *Platinum Resistance Thermometry*, U.S. National Bureau of Standards Monogr. 126, appendices A-F, April 1973.

Quinn, T. J., *Temperature*, Academic Press, London, 1983, chap. 2.

REFERENCES

1. **Eisenhart, C.**, Realistic evaluation of the precision and accuracy of instrument calibration systems, *J. Res. Nat. Bur. Stand.*, 67C, 161, 1963.
2. **Guildner, L. A.**, private communication.
3. **Ku, Harry H.**, Statistical concepts in metrology, in *Handbook of Industrial Metrology*, American Society of Tool and Manufacturing Engineers, Ku, H. H., Ed., Prentice-Hall, New York, 1967, 20; reprinted in *Precision Measurement and Calibration; Statistical Concepts and Procedures*, Ku, H. H., Ed., U.S. National Bureau of Standards Special Publ. 300, Vol. 1, February 1969, 296.
4. Documents diplomatiques de al Conférence du Metre, Paris, Imprimeria Nationale, 1875.
5. **Page, C. H. and Vigoureux, P.**, Eds., *The International Bureau of Weights and Measures 1875-1975*, U.S. National Bureau of Standards Special Publ. 420, May 1975.
6. **de Boer, J.**, *The International Bureau of Weights and Measures 1875—1975*, Page, C. H. and Vigoureux, P., Eds., U.S. National Bureau of Standards Special Publ. 420, May 1975, 9.
7. Comptes Rendus des Séances de la Dix-septième Conférence Générale des Poids et Mesures, October 1983, International Bureau of Weights and Measures (Bureau International des Poids et Mesures), Sèvres, France.
8. *Comité International des Poids et Mesures — Procès-Verbeaux des Séances* 71st session, Bureau International des Poids et Mesures, Sèvres, France, 1984.
9. Comité Consultatif de Thermométrie, 15e Session — 1984, Bureau International des Poids et Mesures, Sèvres, France.
10. **Preston-Thomas, H. and Quinn, T. J.**, working paper submitted to the Comité Consultatif de thermométrie, 15e Session — 1984, Bureau International des Poids et Mesures, Sèvres, France, 1984.
11. **Callendar, H. L.**, On the practical measurement of temperature, *Philos. Trans. R. Soc. London A*, 178, 161, 1887.
11a. **Meuller, E. F.**, Precision resistance thermometry, in *Temperature, Its Measurement and Control in Science and Industry*, Vol. 1, Reinhold, New York, 1941, 162.
12. Comptes Rendus des Séances de la Septiéme Conférence Générale des Poids et Mesures (in French), Bureau International des Poids et Mesures, Sèvres, France, 94—99.
12a. **Burgess, G. K.**, The international temperature scale, *J. Res. Natl. Bur. Stand.*, 1, 635, 1928.
13. **Van Dusen, M. S.**, Platinum-resistance thermometry at low temperatures, *J. Am. Chem. Soc.*, 47, 326, 1925.
14. **Wensel, H. T.**, *Temperature, Its Measurement and Control in Science and Industry*, Vol. 1, Reinhold, New York, 1941, 3.
15. Comptes Rendus des Séances de la Neuviéme Conférence Générale des Poids et Mesures (in French), Bureau International des Poids et Mesures, Sèvres, France, 89—100.
15a. **Stimson, H. F.**, The International Temperature Scale of 1948, *J. Res. Natl. Bur. Stand.*, 42, 209, 1949.
16. **Hall, J. A.**, The international temperature scale, in *Temperature, Its Measurement and Control in Science and Industry*, Wolfe, H. D., Ed., Vol. 2, Reinhold, New York, 1955, 115.
17. Comptes Rendus de la Onziéme Conférence Générale des Poids et Mesures (in French), Bureau International des Poids et Mesures, Sèvres, France, 1960, 124—133.
17a. **Stimson, H. F.**, The International Practical Temperature Scale of 1948; text revision of 1960, *J. Res. Natl. Bur. Stand.*, 65A, 139, 1961; also published as NBS Monogr. 37, September 1961.
18. **Preston-Thomas, H.**, The origin and present status of the IPTS-68, in *Temperature, Its Measurement and Control in Science and Industry*, Vol. 4, Plumb, H. H., Ed.-in-Chief, Instrument Society of America, Pittsburgh, 1972, 3.
19. **Wolfe, H. D.**, Ed., *Temperature, Its Measurement and Control in Science and Industry*, Vol. 2, Reinhold, New York, 1955.
19a. **Herzfeld, C. M.**, Ed.-in-Chief, *Temperature, Its Measurement and Control in Science and Industry*, Vol. 3, Parts 1 and 2, Reinhold, New York, 1962.
20. *Metrologia*, 1, 1965.
21. **Beattie, J. A.**, Gas thermometry, in *Temperature, Its Measurement and Control in Science and Industry*, Wolfe, H. D., Ed., Vol. 2, Reinhold, New York, 1955, 63.
22. **Keller, W. E.**, Thermodynamic temperatures obtained from P-V isotherms of He gas Between 2 and 4°K, in *Temperature, Its Measurement and Control in Science and Industry*, Wolfe, H. D., Ed., Vol. 2, Reinhold, New York, 1955, 99.
23. **Moser, H.**, High temperature gas thermometry, in *Temperature, Its Measurement and Control in Science and Industry*, Wolfe, H. D., Ed., Vol. 2, Reinhold, New York, 1955, 103; also **Moser, H.**, Gasthermometrie bei höheren temperaturen, *Metrologia*, 1, 68, 1965.
24. **Barber, C. R.**, Gas thermometry at low temperatures, in *Temperature, Its Measurement and Control in Science and Industry*, Vol. 3, Part 1, Herzfeld, C. M., Ed.-in-Chief, Reinhold, New York, 1962, 103.

25. **Moessen, G. W., Aston, J. G., and Ascah, R. G.**, The Pennsylvania State University Thermodynamic Temperature Scale below 90°K and the normal boiling points of oxygen and normal hydrogen on the Thermodynamic Scale, in *Temperature, Its Measurement and Control in Science and Industry,* Vol. III, Part 1, Herzfeld, C. M., Ed.-in-Chief, Reinhold, New York, 1962, 90.
26. **Borovik-Romanov, A. C., Strelkov, P. G., Orlova, M. P., and Astrov, D. N.**, The IMPR Temperature Scale for the 10 to 90°K Region, in *Temperature, Its Measurement and Control in Science and Industry,* Vol. 3, Part 1, Herzfeld, C. M., Ed.-in-Chief, Reinhold, New York, 1962, 113.
27. **Plumb, H. and Cataland, G.**, Acoustical thermometer and the NBS Provisional Temperature Scale 2-20 (1965), *Metrologia,* 2, 127, 1966.
28. **Preston-Thomas, H. and Kirby, C. G. M.**, Gas thermometer determinations of the Thermodynamic Temperature Scale in the range $-183°C$ to $100°C$, *Metrologia,* 4, 30, 1968.
29. **Rogers, J. S., Tainsh, R. J., Anderson, M. S., and Swenson, C. A.**, Comparison between gas thermometer, acoustic, and platinum resistance temperature scales between 2 and 20 K, *Metrologia,* 4, 47, 1968.
30. **Moser, H.**, A review of recent determinations of thermodynamic temperatures of fixed points above 419°C, in *Temperature, Its Measurement and Control in Science and Industry,* Vol. 3, Part 1, Herzfeld, C. M., Ed.-in-Chief, Reinhold, New York, 1962, 167.
31. **van Dijk, H.**, Selected values for the thermodynamic temperatures of thermometric fixed points below 0°C, in *Temperature, Its Measurement and Control in Science and Industry,* Vol. 3, Part 1, Herzfeld, C. M., Ed.-in-Chief, Reinhold, New York, 1962, 173.
32. **Harper, A. F. A., Kemp, W. R. G., and Lowenthal, G. C.**, The extension of the International Practical Temperature Scale below 90° K, in *Temperature, Its Measurement and Control in Science and Industry,* Vol. 3, Part 1, Herzfeld,C. M., Ed.-in-Chief, Reinhold, New York, 1962, 339.
33. **Barber, C. R.**, Low-temperature scales 10 to 90° K, in *Temperature, Its Measurement and Control in Science and Industry,* Vol. 3, Part 1, Herzfeld, C. M., Ed.-in-Chief, Reinhold, New York, 1962, 345.
34. **Sharevskaya, D. I., Strelkov, P. G., Borovik-Romanov, A. S., Astrov, D. N., and Orlova, M. P.**, On methods of establishment of a practical scale in the range 10 to 90° K, in *Temperature, Its Measurement and Control in Science and Industry,* Vol. 3, Part 1, Herzfeld, C. M., Ed.-in-Chief, Reinhold, New York, 1962, 351.
35. **Hudson, R. P.**, The helium vapor-pressure scale of temperature, in *Temperature, Its Measurement and Control in Science and Industry,* Wolfe, H. D., Ed., Vol. 2, Reinhold, New York, 1955, 185.
36. **Hudson, R. P.**, The thermodynamic scale of temperature below 1° K, in *Temperature, Its Measurement and Control in Science and Industry,* Vol. 3, Part 1, Herzfeld, C. M., Ed.-in-Chief, Reinhold, New York, 1962, 51.
37. **Clement, J. R.**, The 1958 He^4 temperature scale, in *Temperature, Its Measurement and Control in Science and Industry,* Vol. 3, Part 1, Herzfeld, C. M., Ed.-in-Chief, Reinhold, New York, 1962, 67.
38. **Roberts, T. R., Sydoriak, S. G., and Sherman, R. H.**, Problems in establishment of a He^3 scale of temperature, in *Temperature, Its Measurement and Control in Science and Industry,* Vol. 3, Part, 1, Herzfeld, C. M., Ed.-in-Chief, Reinhold, New York, 1962, 75.
39. **Barber, C. R. and Horsford, A.**, Differences between the thermodynamic scale and the International Practical Scale of Temperature from 0°C to $-183°C$, *Metrologia,* 1, 75, 1965.
40. **Hall, J. A.**, The radiation scale of temperature between 175°C and 1063°C, *Metrologia,* 1, 140, 1965.
41. **Heusinkveld, W. A.**, Determination of the differences between the thermodynamic and the practical Temperature Scale in the range 630 to 1063°C from radiation measurements, *Metrologia,* 2, 61, 1966.
42. **Orlova, M. P., Belyansky, L. B., Astrov, D. N., and Sharevskaya, D. I.**, A new determination of the normal oxgyen boiling temperature, *Metrologia,* 2, 163, 1966.
43. **Kostkowski, H. J.**, Lack of uniqueness in the International Practical Temperature Scale above the gold point, *Metrologia,* 3, 28, 1967.
44. **Snellemann, W.**, A flame as a primary standard of temperature, *Metrologia,* 4, 117, 1968.
45. **Preston-Thomas, H. and Bedford, R. E.**, Practical temperature scales between 11 K and 273 K, *Metrologia,* 4, 14, 1968.
46. **Betteridge, W., Rhys, D. W., and Withers, D. F.**, Laboratory control of production of platinum for thermometry, in *Temperature, Its Measurement and Control in Science and Industry,* Vol. 3, Part 1, Herzfeld, C. M., Ed.-in-Chief, Reinhold, New York, 1962, 263.
47. **Berry, R. J.**, The stability of platinum resistance thermometers up to 630°C, in *Temperature, Its Measurement and Control in Science and Industry,* Vol. 3, Part 1, Herzfeld, C. M., Ed.-in-Chief, Reinhold, New York, 1962, 301.
48. **Evans, J. P. and Burns, G. W.**, A study of stability of high temperature platinum resistance thermometers, in *Temperature, Its Measurement and Control in Science and Industry,* Vol. 3, Part 1, Herzfeld, C. M., Ed.-in-Chief, Reinhold, New York, 1962, 313.

49. **Corruccini, R. J.**, Interpolation of platinum resistance thermometers, 10 to 273.15°K, in *Temperature, Its Measurement and Control in Science and Industry*, Vol. 3, Part 1, Herzfeld, C. M., Ed.-in-Chief, Reinhold, New York, 1962, 329.
50. **van Dijk, H.**, On the use of platinum resistance thermometers for thermometry below 90° K. A study of two possibilities for relating resistance data for platinum thermometers to each other, in *Temperature, Its Measurement and Control in Science and Industry*, Vol. 3, Part 1, Herzfeld, C. M., Ed.-in-Chief, Reinhold, New York, 1962, 365.
51. **Barber, C. R. and Hayes, J. G.**, The derivation of the Provisional Reference Table CCT 64, T = f(W) for platinum resistance thermometers for the range from 12 to 273.15°K, *Metrologia*, 2, 11, 1966.
52. **Berry, R. J.**, Platinum resistance thermometry in the range 630—900°C, *Metrologia*, 2, 80, 1966.
53. **Lee, R. D.**, The NBS photoelectric pyrometer and its use in realizing the International Practical Temperature Scale above 1063°C, *Metrologia*, 2, 150, 1966.
54. **Quinn, T. J. and Barber, C. R.**, A lamp as a reproducible source of near blackbody radiation for precise pyrometry up to 2700°C, *Metrologia*, 2, 19, 1966.
55. Comptes Rendus de Séances de la Treiziéme Conférence Générale des Poids et Mesures (in French), Bureau International des Poids et Mesures, Sèvres, France, A1—A24; English version: *Metrologia*, 5, 35, 1969.
56. 1975 Amendment to the IPTS-68, Comptes Rendus de Séances de la Quinziéme Conférence Générale des Poids et Mesures (in French), Bureau International des Poids et Mesures, Sèvres, France, A1—A21; English version: *Metrologia*, 12, 7, 1976.
57. **Riddle, J. L., Furukawa, G. T., and Plumb, H. H.**, *Platinum Resistance Thermometry*, U.S. National Bureau of Standards, Monogr. 126, April 1973.
58. **Preston-Thomas, H., Bloembergen, P., and Quinn, T.**, Supplementary information for the IPTS-68 and the EPT-76, *Monograph of the International Bureau of Weights and Measures*, 1983. Available from the BIPM, Pavillon de Breteuil, F-92310 Sèvres, France.
59. **Brickwedde, F. G., van Dijk, H., Durieux, M., Clement, J. R., and Logan, J. K.**, The 1958 He^4 scale of temperatures, *J. Res. Natl. Bur. Stand.*, 64A, 1, 1960; also issued as NBS Monograph 10, June 1960.
60. **Sydoriak, S. G., Sherman, R. H., and Roberts, T. R.**, The 1962 He^3 scale of temperatures, parts I to IV, *J. Rev. Natl. Bur. Stand.*, 68A, 547, 1964.
61. **Berry, K. H.**, A low-temperature gas thermometer scale from 2.6 K to 27.1 K, *Metrologia*, 15, 89, 1979.
62. **Durieux, M., Astrov, D. N., Kemp, W. R. G., and Swenson, C. A.**, The derivation and development of the 1976 Provisional 0.5 K to 30 K Temperature Scale, *Metrologia*, 15, 57, 1979.
63. **Durieux, M., Astrov, D. N., Kemp, W. R. G., and Swenson, C. A.**, EPT-76, the 1976 Provisional 0.5 K to 30 K Temperature Scale, *Metrologia*, 15, 65, 1979.
64. **Plumb, H. H. and Cataland, G.**, Acoustical thermometer and the National Bureau of Standards Provisional Temperature Scale 2-20 (1965), *Metrologia*, 2, 127, 1966.
65. **Cetas, T. C. and Swenson, C. A.**, A paramagnetic salt temperature scale, 0.9 to 18 K, *Metrologia*, 8, 46, 1972.
66. **Cetas, T. C.**, A magnetic temperature scale from 1 to 83 K, *Metrologia*, 12, 27, 1976.
67. **van Rijn, C. and Durieux, M.**, A magnetic temperature scale between 1.5 K and 30 K, in *Temperature, Its Measurement and Control in Science and Industry*, Vol. 4, Plumb, H. H., Ed.-in-Chief, Instrument Society of America, Pittsburgh, 1972, 73.
68. **Rusby, R. L.**, Resistance thermometry using rhodium-iron, 0.1 K to 273 K, in *Temperature Measurement 1975*, Billing, B. F. and Quinn, T. J., Eds., Conference Series No. 26, The Institute of Physics, London, 1975, 125.
69. **Schooley, J. F., Soulen, R. J., Jr., and Evans, G. A., Jr.**, *Preparation and Use of Superconductive Fixed Point Devices, SRM 767*, Natl. Bur. Stand. (U.S.) Spec. Publ., 260-44, Dec. 1972, 35 p.
70. **Compton, J. P. and Ward, S. D.**, International comparison of low-temperature platinum resistance thermometers, in *Temperature Measurement 1975*, Billing, B. F. and Quinn, T. J., Eds, Conference Series No. 26, The Institute of Physics, London, 1975, 91. A more extensive discussion of this comparison is given by **Ward, S. D. and Compton, J. P.**, Intercomparison of platinum resistance thermometers and T_{68} calibrations, *Metrologia*, 15, 31, 1979.
71. **Besley, L. M. and Kemp, W. R. G.**, An intercomparison of temperature scales in the range 1 to 30 K using germanium resistance thermometry, *Metrologia*, 13, 35, 1977.
72. **Pfeiffer, E. R. and Kaeser, R. S.**, Realization of the 1976 Provisional 0.5 K to 30 K Temperature Scale at the National Bureau of Standards, in *Temperature, Its Measurement and Control in Science and Industry*, Vol. 5, Schooley, J. F., Ed.-in-Chief, American Institute of Physics, New York, 1982, 159.

Chapter 5

THE MEASUREMENT OF THERMODYNAMIC TEMPERATURES

I. INTRODUCTION

Any temperature-dependent property of matter can be used to measure thermodynamic temperatures, providing that there is an established theoretical relationship between the property and temperature. The uncertainties with which the temperature dependence of most properties can be predicted, however, are quite large. Only a handful of techniques have contributed to our knowledge of thermodynamic temperatures. Of these, we outline in this chapter the following: gas thermometry, radiation thermometry, noise thermometry, and gamma-ray anisotropy thermometry.

II. GAS THERMOMETRY

Throughout the history of scientific thermometry, there has been no other method for thermodynamic temperature determination to compare in effectiveness with gas thermometry. All of the international temperature scales have been based upon thermodynamic measurements of temperature by gas thermometry. The ideal gas law

$$Pv = nRT \qquad (1)$$

is the basis for gas thermometry. After the reciprocal relationship between pressure and volume in an isothermal gas was elucidated by Boyle and his contemporaries and the temperature dependence of the volume of a gas at constant pressure was shown to be essentially linear by Charles, Gay-Lussac, and others, the ideal gas law was given a firm theoretical basis by the development of the sciences of thermodynamics and statistical mechanics. This work was discussed in some detail in Chapter 2.

We shall spend some time presently calling attention to the difficulties inherent in using the ideal gas law to measure with high accuracy the thermodynamic temperatures of real gases. Nevertheless, in comparison with other thermodynamic methods, the straightforward nature of the ideal gas law and the intrinsic simplicity of its adaptation to real-gas measurements have kept gas thermometry first among the techniques that are useful for thermodynamic temperature determinations.

Soon after Kelvin proposed the creation of a thermodynamic scale of temperature based upon the efficiency of an ideal heat engine, the thermal expansion of a gas was considered as a possible thermodynamic thermometer. The regular use of gas thermometers for the purpose of calibrating secondary thermometers commenced with the efforts of Chappius[1] and Callendar.[2] Chappius developed the normal hydrogen scale of temperatures at the International Bureau of Weights and Measures in order to give uniformity to the calibration of mercury-in-glass thermometers used in length metrology. Callendar wished to establish a practical temperature scale based upon the use of platinum resistance thermometers.

Many thermometrists have contributed to further progress in gas thermometry from the beginning of this century. Detailed discussions of the results obtained from their work can be found in articles by Beattie,[3,3a] Hoge and Brickwedde,[4] Day and Sosman,[5] Moser,[6] and Barber.[7]

Methods were developed throughout the years to counter a variety of measurement problems in experimental gas thermometry. These problems include an inaccurate knowledge of the gas constant; deviations from ideality found in real gases; dead-volume, thermal-expan-

Table 1
GAS CONSTANT VALUES

Year	R (J/mol · K)	Ref.
1929	8.3140 ± 0.0005	8
1969	8.31434 ± 0.00035	9
1973	8.31441 ± 0.00026	10

sion, hydrostatic-head, and thermomolecular corrections peculiar to each apparatus; and contamination of the working gas by impurities. We discuss each of these problems briefly in the following sections.

A. Uncertainty of the Ideal Gas Constant

The methods chosen for gas thermometry reflect the uncertain value of R, the ideal gas constant. R and the Boltzmann constant, k, are related by the equation $R = N_o k$, where N_o is Avogadro's constant. Table 1 shows the chronological sequence of values and uncertainties of the gas constant.

The relatively high uncertainty level in the value of the gas constant (~30 ppm) limits the accuracy with which an arbitrary temperature can be measured using the ideal gas law. If, for example, P, v, and n were to be known perfectly well for a certain quantity of an ideal gas at temperature T, then the uncertainty in T could not be less than 30 ppm — say, ±0.009 K at 300.000 K.

Despite this uncertainty in R, gas thermometry measurements can be performed with inaccuracy levels below 30 ppm; this is done by comparing measurements made at the unknown temperature with similar measurements obtained at a well-known reference temperature. In principle, one evaluates the ratio

$$\frac{P_0 v_0}{P_u v_u} = \frac{n_0 R T_0}{n_u R T_u} \quad \text{or} \quad T_u = T_0 \left(\frac{n_0 P_u v_u}{n_u P_0 v_0} \right) \quad (2)$$

where the subscripts 0 and u refer respectively to measurements at the well-known and the unknown temperature in kelvins. Use of this "ratio" technique has been universal since the first days of gas thermometry. Acceptance of the water triple point as defining the principal reference temperature in the KTTS makes the temperature 273.16 K available as a convenient reference state for gas thermometry. But while this technique does allow the elimination of R from the calculation of the gas-thermometer temperature, it introduces some experimental complexities into the measurements, too, as we shall see.

There are several methods of gas thermometry that differ according to the manner in which measurements are made (see, for example, References 11 and 11a). The most prevalent method is the "constant-volume" one, in which a single gas bulb is loaded with n moles of the working gas and thermostated first at the reference temperature T_0 and then at T_u, attaining equilibrium pressures P_0 and P_u, respectively. The unknown temperature T_u can then be calculated from Equation 2

$$T_u = (P_u/P_0) T_0 \quad (3)$$

In this calculation we assume ideal behavior of the gas and of the apparatus (e.g., no thermal expansion of the bulb is assumed to take place upon heating).

A second method is called "constant-pressure" gas thermometry. In this case, a second gas bulb of variable volume, thermostated at the reference temperature T_0, shares the working gas with the main bulb of volume v_1 whenever the latter is thermostated above its lowest temperature. The volume v_2 of the second bulb is varied so that the measured pressure is always constant. The unknown temperature T_u is calculated by equating the number of moles $n = (Pv)/(RT)$ of working gas in the two states. If T_u is higher than T_0, then the second bulb is not used in the reference state and $n_{ref} = (P_0v_1)/(RT_0)$. As the temperature is increased to T_u, the auxiliary bulb, thermostated at T_0, is introduced to maintain the total pressure at P_0. Then $n_{final} = [(P_0v_1)/(RT_u)] + [(P_0v_2)/(RT_0)]$. Equating n_{ref} and n_{final} allows one to calculate T_u. In a similar way, T_u can be evaluated in the case that $T_u < T_0$. In that case, of course, the reference state involves both volumes. In terms of the measured bulb volumes v_1 and v_2, the unknown temperature T_u is

$$T_u = T_0v_1/(v_1 - v_2) \quad \text{or} \quad T_u = T_0v_1/(v_1 + v_2) \tag{4}$$

with the choice of sign depending upon whether the unknown temperature is higher or lower than the reference temperature.

A third method of gas thermometry is called the "constant-bulb-temperature" technique. In this case, the working gas is thermostated in a bulb of volume v_1 at the unknown temperature T_u and a measurement is made of the pressure P_1. Then the gas is shared with a second bulb of volume v_2 that is thermostated at the reference temperature T_0. The pressure is remeasured as P_2. Again writing Equation 1 as $n = (Pv)/(RT)$, one can note, as in the previous case, that the total amount of gas originally in one bulb now occupies both. Thus

$$n = (P_1v_1)/(RT_u) = (P_2v_1)/(RT_u) + (P_2v_2)/(RT_0) \tag{5}$$

and T_u can be calculated readily. If the volume of the second bulb was chosen so that the pressure P_2 is exactly one half its former value, then Equation 5 simplifies to

$$T_u = T_0(v_1/v_2) \tag{6}$$

B. Deviations from the Ideal Gas Law

The methods chosen for gas thermometry also reflect the fact that no real gas can be represented by the ideal gas law with high accuracy. The quantity PV/RT is called the "compressibility factor" [not to be confused with the quantity $(-\partial v/\partial P)/v$ of a gas].[12] The compressibility factor tends to vary both with pressure and with temperature; it is shown in Figures 1 and 2 for several gases.

Several modifications to the ideal gas law have been introduced to account for the non-ideality of real gases. One of these is the virial equation, derived from kinetic theory by Clausius.[13]

$$Pv = nRT\left(1 + \frac{nB_v}{v} + \frac{n^2C_v}{v^2} + \frac{n^3D_v}{v^3} + \ldots\right) \tag{7}$$

In this equation, which really is an expansion in powers of the gas density, the coefficients B_v, C_v, and D_v are called the second, third, and fourth volume virial coefficients. They are constant for a particular gas at a particular temperature. (For tabular values, see, for example, Reference 14.)

The behavior of the second virial coefficient for a number of gases is shown in Figure 3. For each gas, a temperature exists at which $B_v(T) = 0$ so that the gas behaves ideally to second order in the viral expansion; this characteristic temperature is called the Boyle temperature.

Note that, in the limited range of 0 to 100°C, nitrogen behaves very nearly as an ideal gas. For gas thermometry work over an extended range of temperature, however, helium and hydrogen are preferable.

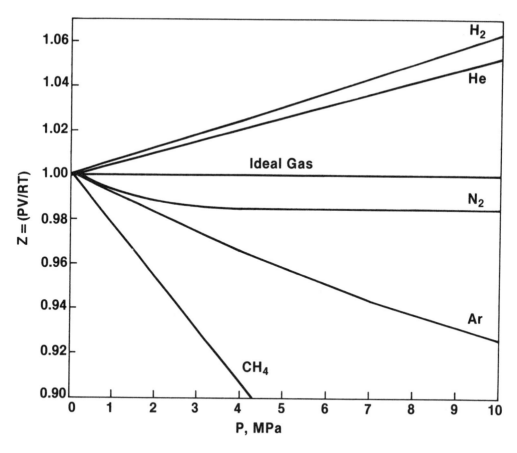

FIGURE 1. "Compressibility factors" $Z = (PV/RT)$ at 273 K for several gases as functions of pressure. Note that both positive and negative deviations from ideality occur.

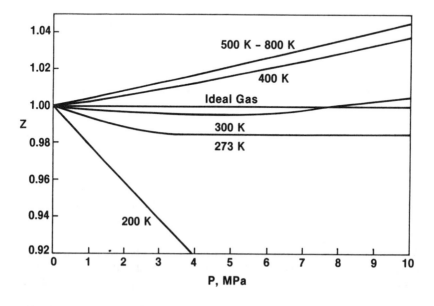

FIGURE 2. "Compressibility factor" $Z = PV/RT$ for nitrogen at several temperatures, as functions of pressure. Note that room-temperature nitrogen behaves nearly as an ideal gas at pressures below 10 MPa.

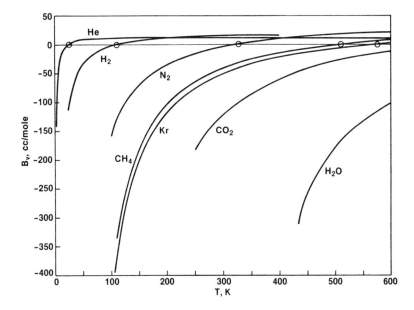

FIGURE 3. Temperature dependences of the second (volume) virial coefficient B_v for several gases. The circled intersection of each curve with $B_v = 0$ is called the Boyle temperature; at that temperature, the gas behaves ideally to second order.

The virial equation also can be employed in the form of a power dependence in pressure

$$Pv = nRT (1 + B_p P + C_p P^2 + \ldots) \quad (8)$$

where

$$B_p \equiv B_v/(RT) \quad \text{and} \quad C_p \equiv (C_v - B_v^2)/(RT)^2$$

For experimental gas thermometry at low to moderate densities, the ideal gas law generally need be corrected only by the use of the second virial coefficient.

Several other modifications to the ideal gas law have been introduced to account for the nonideality of various real gases. These include the Van der Waals equation

$$(P + n^2 a/v^2)(v - nb) = nRT \quad (9)$$

and Berthelot's equation

$$Pv = nRT [1 + (9PT_c)(1 - 6T_c^2/T^2)/(128 \, P_c T)] \quad (10)$$

where the subscripts c refer to critical parameters.[13]

Most thermometrists currently performing gas thermometry measurements employ the virial Equation 7 or 8 in analyzing their results; the needed values of the virial coefficients generally are obtained from the work of others. However, the value of the virial coefficients at any desired temperature also can be determined by measuring the quantity Pv/nRT as a function of pressure or of gas density. Appropriate values of the virial coefficients then are derived by fitting the corresponding equation to the experimental data. This technique is well illustrated by the work of Berry,[15] who applied it to a modification of the "constant-

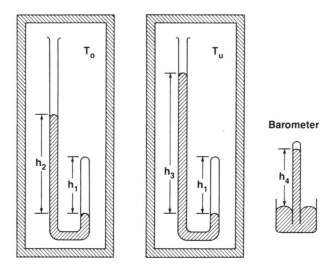

FIGURE 4. A constant-volume gas thermometer with no "dead volume". The unknown temperature, T_u (in Kelvins), is given very approximately by the quantity $T_0(h_3 + h_4)/(h_2 + h_4)$. The test gas is kept at a constant volume indicated by the length of manometer h_1 by increasing the height of the mercury in the open arm of manometer from h_2 to h_3 as the temperature is increased from T_0 to T_u.

bulb-temperature" method of gas thermometry in order to determine not only the value of the unknown temperature, but also the value of the virial coefficients of the working gas.

C. Dead Space Corrections

In its simplest form, a gas thermometer is composed of a bulb that contains the working gas and a pressure-measuring device. Figure 4 shows once again the P-v apparatus used by Boyle (cf. Figure 3 of Chapter 2). Boyle's manometer could suffice, in principle, for the measurement of temperatures over a limited range by gas thermometry. The unknown thermodynamic temperatures T_u would be approximated by the equation

$$T_u = T_0(h_3 + h_4)/(h_2 + h_4) \qquad (11)$$

In practice, one would find it extremely awkward to maintain the manometer temperature at a uniform value and yet measure its pressure. Furthermore, sizeable systematic errors would result from variations in the mercury density and in its vapor pressure between the two temperatures, and from the thermal expansion of the manometer glass.

A common practice in constant-volume gas thermometry is to locate both the single gas bulb and the manometer in separate, carefully controlled, constant-temperature environments, as indicated in Figure 5. This experimental arrangement offers a great improvement over the simpler apparatus of Figure 4 with respect to the convenience and accuracy of pressure measurement, but we must note that no longer is the working gas entirely maintained at the test temperature. Most of the gas is contained in the bulb G, but some of it resides in the capillary C that must penetrate the wall of the experimental thermostat in order to permit the pressure measurement to be made. The size of the capillary is a compromise between the desire on the part of the operator to minimize the so-called "dead space" correction and the need for accuracy in the pressure measurement.

The capillary diameter usually chosen[3] is about 1 mm. If the bulb volume is approximately 1 ℓ, then the ratio of the capillary volume to the bulb volume is about (0.8×10^{-5}) times the length of the capillary in cm. Thus, given a 1-m-long capillary, approximately 0.1% of

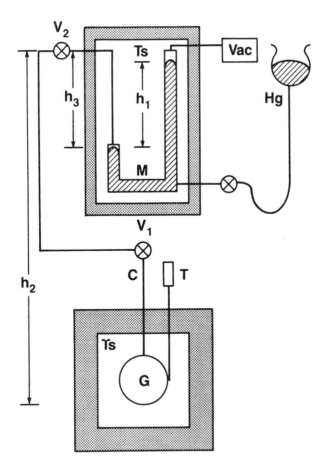

FIGURE 5. Sketch of simplified constant-volume gas thermometer. M, U-tube manometer filled with mercury and enclosed in a thermostat Ts; h_1, vertical distance between the meniscus on the pumped side of the manometer and the meniscus on the lower (measuring) side; h_2, vertical distance between the gas bulb G and the valve V_2; h_3, vertical distance between the lower meniscus and the highest point in the measuring capillary, near V_2; C, capillary for sensing the pressure in the gas bulb G; T, secondary thermometer to transfer gas thermometer temperature determinations.

the working gas is not maintained at the test temperature at all, but is exposed to a range of temperatures. A further complication arises if the dead volume is varied during the measurement process — for example, if the volumes of valves in the system are different when opened than they are when closed, or if the level of the mercury in the short (measuring) arm of the manometer varies during the measurement.

To minimize the sizeable systematic error possible as a result of the existence of a dead volume in gas thermometry, the operator can take four types of precautionary steps. First, he can maximize the volume of the gas bulb. Second, he can minimize the dead volume itself; a small-diameter capillary can be used to measure the pressure in the gas bulb, and the volumes of any valves in the system can be minimized. Third, he can minimize the change in the dead volume by holding the mercury level constant in the measuring side of the manometer, and by using a "constant-volume" valve at V_1. Finally, he can carefully measure the temperature distribution along the capillary so that an appropriate correction can be made to the calculated gas thermometer volume.

Table 2
THERMAL EXPANSION COEFFICIENTS OF COMMON MATERIALS AT 300 K

Substance	$(1/L)(dL/dT) \times 10^6$, 1/K
Aluminum	23
Graphite	7.8
Copper	17
Mercury	182
Platinum	9.2
Invar	1.2
Al_2O_3	5.6
Vitreous silica	0.4

D. Corrections for Thermal Expansion

It is a general property of materials that they expand as the temperature rises. This is a consequence of the general increase of kinetic and vibrational energy with temperature.

There are several important consequences for gas thermometry of the magnitudes of the thermal expansion coefficients of materials used therein. Of course, the foundation of gas thermometry, the ideal gas law, is itself an example of thermal expansion. From the point of view of the simplified gas thermometer shown in Figure 5, potential systematic errors arising from neglected thermal expansion corrections can occur through changes in the volume of the gas bulb G, from the thermal expansion of the connecting capillary tubing C, and from temperature variation in the mercury density.

The coefficients of thermal expansion of materials commonly used in the construction of gas thermometry apparatus are listed in Table 2.

The influence of thermal expansion on the measurement of thermodynamic temperatures by gas thermometry can be illustrated by a simple example based upon the "constant-bulb-volume" technique. Suppose that a copper bulb is used to contain n moles of an ideal gas, that its volume at the triple point of water is v_1 and the pressure of the gas is measured as 27.3160 kPa at the triple-point temperature T_0. Suppose that the copper bulb, with its n moles of gas, then is placed in a bath of water near the boiling point; the pressure of the gas is now remeasured as 37.1228 kPa. Using the ideal gas law, one can calculate the temperature of the hot-water bath by evaluating the ratio

$$\frac{P_2 v_2}{P_1 v_1} = \frac{nRT_2}{nRT_0} \quad \text{or} \quad T_2 = 273.160 \frac{37.1228}{27.3160} \frac{v_2}{v_1} \text{ kelvins} \tag{12}$$

If the thermal expansion of the gas bulb were to be ignored, T_2 would be calculated erroneously as 371.228 K. If proper account is taken of the bulb expansion, however, v_2/v_1 is seen to become

$$(1 + \alpha \Delta t)^3 = 1 + 3\alpha \Delta t + 3\alpha^2 (\Delta t)^2 + \alpha^3 (\Delta t)^3 \tag{13}$$

Taking $\alpha = 17 \times 10^{-6}$/K and $\Delta t \sim 100$ K, then $v_2/v_1 = 1.005109$ and $T_2 = 373.125$ K. The potential error in the calculated temperature arising from neglect of bulb thermal expansion is 1.9 K. From this example, we find that millidegree accuracy in gas thermometry requires careful measurements of the thermal expansion of the material from which the gas bulb is constructed. Although a single number was used for α in the example just given, good practice requires evaluation of $\alpha(t)$ throughout the measured temperature range. Similar care must be used in accounting for changes in the volume of the capillary tube connecting the gas bulb to the pressure-measuring device and in the density of mercury which may be employed in a manometer.

E. Hydrostatic Pressure Head Corrections

Accurate gas thermometry requires also that the measured pressure be corrected for the weight of the gas columns in the manometer and in the lines leading to the gas bulb. Referring again to Figure 5, we caution that the pressure at the position of the lower meniscus of the manometer, P(L), is slightly larger than the pressure existing at the highest point in the capillary near to valve V_2, owing to the "hydrostatic head" of gas in the section of capillary labeled h_3. If we let (Π) denote the pressure correction, then

$$P(L) = P(V_2)[1 + \Pi (V_2 \text{ to } L)] \qquad (14)$$

Likewise the pressure at the bottom of the gas bulb, P(G), must be corrected for the hydrostatic head of gas existing below the position of valve V_2;

$$P(G) = P(V_2)[1 + \Pi (V_2 \text{ to } G)] \qquad (15)$$

The hydrostatic-head pressure correction Π can be evaluated by the relation[16]

$$\Pi = \frac{Mg}{R} \sum_k \left(\frac{l_k}{T_k}\right) \qquad (16)$$

where M is the molecular weight of the gas, g is the local acceleration due to gravity, R is the gas constant, and the sum over k allows for the variation in temperature T_k of the gas in the length increments l_k connecting the two end points of the vertical run of tubing.

F. Thermomolecular Pressure Corrections

In certain cases, a correction for thermal transpiration, or thermomolecular flow, must be applied to the pressure measured at one end of a capillary that traverses a thermal gradient. Such a situation is shown in Figure 5 between the valve V_1 and the gas bulb G.

For a capillary of diameter smaller than the mean free path of the gas, the ratio of pressures at the two ends of the tube is given by[17]

$$P_A/P_B = (T_A/T_B)^{1/2} \qquad (17)$$

where the temperatures T_A and T_B are expressed in kelvins.

If the capillary tubing has a much larger diameter, so that intermolecular collisions are the dominant form of energy exchange, then the thermomolecular pressure correction disappears.

McConville[18] and Guildner and Edsinger[16] have discussed the applicability of thermomolecular corrections in various cases, pointing out the dependence of the correction upon the nature of the inner surface of the capillary tubing; in both presentations, the authors recommend measurement of the correction by the use of an apparatus such as that shown in Figure 6 (cf. Figure 2 of Reference 16). Guildner and Edsinger[19] have reported thermomolecular corrections measured with one such apparatus; the pressure corrections exceeded 400 ppm when a 0.8-mm-diameter platinum-10% rhodium capillary was held at a pressure of 10 kPa with end-point temperatures of 273 and 1000 K. In the case of low-pressure measurements at very low temperatures, the corrections can exceed 50% of the measured pressure.

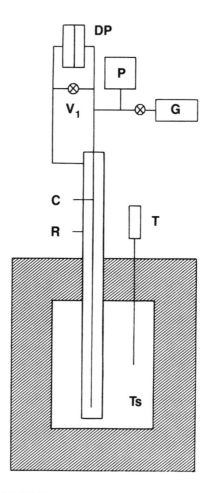

FIGURE 6. Apparatus for direct determination of thermomolecular corrections in gas thermometry. C, Actual capillary tubing from gas thermometer, open at the bottom end; R, reference tube of diameter more than ten times the capillary-tube diameter; Ts, thermostated bath; T, thermometer; G, gas supply to set the pressure P to actual values used in gas thermometry experiment; DP, differential pressure gage; V_1, Valve for measurements. When V_1 is open, the operator checks the DP null reading; when V_1 is closed, DP indicates the difference between the capillary pressure and the reference-tube pressure at temperature T.

G. Effects of Adsorbed Impurities

It appears that the widespread presence of impurities in gas thermometers led to the erroneous selection of 273.16 K as the thermodynamic triple-point temperature of water. We noted in Chapter 4 that the thermodynamic ice-point temperature was assigned the value 273.15 °K in the 1948 International Temperature Scale; that scale also specified 0 and 100°C at the ice-point and steam-point temperatures.

The water triple point was approved by the General Conference of Weights and Measures in 1954 as the primary defining point of the Kelvin thermodynamic scale; its assigned value,

273.16 °K, was consistent with the retention of 273.15 °K as the ice-point temperature. The triple-point value was chosen on the basis of contemporary gas thermometry measurements in an effort to retain the 100-degree interval between the ice and steam points. Gas-thermometry results were available from laboratories at the Physicalische-Technische Reichanstalt in Germany, at the Kamerlingh Onnes Laboratory in Holland, at the Tokyo Institute of Technology in Japan, and at the Massachusetts Institute of Technology in the U.S. The values did not agree very well (the triple-point temperature varied from 0.005 to 0.02°C), but the average of the values was not far from the "traditional" value, 0.01°C.

Similarly conflicting values relating to the choice of 273.16 K for the water triple point appeared in more recent gas thermometry;[19] steam-point temperatures of 99.984 and 99.999°C were obtained prior to the reaffirmation of the water triple-point temperature in the International Practical Temperature Scale of 1968.

In gas thermometry studies at the National Bureau of Standards, Guildner and Edsinger have placed special emphasis upon obtaining pure working gas and upon achieving a clean gas-handling system. They have noted particularly that impurities may be adsorbed on the walls of a gas thermometer bulb at the reference temperature, then may be desorbed at higher temperatures to enhance the measured gas-bulb pressure. For example, if a gas-bulb pressure is measured as 27.3150 kPa at the ice-point reference temperature and is measured again at the steam point as 37.3150 kPa, then the thermodynamic steam-point temperature will be calculated (ignoring, for the moment, other corrections) as 100.000°C. If however, 0.0025 kPa (~67 ppm) of the 37.3150 kPa measured at the steam point were to have been contributed by impurities that desorbed from the walls of the gas bulb, then the correct steam-point temperature would have been 99.975°C.

Beattie[20] and Guildner and Edsinger[19] have called attention to the relatively large variation in ice-point temperatures determined by gas thermometry. The latter experimenters have sought to minimize sorption as a contributing factor to that variation by use of helium as the working gas, by careful purification of the helium gas, by pumping the gas-handling system for extended periods of time while heating its components, and by examining the progress of the gas clean-up by use of a residual gas analyzer. In the next section, we note that their results have reflected this progress. According to the results obtained up to the present time — not only by Guildner and Edsinger but also by Quinn and Martin[21] — if it were desired to maintain 100 degrees exactly between the ice and steam points, then the ice-point temperature should be assigned the value 273.22 K on the thermodynamic scale.

H. Current Gas Thermometry Experiments

It is instructive to review current gas thermometry experiments that have provided improved accuracy in the determination of Kelvin thermodynamic temperatures both above the triple point of water and below. Each experiment illustrates an attempt to minimize systematic errors characteristic of the temperature range involved. In an effort to make the discussion as clear as possible for readers who wish to consult them, we generally have used the notation from the original papers.

1. Guildner and Edsinger's Measurements from 273 to 730 K

Working over a period of more than twenty years at the National Bureau of Standards, Guildner and Edsinger brought classical constant-volume gas thermometry to a new level of accuracy. Their efforts were marked by successful attempts to minimize the systematic errors we have just discussed. Figure 7 shows the NBS experiment in schematic form.

The 430-cc gas thermometer bulb GB was made from an alloy of platinum and rhodium (melting point above 1700°C). Errors arising from uncertainty in its thermal expansion were minimized by careful measurements of the thermal expansion coefficient of a sample of the bulb material over the range −55 to 550°C. The measurements were fitted by the equation

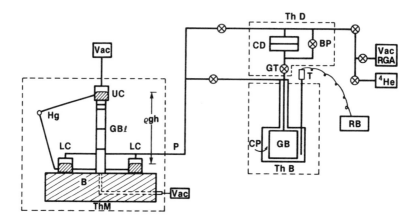

FIGURE 7. Schematic drawing of the NBS constant-volume gas thermometer. At the left of the drawing is the mercury manometer, with the upper cell UC raised above the two lower cells LC by gage blocks GBl. On the right are the gas thermometer bulb GB and the capacitance diaphragm null detector CD. Each component is contained in its own thermostated environment. See text for further details.

$$L/L_0 = 1 + C_1 t + C_2 t^2 + C_3 t^3 + C_4 t^4 + C_5 t^5 \qquad (18)$$

The equation fitted the measured data within less than 0.14 ppm throughout the measuring range of temperature.[16]

During gas thermometer measurements, ^{4}He gas was introduced into the annular space CP that lies between an Inconel® protective case and the outside of the gas bulb GB. The pressure of this ^{4}He gas was kept the same as the gas-bulb pressure within about 150 Pa to avoid distortions of the gas bulb resulting from inside-outside pressure differences. The thin layer (about 0.25 mm) of ^{4}He gas served another purpose as well. This was to enhance the thermal contact between the working gas in the bulb GB and the platinum resistance thermometers T (actually four thermometers were used to minimize the uncertainty in the IPTS-68 temperatures) that were inserted into holes in the Inconel® case.

Guildner and Edsinger realized the need for a pure, one-component gas in the thermometer bulb. The systematic error caused by nonideality of the gas becomes much larger if sorbable contaminants enter the bulb. Traces of water, carbon dioxide, hydrogen, nitrogen and other gases substantially degrade the accuracy of gas thermometer temperature determinations, as we already have noted. Figure 3 shows, for example, the large departures from ideality above room temperature found in water and in CO_2. To maximize the purity of the ^{4}He working gas in their experiment, Guildner and Edsinger baked the metal gas bulb at temperatures well above the measuring temperatures, pumping all the while with an ion pump; a residual gas analyzer was used to measure both the amount and composition of the gas remaining in the bulb. By encasing the gas bulb in a thin layer of purified He gas and surrounding the Inconel® case with an inert gas, they further reduced the possibility that impurity gases — particularly hydrogen, which penetrates metal walls at high temperatures — would contaminate the working gas.

The capillary tubing, also of Pt-Rh alloy, was carefully prepared before its installation to connect the gas bulb to a constant-volume valve labeled GT in Figure 7. The inner diameter of the tubing was measured throughout its length in order to evaluate its volume.[22] As many as 20 Pt wires were welded to the tubing over its 1/2-m length to form a set of platinum-rhodium vs. platinum thermocouple thermometers. These allowed measurements of the capillary temperature profile to be made and thus its contribution to the dead space correction could be evaluated. In a separate experiment, the thermomolecular effect arising from the use of the capillary in a temperature gradient was measured for a variety of end-point temperatures; this correction is difficult to calculate accurately, as we noted earlier.[19]

Small constant-volume valves GT and BP were designed and built in order to minimize the dead space that necessarily is added to the system by the GT valve and in order to avoid any change in the volume of the gas thermometer system when the valves were operated.

A low-volume capacitance diaphragm gage CD was installed between the gas thermometer and the manometer system to avoid contamination of the working gas by mercury vapor. The gage was used as a null detector to determine that the pressure at GT was equal to that in the manometer system.

The capacitance diaphragm gage and the two constant-volume valves are enclosed by a dashed line in Figure 7. This is done to indicate the extent of a thermostat designed to minimize drifts in the null-point reading of the gage CD.

A more elaborate thermostat enclosed the manometer system shown to the left in the figure. In this case the temperature of an entire room some 5 m below the basement laboratory level was carefully regulated so as to maintain the mercury in the manometer at a constant density and to allow accurate hydrostatic-head corrections to be made.

The manometer itself is a unique facility designed to provide steady reference pressures in the range 4 to 125 kPa with an accuracy[23] of about 2 ppm. The principal features of the manometer are shown in the figure. They include one upper (UC) and two lower (LC) mercury cells resting on a rigid base B and connected by leak-tight, articulating tubes; a set of calibrated gage blocks GBℓ to provide the desired elevation of the upper cell; and two vacuum systems placed so as to evacuate the upper cell and to monitor the proper assembly of the stack of gage blocks.

The base B is a copper-clad block of invar resting upon a large concrete pad that in turn rests upon bedrock beneath the building. The pad is separated from the building supports so as to free it from many of the vibrations arising from rotating machinery and other mechanical disturbances. The use of two lower cells allows direct measurement of tilt in the base in the plane of the manometer cells. The level of the mercury in each cell is monitored by the guarded capacitance plates connected to a bridge circuit. Observation of the capacitance between the plate and the mercury in each cell allows the operator to maintain the mercury levels reproducibly[22] within about 0.1 μm.

The use of gage blocks to support the upper cell limits the manometer pressure settings to discrete, predetermined values that depend upon the variety of available blocks. Though cumbersome to operate, the blocks provide part-per-million accuracy in the manometer setting; their use thus removes a major source of error in the gas thermometry measurements. Contributing to the preservation of the calibration accuracy of the gage blocks is the practice by Guildner and Edsinger of wringing the blocks to each other and to the cell and base. The quality of the wring is checked by making use of axial holes in all of the blocks; pumping through a matching hole in the base allows the operator to measure the leak rate at the wrung joints, thus ensuring reproducible contact throughout the stack.

In operation, the gas bulb/Inconel® case/platinum resistance thermometer system first was placed in a stirred liquid bath. The bath was thermostated at the highest temperature to be measured. Then the gas bulb and the counter-pressure space were filled with purified ^{4}He gas to the pressure that was predetermined by the height of the gage blocks. Valve GT was not closed until the capacitance bridge reading indicated that the lower manometer cells were filled to the desired pressure. After pressure and temperature equilibrium were reached — usually within a few hours — the capillary temperature profile was measured via the thermocouples, the platinum resistance thermometer resistances were recorded, valve GT was closed, and the bath thermostat was reset for a lower temperature. Ordinarily, the system was allowed to equilibrate overnight before measurement at the new temperature was attempted. This time interval allowed the manometer room to return to its thermostated temperature following the disturbance caused by the operator's entry to change the gage blocks.

The new and each succeeding temperature was determined by the choice of gage blocks selected to elevate the center cell of the manometer; having guessed the height of gage-block stack that would be appropriate for a given temperature, the operator would close valve BP and open valve GT after equilibrium at that temperature was reached. If, as was commonly the case on a first trial at a particular temperature, the capacitance diaphragm CD indicated too high or too low a pressure in the gas bulb, then the thermostat was reset to a temperature such that the null capacitance reading was regained. Using this technique, Guildner and Edsinger avoided loss or gain of gas in the gas bulb/capillary system.

The reference temperature setting was achieved by immersing the gas thermometer in an ice bath. Although the resulting temperature was not exactly 273.16 K, it was sufficiently close that all measurements could be corrected to the triple-point temperature with negligible error.

Calculation of each gas thermometer temperature involved evaluation of the corresponding ratio of gas-bulb pressures adjusted for all of the systematic corrections that we have outlined heretofore. To calculate the thermodynamic temperature of a particular state, T_i, in terms of the reference state temperature T_0, one can use the ideal gas law modified by the second virial correction:

$$P_i v_i = n_0 (RT_i + B_{vi} P_i) \tag{19}$$

Here the use of n_0 denotes the fact that the gas thermometer contains the same quantity of working gas in both states. The gas always is divided between the bulb, held at temperature T_i, and the dead space, which varies in temperature between T_i and room temperature. In this experiment, the dead space consists of the capillary tube connecting the bulb to valve GT and the volume of GT itself. Thus

$$n_0 = n_b + n_{ds} \quad \text{or} \quad n_0 = \frac{P_b v_b}{(RT_b + B_{vb} P_b)} + \frac{P_{ds} v_{ds}}{(RT_{ds} + B_{vds} P_{ds})} \tag{20}$$

In Equation 20, the subscripts b and ds refer respectively to quantities measured in the bulb and in the dead space. B_v is the second volume virial coefficient. The pressure set by the gage blocks at the manometer is $\rho_i g h$, where ρ_i is the density of the mercury in the manometer during the measurement of state i, g is the acceleration due to gravity at the manometer elevation, and h is the height of the gage blocks. Accounting for the manometer temperature and for the hydrostatic head correction Π_i occasioned by the difference in elevation between the manometer and the capacitance diaphragm, Guildner and Edsinger could evaluate

$$P_{0i} = \rho_i g h_i (1 + \Pi_i) \tag{21}$$

the pressure generated at the diaphragm CD by the manometer. No correction was necessary for temperature variation within the manometer line. After the bulb temperature was adjusted until a null was realized at CD with the by-pass valve BP closed and the gas thermometer valve GT open, the following equation could account for the disposition of the n_0 moles of working gas in each state i.

$$n_0 R = P_{0i} \left\{ \left[\frac{(1 + \Pi_i' + \theta_i) v_{0b} (1 + \alpha \Delta t_{bi})^3}{T_i + B_{vi} P_i / R} \right] + \sum_k \left[\frac{v_{ki} (1 + \alpha' \Delta t_{ki})^3}{T_{ki}} \right] \right\} \tag{22}$$

Table 3
SECOND VIRIAL COEFFICIENTS OF ^{4}He DERIVED BY GUILDNER AND EDSINGER[16,24]

t (°C)	B (cm^3/mol)
0	12.00
25	11.89
50	11.77
75	11.67
100	11.56
125	11.46
150	11.36
300	10.76
400	10.45
500	10.14
600	9.82

In Equation 22, Π_i' and θ_i refer respectively to the hydrostatic head and the thermomolecular corrections between the capacitance diaphragm and the gas bulb; α and α' refer to the linear thermal expansion coefficients of the gas bulb volume, v_{0b}, and the capillary volume elements, v_{ki}, respectively; and the sum over k permits a detailed accounting of the temperature profile of the dead space volume elements. Note that we have neglected the virial correction and the hydrostatic head correction in the relatively small dead space term.

Equation 22 holds true at each state i. The total quantity of gas n_0, which remains unchanged throughout a given series of measurements, always is divided between the gas bulb and the capillary as indicated by Equation 22. By equating the quantity n_0R for each state of unknown temperature with n_0R for the reference state, Guildner and Edsinger could determine the unknown temperatures from the known reference temperature and the measured quantities in each state.

Over a period of several years, Guildner and Edsinger measured more than eighty gas thermometer pressure ratios at some twenty temperatures ranging from 20 to 457°C. These measurements were converted into values of Kelvin thermodynamic temperature by the use of virial coefficients for helium derived from compressibility data.* The coefficients used are given in Table 3.

The values of Kelvin thermodynamic temperature T_i that Guildner and Edsinger obtained in this way were compared with measured values of temperature on the IPTS-68. These temperatures were obtained at the same time from the four calibrated platinum resistance thermometers that were incorporated into the gas thermometer apparatus. The resulting determinations of $(T_i - T_{68})$, converted to the corresponding Celsius values $(t_{th} - t_{68})$, are plotted against t_{68} in Figure 8.

Guildner and Edsinger carefully examined both the random and systematic uncertainties of each of their temperature determinations. The data points plotted in Figure 8 actually do not include all of the determinations of T_i that were presented in Tables 2 to 4 of Reference 16, although this fact is not clear from reading their paper. Data from two series of experiments were discarded because they appeared to reflect large systematic errors. One series,

---

* See Reference 16. From 0 to 150°C, the values of the helium second-virial coefficients used were based upon measurements made by M. Waxman at the National Bureau of Standards; the higher-temperature values were derived from Reference 24. The data were interpolated and smoothed so as to be consistent, so far as possible, with the acoustic measurements of Reference 24a.

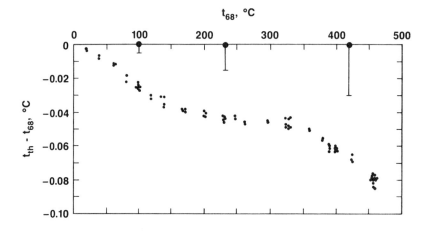

FIGURE 8. Differences between Kelvin thermodynamic temperatures as determined from gas thermometer measurements by Guildner and Edsinger[16] and temperatures measured on the IPTS-68. The error bars show the KTTS uncertainty values estimated in 1968 in conjunction with the assignments of temperatures to the steam point and the tin and zinc freezing points.[25]

Table 4
ANALYSIS OF NBS GAS THERMOMETRY DATA

t (°C)	No. of measurements	Range of measurements (°C)	$t_{th} - t_{68}$ (°C)	SD[a] (°C)	$(SD)/\sqrt{N}$[b] (°C)	U[c] (°C)
100	7	0.005	−0.025	0.0016	0.0006	0.0019
325	6	0.007	−0.048	0.0025	0.0010	0.0029
457	10	0.006	−0.079	0.0017	0.0006	0.0041

[a] Calculated standard deviation of one measurement (see Equation 1, Chapter 4).
[b] Calculated standard deviation of the mean of the set of measurements.
[c] Estimated overall uncertainty at the 99% confidence level.

set "A", the first series of (14 in number) measurements listed in Table 2 of Reference 16, was discarded on account of an apparent distortion of the gas bulb arising from an accidental inequality in the pressures maintained in the gas-bulb and counter-pressure spaces. Another series, set "G", the third series of (11) measurements listed in Table 3 of Reference 16, was discarded because of suspected contamination of the working gas.

A simple analysis of the determinations of t_{th} for three temperatures that are representative of the NBS gas thermometry experiment is presented in Table 4. This table illustrates the high quality of the results obtained by Guildner and Edsinger and further demonstrates useful statistical methods for treating experimental data. Note that, whereas the standard deviations of the means of the determination of the KTTS at 100°C, 325°C, and 457°C do not exceed 0.001°C, the increasing systematic uncertainty raises the 99%-confidence total uncertainty to ±0.004°C at 457°C.

Examining Figure 8, we see that the temperature value on the IPTS-68 appears to exceed that on the Kelvin thermodynamic scale as realized by the NBS gas thermometry by some 0.08°C at 450°C. Also shown in the figure are the 1968 estimates of the uncertainty of the KTTS at three fixed-point temperatures; 100°C (the temperature of the water liquid-vapor phase-equilibrium state), 232°C (the tin solid-liquid equilibrium temperature), and 420°C (the zinc solid-liquid equilibrium temperature). The indication by Guildner and Edsinger

Table 5
KELVIN THERMODYNAMIC TEMPERATURES OF THE STEAM, TIN, AND ZINC POINTS FROM GAS THERMOMETRY[16]

	Steam point		Tin point		Zinc point	
	K	°C	K	°C	K	°C
T_{th}	373.125	99.975	505.074	231.924	692.664	419.514
T_{68}	373.150	100.000	505.1181	231.9681	692.73	419.58
Estimated errors						
Random	±0.0018		±0.0022		±0.0028	
Systematic	±0.00054		±0.0015		±0.0028	
$T_{th} - T_{68}$	−0.0252		−0.0439		−0.0658	
Uncertainty of KTTS in 1968[a]	±0.005		±0.015		±0.03	

[a] Estimated uncertainty of the KTTS; cf. Table 7, Reference 25.

that the IPTS-68 did not agree precisely with the Kelvin thermodynamic scale might have been expected; after all, one knows that thermodynamic temperatures are not easily realized. However, finding the IPTS-68 to deviate from the KTTS by amounts varying from two to five times the estimated uncertainties of the Kelvin thermodynamic scale, as shown by the summary in Table 5, is surprising indeed. Clearly, the 1968 estimates were optimistic.

Since the publication of the NBS gas thermometry results, other thermometrists, most notably Quinn and Martin,[21] whose work we discuss below, have obtained convincing evidence of their accuracy. Thus one must suspect the accuracy of the IPTS-68 with respect to the KTTS at all temperatures above the triple point of water.

2. Berry's Measurements from 2.6 to 27.1 K

Berry of the National Physical Laboratory, Teddington, England, utilized gas thermometry to determine the Kelvin thermodynamic temperature of the normal (1-atm) boiling point of neon (27.1 K), of the normal boiling point of equilibrium hydrogen (e-H_2) (20.3 K), of the triple point of e-H_2 (13.8 K), and of the normal boiling point of ^{4}He (4.22 K).[15,26] In addition to these fixed-point temperatures, more than 30 other thermodynamic temperatures corresponding to the resistances of a set of stable low-temperature thermometers made of an iron-doped rhodium alloy were evaluated.

The work of Berry was distinguished by his imaginative use of special techniques to avoid the difficulties raised by the low temperatures that he wished to measure, by his exhaustive and careful measurements, and by his extremely thorough data analysis. Furthermore, Berry's written discussion of his work is very lucid; it is worth reading simply as an example of metrological exposition.

The major problems faced by Berry in his desire to establish thermodynamic temperatures in the range 2.5 to 30 K with an inaccuracy not larger than 70 ppm were the following:

- A sizeable uncertainty in the value of R, the gas constant
- The difficulty of using constant-volume gas thermometry (CVGT) below 30 K in conjunction with a 273-K reference temperature
- Uncertainty in the values of the low-temperature virial coefficients of ^{4}He

Berry circumvented these problems by using a combination of three experimental techniques: absolute P-V isotherm thermometry, relative isotherm thermometry, and constant-volume gas thermometry.

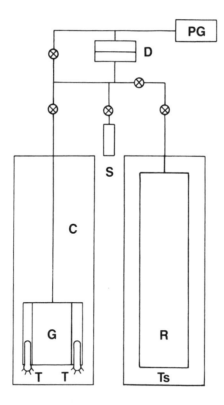

FIGURE 9. Schematic drawing of the gas thermometer used by Berry.[26] C, Liquid helium cryostat to maintain the 1-ℓ gas bulb G at the test temperatures; Ts, thermostat to maintain a 6-ℓ reference volume R at 273.15 K. Pressure ratios were measured by a piston gage PG that was isolated from the gas volumes by a null-reading capacitance diaphragm gage D. S, Small, calibrated volume for evaluating dead-space volumes; T, rhodium-iron resistance thermometers.

a. Absolute P-V Isotherm Thermometry

Both absolute and relative P-V isotherm thermometry have been in use for many years. The absolute isotherm technique was used to derive low-temperature values of the virial coefficients for ^{4}He in the equation

$$Pv = nRT \left[1 + \left(\frac{n}{v}\right) B_v + \left(\frac{n}{v}\right)^2 C_v + ... \right] \qquad (23)$$

early in the twentieth century.[27,28] For that purpose, a measured volume is filled repeatedly at a single temperature with various measured quantities of gas; measurements of the pressure resulting from each filling permits the operator to derive the coefficients $B_v(T)$ and $C_v(T)$ for each temperature (each "isotherm") investigated.

Use of the technique to simultaneously obtain thermodynamic values of particular temperatures presented Berry with an added measure of difficulty. Berry addressed the problem by use of a variation of the "constant-bulb-temperature" method described in Section II.A. His apparatus is shown schematically in Figure 9. In this case, Berry used two volumes, one (G, Figure 9) held at the temperature to be measured (the normal boiling point, nbp, of e-H_2, ~20.3 K), and the other (R, Figure 9) at the ice point, from which temperature nearly error-free corrections can be made to the triple point of water. Berry measured the ratio of the two volumes carefully (accurately within about 20 ppm), then corrected the ratio according to the expected thermal contraction of the low-temperature gas bulb between 273 and 20 K and for pressure dilation upon filling the bulb with the experimental sample of ^{4}He gas. All pressures were measured using a piston gage PG coupled with a capacitance diaphragm gage D.

By sequentially adding ^{4}He gas quantities from the thermostated ice-point reference volume to the gas bulb while it was held within 0.01 K of T_h, the e-H_2 normal boiling point, Berry was able to derive values of the desired temperature T_h as a function of the gas density (n_B/v_B), using the equation

$$T_h = \frac{P_B v_B}{R n_B} = \frac{(T_0 P_B v_B)/(P_i v_r)}{\dfrac{1}{1 + \dfrac{B_0 P_i}{RT_0}} - \dfrac{P_f/P_i}{1 + \dfrac{B_0 P_f}{RT_0}} + \dfrac{T_0}{v_r}\sum_i \left(\dfrac{v}{T}\right)_i - \dfrac{T_0 P_f}{v_r P_i}\sum_f \left(\dfrac{v}{T}\right)_f} \tag{24}$$

In Equation 24, the subscripts B refer to quantities measured on the gas bulb; v_r is the reference volume; P_i and P_f refer to the initial and final reference-volume pressures; T_0 is the Kelvin temperature of the reference volume and B_0 is the second (volume) virial coefficient of ^{4}He, the working gas, at T_0; and the sums over (v/T) refer to the dead volume in the apparatus. All of the quantities on the right-hand side of the equation were measured or were sufficiently well-known to preserve the goal of 50-ppm accuracy for T_h.

Berry measured twelve isotherm points with ^{4}He densities ranging from about 0.8×10^{-4} to 6×10^{-4} moles/cc. From these data he derived T_h, the KTTS equilibrium hydrogen normal boiling-point temperature, as

$$T_h = 20.2712 \pm 0.0004 \text{ K} \tag{25}$$

and at the same time he obtained the value of the helium second virial coefficient at that temperature as

$$B_v(T_h) = -2.4 \pm 0.2 \text{ cc/mol} \tag{26}$$

b. Relative P-V Isotherm Thermometry

Having established the e-H_2 nbp, Berry then could use that temperature as a convenient reference point for other isotherms. The number of moles of gas introduced into the gas bulb for these measurements, n_B, could be established by a pressure measurement at the reference temperature, 20.271 K;

$$n_B = \frac{P_r v_B/(RT_r)}{1 + B_v\left(\dfrac{P_r}{RT_r}\right) + (C_v - B_v^2)\left(\dfrac{P_r}{RT_r}\right)^2} + \frac{P_r}{R}\sum_i \left(\dfrac{v_i}{T_i}\right) - \frac{P_B}{R}\sum_f \left(\dfrac{v_f}{T_f}\right) \tag{27}$$

In Equation 27, B_v and C_v are respectively the second and third volume virial coefficients for ^{4}He at the gas-bulb reference temperature T_r (the e-H_2 nbp); P_r is the gas-bulb pressure at T_r; P_B is the gas-bulb pressure at the new, relative-isotherm temperature T; and the sums over i and f refer again to the initial and final states of the dead volume. Similarly, the relative-isotherm temperatures T_j could be evaluated as functions of the measured gas densities;

Table 6
BERRY'S LOW-TEMPERATURE GAS THERMOMETRY FIXED-POINT RESULTS

Point	T("NPL-75") (K)	T_{58} or T_{68} value (K)	$(T_{th} - T_{58,68})$ (mK)
^{4}He nbp	4.2221 ± 0.0005	4.2150	7.1
e-H$_2$ tp	13.8035 ± 0.0008	13.81	−6.5
"Reduced-pressure bp" of e-H$_2$	17.0356 ± 0.0009	17.042	−6.4
e-H$_2$ nbp	20.2712 ± 0.0009	20.28	−8.8
Ne nbp	27.0979 ± 0.0014	27.102	−4.1

$$T_j = \left(\frac{P_B v_B}{R n_B}\right)_j = \frac{T_r P_B / P_r}{\left[1 + B_v\left(\frac{P_r}{RT_r}\right) + (C_v - B_v^2)\left(\frac{P_r}{RT_r}\right)^2\right]^{-1} + \frac{T_r}{v_B}\sum_i\left(\frac{v_i}{T_i}\right) - \frac{P_B}{P_r}\sum_f\left(\frac{v_f}{T_f}\right)} \quad (28)$$

Berry selected his isotherm temperatures in order to best define the curve of $B_v(T)$, the temperature-dependent second virial coefficient for ^{4}He. Expecting $B_v(T)$ to follow the form $B \sim (20 - 400/T)$ cc/mol, he chose the temperatures 2.6 K, 2.75 K, 3.33 K, 4.22 K, 7.2 K, 13.8 K, and 27.1 K for relative isotherm measurements. The highest four points have the added significance of approximating respectively the nbp of ^{4}He, the superconductive transition temperature (T_c) of Pb, the triple-point temperature of e-H$_2$, and the nbp of Ne.

In three of the isotherms, Berry extended the range of measured densities to obtain information on the third virial coefficient as well as the second. An exhaustive set of measurements provided isotherm temperatures that appeared to be accurate within 10 to 60 ppm, and new values of B_v and C_v were derived for ^{4}He as well.

c. Constant-Volume Gas Thermometry

In order to construct a reasonably complete thermodynamic scale of temperatures in the range 2.6 to 27 K, Berry utilized the reference temperatures as calibration points for constant-volume gas thermometer measurements at some 28 temperatures in that range.

The measurements were recorded on a set of three rhodium-iron resistance thermometers and one capsule platinum resistance thermometer.

The results obtained at the fixed points measured by Berry in his lengthy experiment are given in Table 6. Berry's work provides a new level of precision for the earlier finding of Plumb and Cataland[29] that the ^{4}He boiling point temperature as given by the T-58 vapor-pressure scale[30] is lower than the thermodynamic scale by about 7 mK. In addition, the new work indicates discrepancies in the IPTS-68 assignments of temperature to the hydrogen and neon fixed points.

Berry's values for the virial coefficients of ^{4}He are compared in Table 7 with those determined by previous experimenters.[31,32] Note that Berry evaluated both random and systematic errors associated with the determinations of B_v.

Berry's gas thermometry results have been corroborated by acoustic thermometry measurements made by Plumb and Cataland[29] and by Colclough,[33,33a] and by noise thermometry performed by Klein, Klempt, and Storm;[34] these techniques are discussed later. Therefore, the NPL-75 temperature scale that was derived from Berry's work has been generally accepted as representing the thermodynamic temperature scale within about 1 mK in the range 2.6 to 27 K.

Table 7
⁴He VOLUME VIRIAL COEFFICIENTS

	B_v (cm³/mol)					C_v [(cm³/mol)²]
		Uncertainties (cm³/mol)				
T (K)	Berry[a]	Random	Systematic	Keesom (adj)[b]	KKHM[c]	Berry[a]
2.6014	−142.5	0.8	0.2	−139.4	−136.9	(4000)
2.7479	−133.2	0.5	0.2	−130.9	−128.5	(1500)
3.3303	−105.8	0.5	0.2	−104.7	−102.5	1000 ± 500
4.2201	−79.5	0.3	0.2	−78.6	−76.5	1200 ± 300
7.1992	−39.0	0.5	0.2	−38.2	−36.4	900 ± 300
13.8036	−11.7	0.2	0.2	−10.8	−9.2	(200)
20.2712	−2.4	0.2	0.0	−1.3	+0.3	(300)
27.0979	+2.5	0.1	0.2	+3.8	+5.4	(−100)

[a] Data from Reference 26.
[b] Virial-coefficient data of W. H. Keesom and W. K. Walstra,[31] as reanalyzed and adjusted by W. E. Keller.[31a]
[c] Calculations of J. E. Kilpatrick, W. E. Keller, and E. F. Hammel,[32] interpolated by Berry using the equation B = 20.5 − 409.5/T in units cm³/mol.

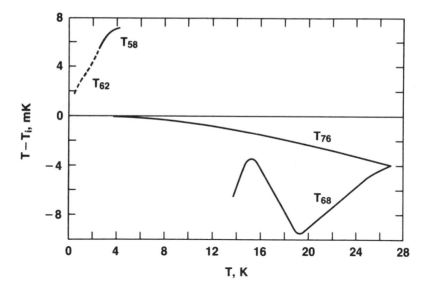

FIGURE 10. Differences between thermodynamic temperatures, represented within about 1 mK by the NPL-75 gas thermometer scale[26] as extended by paramagnetic salt thermometry,[36] and four current practical temperature scales.

Both paramagnetic salt thermometry and helium vapor pressure thermometry have been accomplished in the range 0.5 to 30 K, using the NPL-75 scale as a reference; from these and other studies,[35,36] the 1976 Provisional 0.5 K to 30 K Temperature Scale (EPT-76) was developed.[37]

In Figure 10 we compare with the NPL-75 scale four low-temperature scales that are in current use below 30 K; the 1958 ⁴He Vapor-Pressure scale (T_{58}), the 1962 ³He Vapor-Pressure scale (T_{62}), the IPTS-68, and the EPT-76. The differences between the NPL-75 and these scales have been established through careful laboratory comparisons and included

136 *Thermometry*

as tables in the EPT-76. Note that the thermodynamic discrepancy in the IPTS-68 at 27 K, about +4 mK, has been made a part of the EPT-76 so as to make those two scales consistent.

3. Acoustic Thermometry

The determination of thermodynamic temperatures by measuring the velocity of sound in a gas differs considerably from ordinary gas thermometry. However, we discuss such measurements here in order to stress the similarity of temperatures determined by acoustic thermometry — as practiced earlier by Plumb and Cataland[29] and more recently by Colclough[33,33a] — and those obtained by traditional gas thermometry in the hands of Berry.

The speed of sound in a gas at a vanishingly small pressure, V_0, is related to KTTS temperature by the equation

$$V_0^2 = (C_p/C_v) RT/M \qquad (29)$$

where the ratio of the heat capacity at constant pressure to that at constant volume (C_p/C_v) can be set equal to the fraction 5/3 in the case of an ideal monatomic gas. In Equation 29, R is the gas constant and M is the atomic weight of the gas. By exciting a gas column with acoustic waves and mechanically varying the length of the column, one can observe many acoustic resonances in the gas as the column length coincides with an integral number of acoustic wavelengths. If one can measure accurately the column-length increments that correspond to the resonances, then one can calculate the speed of sound. Conducting isothermal measurements of the sound velocity at several pressures permits an extrapolation to P = 0 to obtain V_0 for the isotherm temperature.

Plumb and Cataland[29] used the acoustic method in the mid-1960s in conjunction with a quartz oscillator operating at 1 MHz, obtaining Kelvin thermodynamic temperatures over the range 2 to 20 K in ^{4}He gas. A provisional temperature scale resulted from that work (see Chapter 4, Section IV.E).

The isotherm results given by Plumb and Cataland show considerable scatter (approximately ±3 mK) when compared with the precise gas thermometry of Berry; however, a smoothed version of the NBS 2 to 20 K scale deviates from the NPL-75 scale by no more than 3 mK.

Colclough[33,33a] has reported Kelvin thermodynamic temperatures for several fixed points based upon acoustic thermometry in ^{4}He gas, using oscillator frequencies of 3.3 to 7.25 kHz. In an earlier publication, thermodynamic temperatures were reported for the ^{4}He normal boiling point and the equilibrium H_2 normal boiling point; more recently, those points were remeasured, along with the superconductive transition temperature of Pb (7.2 K), the triple point of equilibrium H_2 (13.8 K), and the reduced-pressure boiling point of equilibrium H_2 (17 K).

As shown in Table 8, Colclough's results agree with Berry's NPL-75 measurements within ±2 mK (excepting his earlier determination of the equilibrium hydrogen normal boiling point, which, like an early determination of this point by Plumb and Cataland, appears simply to be incorrect). Although not of such excellent precision as Berry's work, the acoustic temperature determinations agree closely with Berry's results, providing evidence that one knows Kelvin thermodynamic temperatures within a few millikelvins in the range 4 to 20 K. One's confidence in the accuracy of the NPL-75 scale is further bolstered by other results that are discussed in subsequent sections of this chapter.

4. Dielectric-Constant Gas Thermometry

The internal consistency of Berry's results has been corroborated by a rather different variation of gas thermometry known as dielectric-constant gas thermometry. Gugan and Michel proposed using the Clausius-Mossotti equation as the basis for precise measurement

Table 8
COMPARISON OF CONSTANT-VOLUME GAS THERMOMETER[26,40] AND ACOUSTIC THERMOMETER[33] FIXED-POINT DETERMINATIONS BETWEEN 4 AND 27 K

Fixed point	NPL-75 T (K)[a]	Kemp et al. T (K)[b]	T(NPL-75) −T(Kemp) (mK)	Colclough T (K)[c]
^{4}He bp	4.2221	—	—	4.2212
e-H$_2$ tp	13.8035	13.8042	+0.7	13.8032
e-H$_2$ 25/76 atm bp	17.0356	17.0354	−0.2	17.0378
e-H$_2$ bp	20.2712	Reference T	—	20.2724
Ne bp	27.0979	27.0968	−1.1	—

[a] Data from Reference 26.
[b] Data from Reference 40.
[c] Data from Reference 33.

of thermodynamic temperatures. The Clausius-Mossotti equation relates the dielectric constant of the working gas, $\epsilon(P,T)$ to the polarizability A_ϵ through the equation[38-39a] $[(\epsilon - 1)/(\epsilon + 2)] = A_\epsilon/V$.

Strictly applicable only for $P \sim 0$, the Clausius-Mossotti equation can be written as a virial expansion in the gas density in a manner similar to that used for the equation of state of nonideal gases.

The dielectric constant method, measuring an *intensive* property of the gas rather than an extensive one, can be used to eliminate the need to know the gas bulb volume as well as dead space corrections. In place of these advantages, the operator now must know or measure the compressibility of the capacitance cell, the gas polarizability, and the dielectric constant. It still is necessary to measure the cell gas pressure with high accuracy.

Gugan and Michel[38] found Berry's results to be internally consistent at the 1-mK level, and they verified Berry's values for the ^{4}He second virial coefficient. In addition, they derived a relation for the third virial coefficient for helium

$$C_v(T) = (5420 \pm 225)/T \; (cc/mol)^2 \qquad (30)$$

They also derived a value of $517.257 \pm 0.025 \times 10^{-3}$ cc/mol for the ^{4}He polarizability.

5. Constant-Volume Gas Thermometry of Kemp, Besley, and Kemp

Evaluations of Kelvin thermodynamic temperatures below 0°C by constant-volume gas thermometry have been undertaken in at least three laboratories besides the NPL. One of these experiments was performed by Kemp, Besley, and Kemp at the NML in Australia.[40]

Kemp et al. made several sets of measurements. One set consisted of gas thermometry in the range 13 to 27 K, using as a reference the equilibrium hydrogen boiling-point temperature obtained by Berry. Table 8 shows that their measurements agree with those of Berry within ±1 mK.

A second set of measurements reported by Kemp et al. was a group of four relative isotherms based upon Berry's value of the e-H$_2$ boiling point. From four to ten different pressures were measured at each of the temperatures, permitting evaluation of the second virial coefficient as well as the temperature. Table 9 contains a summary of these results.

A third set of measurements was reported informally by Kemp during the (1984) 15th Meeting of the Consultative Committee for Thermometry at the International Bureau of Weights and Measures.[41] These measurements included a relative isotherm at 172 K based upon the ice point at 273.15 K — indicating that the IPTS-68 is about 11 mK colder than

Table 9
RELATIVE ISOTHERMS MEASURED BY KEMP ET AL.[40]

T(IPTS-68) (K)	N	B cm³/mol	T(isotherm) (K)	T(Kemp) − T(IPTS-68) (mK)
27.0922	10	2.43	27.0870	−5.2
43.7348	4	7.59	43.7290	−5.8
54.3371	6	9.29	54.3348	−2.3
83.8372	6	11.08	83.8463	+9.1

Table 10
SELECTED KOL GAS-THERMOMETRY RESULTS[43]

T (K)	T(KOL) − T(IPTS-68) (mK)
13.8	−7.3
20.3[a]	−8.8
54.4	−2.4
83.8	+9.2
90.2	+9.0

[a] Reference temperature as determined by Berry.[26]

the KTTS at 172 K — in addition to gas thermometry referred to the ice point in the range from 83 to 273 K. The measurements were precise within less than 1 mK. We refer to these measurements again in Section V.

6. Constant-Volume Gas Thermometry of Steur et al.

Working in the Kamerlingh Onnes Laboratory at Leiden University in the Netherlands, Steur, van Dijk, Mars, ter Harmsel, and Durieux performed constant-volume gas thermometry in the range 4.2 to 100 K.[42,43] All of their Kelvin thermodynamic temperatures were obtained by reference to Berry's measurements at 20.3 or 27.1 K, using calibrated rhodium-iron resistance thermometers as temperature-transfer instruments.

Steur and his colleagues made nearly 150 determinations of gas-thermometer temperatures; their thermometric precision was found to be somewhat better than 1 mK. The scale that resulted from their measurements agrees well with Berry's NPL-75 scale in the range 4.2 to 27 K and with the measurements of Kemp, Besley, and Kemp in the range up to 100 K. In Table 10, we provide some of their results — others will be presented later in Section V.

III. NOISE THERMOMETRY

Noise thermometry derives its name from the fact that a measurable but variable voltage arises in an electrical conductor of resistance R as a result of random thermal excitation of the conduction electrons. An experimental study of the phenomenon was undertaken by Johnson in 1928.[44] He showed that the noise voltage detected across the resistor varies as the square root of the product of R and T, the thermodynamic temperature.

Nyquist, a theorist colleague, derived an expression for the noise voltage E in terms of the energy distribution among the possible modes of excitation of the conductor:[45]

$$E^2 d\nu = \frac{4R(\nu)h d\nu}{e^{(h\nu/kT)} - 1} \tag{31}$$

Here, ν is the frequency of the measurement, h is Planck's constant, and k is Boltzmann's constant. For the case of classical equipartition of energy, this expression reduces to

$$V^2 = \int 4RkT d\nu \tag{32}$$

which is known as the Nyquist equation for the voltage generated by Johnson noise in a resistor. Note that the symbol R in Equation 32 does not refer to the gas constant, but to an electrical resistance which may vary with frequency.

Utilizing the Nyquist relation, Johnson was able to derive values of the Boltzmann constant that agreed well with the then-current value. We note that k is uncertain at the level of ~40 ppm, so that — like gas thermometry — noise thermometry intended to produce accuracy better than 40 ppm must involve two measurements made at different temperatures, thus eliminating k from the resulting ratio calculation. The relatively small magnitude of the Johnson noise voltage requires the use of broad-band voltage amplifiers for its measurement, however; the lack of stability in the amplifiers available until relatively recently precluded the achievement of temperatures that were accurate within less than a few tenths of a percent.

A. Thermometry Using Standard Noise Circuitry

Garrison and Lawson achieved 0.1% accuracy levels in noise thermometry by utilizing a switching technique and a carefully-designed voltage amplifier-filter circuit.[46] They prepared two resistors of resistances R_1 and R_2 and placed them in thermostats at separate temperatures T_1 and T_2, chosen so that $R_1T_1 = R_2T_2$. Furthermore, they adjusted the input capacitances C_1 and C_2 so that $R_1C_1 = R_2C_2$. Then, having achieved balanced voltages across the two resistors, they alternately measured the two voltages while adjusting the unknown temperature T_2 until the voltages became equal. At that point, they found that the ratio R_1/R_2 was equal to T_2/T_1 within 0.1%. In their experiment, they measured a frequency band between 10 and 100 kHz and used sensing resistors of somewhat less than 5 kΩ resistance.

More recently, Pickup of the NML in Australia used a similar switching technique with improved voltage amplifiers and voltage-to-frequency conversion.[47,47a] His interest was in measuring Kelvin thermodynamic temperatures in the range 90 to 425 K by Johnson noise thermometry. Pickup used essentially the same circuit conditions chosen by Garrison and Lawson; $R_1T_1 = R_2T_2$ and $R_1C_1 = R_2C_2$ and a bandwidth from 10 to 200 kHz. His resistances were chosen to be 2 to 10 kΩ. He paid careful attention to minimizing nonthermal circuit noise, incorporating quiet switches and shielding against electromagnetic interference. His results are given in Table 11; although carefully performed, Pickup's temperature determinations are less precise than the best recent thermodynamic results.

Crovini and Actis used a different technique to measure the difference between the IPTS-68 and Kelvin thermodynamic temperatures in the range 900 to 1235 K.[48,48a] They employed a modified Garrison-Lawson method for a few of their measurements, with the equation

$$T_1 = a^2 \frac{R_0}{R_1} T_0 + (a^2 - 1) \frac{R_e}{R_1} T_0 \tag{33}$$

where the parameter a is close to unity. It accounted for a slight mismatch in the quantities R_0T_0 and R_1T_1. R_e is the symbol for the equivalent amplifier noise resistance.

Crovini and Actis also performed measurements with what they called an "attenuation method." In this case, $R_1 = R_2$ and a balance between the two noise voltages was achieved

Table 11 THERMODYNAMIC TEMPERATURES DETERMINED BY NOISE THERMOMETRY[a]		
T_{68} (K)	T_{Ref} (K)	$T_N - T_{68}$ (mK)
90.17	298	+3.5 ± 3
97.14	298	+8 ± 4
97.45	298	+8 ± 4
408	273	−12 ± 2.4

[a] Data from References 47 and 47a.

Table 12 COMPARISON OF T_{68} AND THERMODYNAMIC NOISE TEMPERATURES[a]		
T_{68} (K)	T_{Ref} (K)	$T_N - T_{68}$ (mK)
902	273	−90 ± 180
902	504	−40 ± 180
953	273	+370 ± 290
1053	273	+470 ± 240
1123	273	+670 ± 360
1235	273	+70 ± 340

[a] Data from References 48 and 48a.

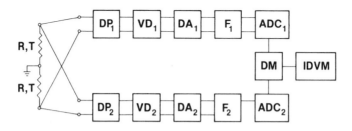

FIGURE 11. Block diagram of correlation noise thermometer.[34] Two temperature-sensing resistors of resistance R are held at the test temperature T. Equivalent channels composed of differential preamplifiers DP_1 and DP_2, symmetric ohmic voltage dividers VD_1 and VD_2, differential amplifiers DA_1 and DA_2, filters F_1 and F_2, and 16-bit analog-to-digital converters ADC_1 and ADC_2 carry their signals to a digital multiplier DM and an integrating digital voltmeter IDVM. The circuitry detects the voltage signal, amplifying correlated noise voltages that arise in the sensing resistors but rejecting uncorrelated circuit noise.

by attenuating the amplifier gain for the high-temperature measurements. Independent measurements of the amplifier noise then permitted calculation of the thermodynamic temperature ratio.

It is interesting that Crovini and Actis used a high-temperature platinum resistance thermometer to provide values of the IPTS-68 in the range that is defined by the platinum-10 wt% rhodium vs. platinum thermocouple thermometer, preferring the improved precision of the resistance thermometer to the official — but relatively imprecise — defining instrument. (See Chapter 4, Section V and Chapter 6, Section III.C.)

The results obtained by Crovini and Actis showed a large level of imprecision. Nevertheless, they indicate that the IPTS-68 deviates by as much as 0.5 K from the thermodynamic scale in the range 900 to 1300 K. Table 12 contains a summary of their results; we include them also in a discussion in Section V.

B. Cross-Correlation Methods

Klein, Klempt, and Storm have made substantial improvements in the thermometric precision available in Johnson noise thermometry.[34,49,50] They have accomplished this advance by utilizing the cross-correlation technique suggested originally by van der Ziel.[51]

Figure 11 shows a block diagram of the measuring circuit used by Klein, Klempt, and Storm in their noise studies. The sensing resistor was grounded at its center, providing

Table 13
JOHNSON NOISE TEMPERATURES ON THE KTTS AS DETERMINED BY KLEIN ET AL.[34,50]

^{4}He pressure (Pa)	T_N (K)	T_{Ref} (K)	$T_N - T_{76}$ (mK)
4659.9	2.1455 ± 0.0003	273	+0.4
5612.2	2.2213 ± 0.0003	273	+0.3
12680	2.6137 ± 0.0004	273	+0.4
25049	3.0274 ± 0.0005	273	0
51784	3.5809 ± 0.0004	273	+0.2
104274	4.2527 ± 0.0005	273	−0.1
104325	4.2534 ± 0.0005	273	+0.1
^{4}He bp (calc)	4.2221 ± 0.0005	—	0
Ar tp	83.8045 ± 0.0011	273	+6.5[a]

[a] $T_N - T_{68}$ (mK).

balanced voltage inputs to two identical differential-amplifier/voltage-divider/filter channels whose outputs are multiplied and then measured with an integrating digital voltmeter. The circuit amplifies the correlated sensing-resistor noise voltage that occurs simultaneously in both channels, but discriminates strongly against uncorrelated noise arising in one or the other amplifier channel.

Klein, Klempt, and Storm developed a functional ratio equation to calculate the unknown sensor temperature from voltage measurements at that temperature (2 to 4 K and the argon triple point temperature) and at the triple point of water:

$$F(R_x) = \frac{V_x}{V_0} \frac{R_0}{R_x} \frac{\alpha_0^a \alpha_0^b}{\alpha_x^a \alpha_x^b} T_0 (1 - \gamma_R - \gamma_c) \quad (34)$$

In Equation 34, the first term contains the ratio of noise voltage measured at the unknown temperature T_x and at the reference temperature T_0 (273.16 K); the ratio of the resistances of the sensing resistor R, measured at the same temperatures; the ratio of the product of the attenuation factors α^a and α^b for the voltage dividers used in the two channels; and the reference temperature. The second and third terms contain also γ_R and γ_c, frequency-dependent terms involving the input capacitance and resistance.

Setting the first term in Equation 34 to be T_x, the test temperature of the sensing resistor, the authors approximated Equation 34 as $F(R_x) \approx T_x[1 + K(T)R_x]$, where K(T) is a small, temperature-dependent coefficient of R_x. F(0) and hence T_x are obtained by plotting experimental values of $F(R_x)$ vs. R_x and extrapolating to $R_x = 0$. Klein and his colleagues evaluated T_N for seven temperatures that had been measured on the EPT-76 temperature scale by ^{4}He vapor-pressure measurements and, later, for the argon triple point. Their results are shown in Table 13.

The experiments of Klein, Klempt, and Storm have provided excellent confirmation of the accuracy of the thermodynamic measurements by gas thermometry in the range of temperature below 273 K.

C. Noise-Voltage Detection by Superconductive Josephson Junctions

Yet another approach to Johnson noise voltage thermometry involves the use of a superconducting device as the sensor of the noise voltage. Soulen, of the NBS, following a technique developed by his colleague Kamper,[52] has studied the measurement of thermodynamic temperatures in the deep cryogenic range 0.01 to 0.5 K with a measuring instrument called a resistively shunted superconducting quantum interference device (R-SQUID).[53] This

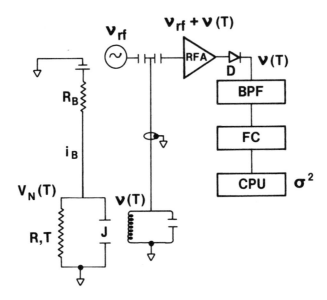

FIGURE 12. Block diagram of Soulen's Josephson-junction noise thermometer (JJNT). The temperature-sensing resistor R is maintained at the test temperature T in a cryogenic refrigerator. Also refrigerated is the Josephson junction J and the adjacent oscillator circuit. Johnson noise voltage $V_N(T)$ is generated in R as a consequence of its temperature. When J is biased by a steady current i_B, it responds to $V_N(T)$ by oscillating at a frequency $\nu(T)$, typically in the audio range. This frequency modulates the carrier frequency ν_{rf}. The modulated radio frequency is amplified by the radio-frequency amplifier RFA, demodulated, and filtered by a band-pass filter BPF. The filtered signal $\nu(T)$ is measured repeatedly by a frequency counter FC. The variance σ^2 is calculated by a dedicated computer CPU to obtain the noise temperature.

device has the unique property of responding to very small voltage variations by changing its characteristic frequency; thus it constitutes a very effective voltage-to-frequency converter. Since frequency measurement can be accomplished with great sensitivity using modern counting equipment, and since the intrinsic noise of Josephson junctions is exceedingly small, voltages as small as 10^{-15} V can be determined in this type of measurement.

Figure 12 contains a schematic drawing of Soulen's noise thermometer. It includes a block of niobium metal in which the point of a niobium screw can be adjusted by a mechanical linkage from room temperature to form a so-called "Josephson point-contact junction." We may note that there are other methods of creating a Josephson junction, among them the introduction of a thin insulating barrier between two superconducting metal strips and also the cutting away of most of a single superconducting metal strip, leaving a narrow bridge to couple the two sides weakly. Each method of forming the Josephson junction results in a device with somewhat different operating characteristics, including internal noise, voltage sensitivity, and response to external excitation. The same voltage-to-frequency relation, however, always holds.

The R-SQUID circuit used by Soulen also is shown in Figure 12. The point-contact Josephson junction J is shunted by a resistor made of CuSi alloy whose resistance is about 10^{-5} Ω. A steady bias current I_B moves the center of the Josephson junction fluctuations away from zero frequency. It then oscillates at a frequency $\nu(t,T)$ given by the equation

$$\phi_0 \nu(t,T) = i_B R + V_N(t,T) + V_T + V_J + V_p \qquad (35)$$

In Equation 35, ϕ_0 is the quantized unit of magnetic flux given by ($\phi_0 \equiv h/2e$), where h is Planck's constant and e is the charge of an electron. The time-varying voltage $V_N(t,T)$ is the temperature-dependent Johnson noise voltage generated in the shunt resistor R, and V_T, V_J, and V_p represent unwanted sources of voltage arising respectively from thermal emfs, from noise in the Josephson junction, and from vibrations or other external parasitic disturbances of the circuit.

The R-SQUID frequency, typically 1 to 100 kHz, is detected by coupling the device to a resonant LC circuit that also is shown in Figure 12. The R-SQUID signal is carried as an amplitude modulation of a 30-MHz rf carrier frequency generated by an oscillator at room temperature. After demodulation, the R-SQUID audio-frequency signal is processed by a filter-counter system containing a dedicated computer. The variances of the detected frequency spectra are related to the temperature through the equation

$$\sigma_N^2 = \sum_i \frac{(\nu_i - \nu_{i+n})^2}{2N} \tag{36}$$

where the index n can take positive integral values (in practice $1 \leq n \leq 4$) and the other parameters are defined as follows: N is the total number of frequency counts; ν_i is the frequency observed during the ith measurement; ν_{i+n} is the frequency measured at a time $\nu\tau$ later, where τ is the measurement gate time.

When the variances for all n are equal, the noise in the measuring circuit can be described as "white" or as independent of frequency. In that case only the V_N term in Equation 35 need be considered, and the temperature of the sensing resistor can be determined by

$$\sigma_N^2 = \frac{2kRT}{\phi_0^2 \tau} \tag{37}$$

In comparison with noise thermometry used at higher temperatures, the Josephson-junction thermometer is not particularly accurate. Soulen estimates its thermodynamic uncertainty as $\pm 0.04\%$ at best, and $\sim \pm 0.5\%$ at 0.01 K. Clearly, the uncertainty of the Boltzmann constant is not a factor in the accuracy of thermodynamic temperatures determined in this fashion. Nevertheless, the method can be used at temperatures below 1 K, where other noise thermometers cannot, and furthermore, its accuracy has proved quite sufficient to achieve new levels of understanding of thermodynamic temperature there.

In conjunction with Marshak, Soulen has used his results to create a new laboratory temperature scale (for a discussion of this scale, see the following section). In addition, he and his associate Dove have preserved his measurements by recording the thermodynamic temperatures of the superconductive transitions in several materials — those of the elements W, Be, Ir, and Cd and of the alloys $AuIn_2$ and $AuAl_2$ (see Chapter 3, Section VIII). Finally, Soulen has measured the resistance vs. temperature relations for several low-temperature resistance thermometers.

IV. NUCLEAR ORIENTATION THERMOMETRY

We already have discussed the use of quantum-level population differences in thermometry. One of the more straightforward applications of this principle is nuclear orientation thermometry.

The basic idea in nuclear orientation thermometry is that the spatial distribution of nuclear radiation emitted from a radioactive source can be related to an identifiable laboratory axis and to the temperature of the source atoms. In the work we discuss here, the laboratory axis is defined by the symmetry of a single crystal of cobalt metal; the nuclear radiation is that

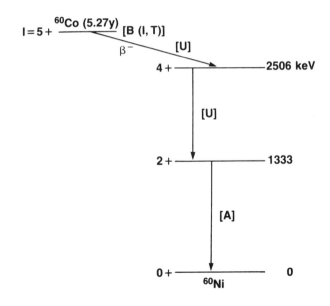

FIGURE 13. Simplified decay scheme of ^{60}Co. The major radiation modes are shown, along with the level spin assignments and energies. The parameters in brackets refer to quantities appearing in Equation 41.

arising in the decay of ^{60}Co nuclei. There are many other systems that also can be used to derive values of temperature below 2 K. Details of nuclear orientation theory and summaries of results obtained by many experimenters have been described in several review papers.[54-57]

Marshak[57,58] has used ^{60}Co gamma-radiation to evaluate Kelvin thermodynamic temperatures in the range 0.01 to 0.05 K. The study is especially interesting from the point of view of thermodynamic thermometry because the radioactive sample used by Marshak was in good thermal contact with a Johnson-noise thermometer operated by Soulen. Thus the two experimenters sought to derive thermodynamic temperatures at the same time by techniques that differ considerably in their principles and in their analyses. We discussed briefly the work of Soulen in the previous section.

A slightly simplified decay scheme for the radioactive nucleus ^{60}Co is shown in Figure 13. ^{60}Co decays with a half-life of about 5.3 years by beta-emission, transforming in the process into ^{60}Ni, a stable nucleus. In Figure 13, we show only the most prominent of the three known beta transitions, that populates the spin 4+ state of ^{60}Ni at 2506 keV energy. The major gamma-rays that depopulate the 4+ level in cascade have energies of 1173 and 1333 keV; it is these gamma-rays that Marshak observed in his experiment.

In order to determine accurately the temperature of the radiating system by measuring the angular distribution of the gamma-rays, it is necessary to know in advance how the distribution varies with temperature. We now describe briefly how one can calculate the temperature-dependent gamma-ray distribution for ^{60}Co in a cobalt single crystal.

Cobalt crystallizes in a hexagonal close-packed lattice, providing the single-crystal sample with a cylindrically-symmetric axis. The metal is ferromagnetic at temperatures below 1400 K, with the electronic magnetic moments aligned parallel to the hexagonal axis. (Note, however, that imperfections in the crystal and in the pattern of the ferromagnetism can cause local variances from this ideal description.) The energies of the (2I + 1) nuclear substates of ^{60}Co are determined by the interaction of the nuclear magnetic moment and the nuclear electric quadrupole moment with the effective magnetic field and the electric field gradient at the nucleus — the so-called "hyperfine" interaction. At the low temperatures involved in this experiment, only the lowest electronic level is involved in the hyperfine interaction. The energies of the (2I + 1 = 11) nuclear substates E_m are given by the relation

$$E_m = \frac{\mu}{I} B_{eff} m + P\left[m^2 - \frac{1}{3}I(I+1)\right]$$

$$= (\mu B_{eff} \, m/5) + P(m^2 - 10) \qquad (38)$$

where μ is the magnitude of the nuclear magnetic dipole moment, B_{eff} is the effective magnetic field at the nucleus, m is the nuclear magnetic quantum number, and P is the quadrupole coupling constant.

If the system is adequately described by the assumptions made so far and if the system is in thermal equilibrium throughout, then the relative populations of the m substates, a_m, can be calculated from the Boltzmann distribution function involving the level energies E_m

$$a_m = \frac{\exp(-E_m/kT)}{\sum_m \exp(-E_m/kT)} \qquad (39)$$

From the appearance of the Boltzmann distribution function, one can see that the temperature range over which the substate populations vary is dependent upon the strength of the hyperfine interaction. At relatively high temperatures, all substates have the same population; at relatively low temperatures, only the lowest substate is populated. In either case, slight changes in temperature will not affect the level populations, so that the gamma-ray distribution will not be a measure of temperature. Thus we can see that, for any particular nuclear orientation thermometer, thermometry is possible only within a specific range of temperatures.

If the coupling constants specified in the hyperfine interaction are accurately known and the crystal lattice is well behaved, then one can evaluate the relative substate populations and describe the degree of spatial orientation of the ensemble of ^{60}Co nuclei by a statistical tensor $B_\lambda(I,T)$

$$B_\lambda(I,T) = \sum_{m=-I}^{I} (-1)^{I-m} [(2\lambda + 1)(2I + 1)]^{1/2} \begin{pmatrix} I & I & \lambda \\ m & -m & 0 \end{pmatrix} a_m(T) \qquad (40)$$

where the index λ takes all integral values from 0 to 2I, with $B_0(I,T) = 1$. If B is nonzero for any odd values of λ then the ensemble of spins is said to be "polarized"; if only even-lambda B are nonzero, then the spin system is said to be "aligned." Note that all of the temperature dependence of the tensor is contained in the substate populations $a_m(T)$.

In order to relate the degree of orientation of the assembly of ^{60}Co nuclei to the observed anisotropy of the 1172 and 1333-keV gamma-rays in ^{60}Ni, one must consider the effect of the beta emission and the gamma radiation upon the original level of orientation. This is accomplished through the use of the following relation:

$$W(\theta,T) = \sum_{\lambda=0}^{\lambda max} B_\lambda(I,T) \, U_\lambda A_\lambda Q_\lambda P_\lambda(\cos\theta) \qquad (41)$$

$W(\theta,T)$ is the normalized directional distribution of the observed gamma radiation from the ^{60}Co nuclei, where θ is the angle between the direction of emission and the crystal symmetry axis. The quantities B_λ are given by Equation 40. The quantities U_λ are called angular

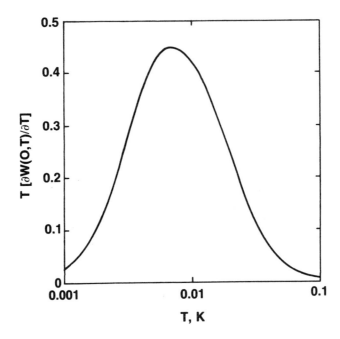

FIGURE 14. The "sensitivity function," $T[\partial W(\theta,T)/\partial T]$, vs. temperature for the thermometer ^{60}Co in a cobalt single crystal, evaluated for $\theta = 0$.

momentum deorientation coefficients and the quantities A_λ are called angular distribution coefficients; both depend upon the spins and multipole amplitudes characterizing the nuclear decay scheme. Thus, although neither U nor A depends upon temperature, their accurate evaluation demands a secure knowledge of the details of the nuclear states and radiations involved. The actual angular dependence of the radiation is dictated by the Legendre polynomials P_λ; the very definite radiation patterns that are predicted by these polynomials are blurred by the "solid-angle" functions Q_λ however.

In Figure 13, we draw attention to the quantities in Equation 41 as they pertain to the various states and radiations. Note that the degree of orientation B is determined by the hyperfine interaction of the spin 5+ level of ^{60}Co. We have treated the case as if only the lower, 1333-keV gamma ray were being monitored; therefore the upper gamma ray is associated in Figure 13 with the U parameter.

For ^{60}Co in a cobalt single crystal with no applied field, only alignment (not polarization) of the nuclei takes place; thus only even values of λ are nonzero. Marshak[57] discusses the evaluation of the several quantities in Equation 41, including the calculated overall temperature sensitivity of the thermometer. As it happens, the most favorable geometry for observing the gamma-ray distribution is achieved by placing the detectors along the crystal axis, so that θ is 0 or 180 degrees. The temperature sensitivity can be defined as

$$\frac{\Delta T}{T} = \frac{\Delta W}{T}\left(\frac{\partial T}{\partial W}\right) = \Delta W / \left[T\left(\frac{\partial W}{\partial T}\right)\right] \qquad (42)$$

Here ΔW reflects a relatively fixed capability of the gamma-ray detection equipment to resolve differences in the gamma-ray anisotropy; the quantity $T(\partial W/\partial T)$ dictates the variation in temperature sensitivity of the nuclear orientation thermometer. Figure 14 shows the temperature dependence of this "sensitivity function" for ^{60}Co in a cobalt single crystal as

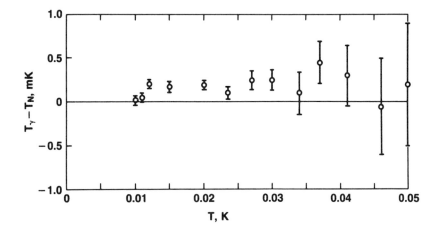

FIGURE 15. The differences in Kelvin thermodynamic temperatures in the range 0.01 to 0.05 K as determined by a ^{60}Co-in-cobalt-single-crystal thermometer (T_γ) and by a Josephson-junction Johnson noise thermometer (T_N). The error bars denote calculations of the combined experimental standard deviations of the two sets of thermodynamic determinations at each temperature.

calculated by Marshak. One can readily see that the thermometer is most sensitive in the range 0.003 to 0.03 K.

In accomplishing the comparison with Soulen's noise thermometer, Marshak was limited to temperature measurements above 0.01 K by the cryogenic apparatus used. The upper limit was set by the declining thermometer sensitivity above 0.05 K.

Figure 15 shows the results of the comparison experiment. The individual data points indicate temperatures at which both gamma-ray and noise measurements were performed; the error bars denote calculations of the combined standard deviations of both measurements.

The satisfactory level of agreement between the two very different thermodynamic techniques (approximately ±0.5%) indicates that thermodynamic temperatures in this very low temperature regime can be provided by either of the thermometers involved in the comparison. A laboratory temperature scale, the "NBS-CTS-1" (National Bureau of Standards Cryogenic Temperature Scale No. 1) has been based upon this comparison experiment; it is disseminated through the superconductive reference-temperature device SRM 768, through ^{60}Co-in-cobalt-single-crystal thermometers, and through various resistance thermometer calibrations.

V. KELVIN THERMODYNAMIC TEMPERATURES BELOW 273 K

Based upon the measurements discussed in the previous sections of this chapter, it appears that the KTTS now is known with an uncertainty not exceeding ±2 mK over the range 0.01 to 273 K.

We have shown in Figure 15 the close agreement obtained by Soulen and Marshak from thermodynamic temperature measurements in the range 0.01 to 0.05 K, and we have noted that Soulen has provided superconductive fixed points and thermometer calibrations to temperatures as high as 0.5 K, based upon continuing Josephson-junction noise thermometer measurements.

In Figure 10 we have shown the relation between the EPT-76 scale and Berry's NPL-75 scale (extended to 0.5 K by paramagnetic salt thermometry and helium vapor-pressure thermometry) in the range 0.5 to 27 K.

In Figure 16, we summarize the relation that has been developed between the KTTS and the IPTS-68 below 273 K. It appears that the IPTS-68 is colder than the KTTS by as much

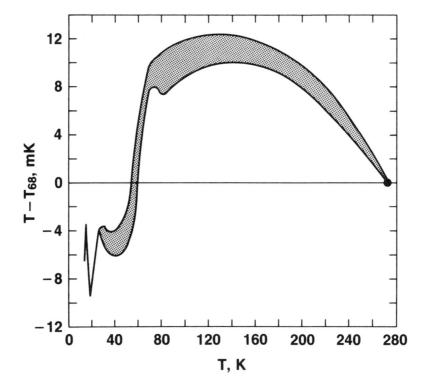

FIGURE 16. Differences between Kelvin thermodynamic temperatures as recently determined, and temperatures on the IPTS-68, in the range below 273 K.

as 10 mK in the range 60 to 273 K, and that it is hotter than the KTTS by about the same amount near 20 K.

Many of the results shown in Figure 16 must be regarded as preliminary at the time this book is written. It is possible that the careful rechecking of calculations, measurements, and calibrations that accompanies the preparation of scientific results for publication may entail small modifications of the thermodynamic temperature differences shown in the figure. However, barring the appearance of conflicting results from new experiments, the measurements on which Figure 16 is based will provide more accurate assignments of temperature to many fixed points below room temperature. Table 14 contains a summary of the new values for comparison with those assigned on the IPTS-68 and EPT-76 scales.

VI. RADIATION THERMOMETRY

Thermodynamic determinations of temperature may be accomplished by observing the radiation that is emitted by all objects. The Stefan-Boltzmann law

$$P = \epsilon \sigma T^4 \tag{43}$$

shows that the rate at which energy is radiated varies enormously with temperature. P is the radiated power, ϵ is the emissivity of the radiator, and σ is the Stefan-Boltzmann constant[59,59a] [$\sigma = 5.67032 \pm 0.00071 \times 10^{-8}$ W/(m^{2}K^4)].

The requirement of detection sensitivity alone usually restricts radiation thermometry to temperatures well above the water triple point, where spectral radiation thermometry can comfortably be practiced by means of Planck-law ratios.[59]

Table 14
SUMMARY OF RECENT DETERMINATIONS OF THE KELVIN THERMODYNAMIC TEMPERATURES OF SOME LOW-TEMPERATURE FIXED POINTS

Fixed point	T(practical) (K)	T(thermodynamic) (K)	Ref.
H_2O tp	273.16	Reference temp	—
(172)	172.000	172.011 ± 0.0015	41
O_2 bp	90.188	90.196 ± 0.002	41
		90.195	42
Ar tp	83.798	83.804 ± 0.004	41
		83.804 ± 0.001	41
N_2 tp	63.146	63.150 ± 0.003	41
O_2 tp	54.361	54.358 ± 0.002	41
		54.3585 ± 0.001	41
		54.357	42
Ne bp	27.102	27.097 ± 0.001	41
		27.098 ± 0.001	41
		27.098 ± 0.001	26
		27.096	42
H_2 bp	20.28	20.271 ± 0.001	26
H_2 17 K pt	17.042	17.0356 ± 0.001	26
H_2 tp	13.81	13.8035 ± 0.001	26
^{4}He bp		4.2221 ± 0.0005	26
		4.2221 ± 0.0005	34

$$\frac{L(\lambda,T)}{L(\lambda,T_{ref})} = \frac{\epsilon(\lambda,T)[\exp(c_2/\lambda T_{ref}) - 1]}{\epsilon(\lambda,T_{ref})[\exp(c_2/\lambda T) - 1]} \qquad (44)$$

The Planck law is used to define the IPTS-68 above the freezing temperature of gold (see Equation 28 in Chapter 4). $L(\lambda,T)$ is the spectral radiance at wavelength λ and temperature T. The emissivity $\epsilon(\lambda,T)$ of the object may vary with temperature and wavelength. The second radiation constant $c_2 = 1.438786 \pm 0.000045 \times 10^{-2}$(m·K). The wavelength is given by λ, and the temperatures are T and T_{ref}. Radiation thermometry commonly is performed at temperatures above 1000 K. However, the uncertainty associated with the Kelvin thermodynamic temperatures of high-temperature fixed points has driven radiation thermometrists to perform experiments based upon both the Stefan-Boltzmann law and the Planck law at temperatures well below "normal." We discuss several of these in this section.

A. Total Radiation Thermometry

The deviations observed by Guildner and Edsinger in the IPTS-68 above 0°C (Section II.H.1) have been observed also by the method of total radiation thermometry. Quinn and Martin of the NPL in England compared the radiation emitted by a black body held at various test temperatures with that emitted by the black body when it is held at the triple point of water.[60] Quinn and Martin based their experiment upon that of Ginnings and Reilly,[61] employing a cryogenic total radiation detector in conjunction with a black body furnace. A schematic drawing of the apparatus is shown in Figure 17. In this experiment, radiation from a black body B, heated uniformly by a heater H_2 to a temperature T in the range 233 to 383 K, is allowed to enter a collector C through a shutter S and 4.2-K radiation trap RT. A weak thermal link L connects C to a 2-K bath of pumped ^{4}He, so that the collector temperature rises to a temperature T_1. By closing the shutter and adjusting the current in heater H_1, the operator can duplicate the temperature rise, thus establishing the rate at which

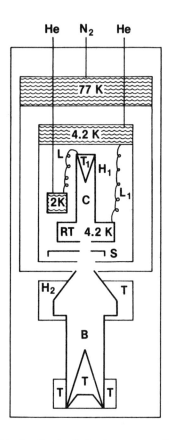

FIGURE 17. Total radiation thermometer of Quinn and Martin.[60] The temperature of radiating black body B is regulated by a heater H_2 in the range -40 to $+110°C$. Its temperature uniformity is evaluated by thermometers at positions T. A calorimeter C is connected by a weak thermal link L to a 2-K pumped ^{4}He bath. The radiation from B passes an aperture system and a shutter S cooled to 77 K, as well as a radiation trap RT cooled to 4.2 K. Thermometer T_1 measures the temperature rise in the calorimeter while the shutter is open; after the shutter is closed, the electrical heater H_1 is energized to provide the same temperature rise in the calorimeter. The electrical energy dissipated by H_1 provides a measure of the power previously radiated by B at its measured IPTS-68 temperature.

power is radiated from the high-temperature black body. The black body temperature T is determined from the ratio

$$T = T_{ref} \left[\frac{P(T)}{P(T_{ref})} \right]^{1/4} \tag{45}$$

There are several potential sources of systematic error in this experiment. One of these is the uncertainty in the effect of diffraction at the limiting aperture, which is wavelength dependent and therefore temperature dependent. A second source is the scattering of radiation into the collector from sources other than the high-temperature black body. It similarly may vary with temperature. A third possible systematic error arises from the uncertainty in the

Table 15
KTTS-T$_{68}$ DIFFERENCES OBTAINED BY TOTAL RADIATION AND GAS THERMOMETRY

t$_{68}$ (°C)	t(Rad) − t$_{68}$ (°C)[a]	t(Gas) − t$_{68}$ (°C)[b]
0.01	Reference	Reference
10	−0.003	—
20	−0.006	−0.003
40	−0.009	−0.007
62	−0.016	−0.012
82	−0.023	−0.020
100	−0.029	−0.025

[a] Data from References 60, 62, and 62a.
[b] Data from Reference 16.

amount of energy carried from place to place within the apparatus by residual molecules of gas. Yet another problem consists of the uncertainty in the value of the total emissivities of the radiator and the collector, either of which may differ from unity and may vary with temperature. Temperature nonuniformity of the radiator also may contribute to the uncertainty of this experiment.

Quinn and Martin sought to minimize the many uncertainties in their total-radiation measurements by increasing the size of their apparatus relative to the earlier but quite similar equipment built by Ginnings and Reilly. By careful measurement, Quinn and Martin have been able to determine values of Kelvin thermodynamic temperature in the range −40 to 110°C (233 to 383 K) with an overall uncertainty at the 99% confidence level (three times the estimated standard deviation) ranging from 6 to 18 mK, depending principally upon the size of the apertures used in the experiment.[60,62,62a] The values of Kelvin thermodynamic temperature thus obtained are shown in Table 15 in comparison with the earlier data of Guildner and Edsinger. The differences between the two sets of results lie well within their combined uncertainty limits.

B. Spectral Radiation Thermometry

We have noted already that use of the Planck law (Equation 44) permits the evaluation of any temperature in terms of any other temperature by a method that is soundly based in statistical mechanics and thermodynamics. As the accuracy of spectral radiation thermometry has improved, there has been a corresponding increase in interest in using this method to evaluate the thermodynamic accuracy of the IPTS-68.

In principle, the Planck law can be employed to measure the ratio of any desired temperature to the temperature of the triple point of water. However, the difficulties in such an undertaking are, up to the present time, insurmountable. The comment of Hall[63] in this respect is illuminating: "Ideally, to establish the radiation scale (by means of Planck-law ratios) absolutely, one should take as T$_1$ the triple point of water, where the numerical value of the thermodynamic temperature is fixed by definition as 273.16 °K (+0.01°C), but this would call for higher sensitivity than is at present available and would introduce many experimental difficulties."

Hall performed spectral radiation thermometry over the range 175 to 1064°C, using as his reference temperature the zinc freezing point. The magnitudes of Hall's measurement uncertainties are larger than one would like to see today (they range from ±0.2°C overall at 630°C to ±0.6°C at 1064°C), and his experimental apparatus is no longer the best available for this purpose. Nevertheless, Hall's lucid exposition of his measurement procedures and analytical techniques still provides a useful starting point for the study of Planck-law thermometry.

Hall's contemporary, Heusinkveld,[64] measured the thermodynamic temperatures of the aluminum, silver, and gold freezing points, using 630.47 °C for the antimony freezing point.

More recent experiments designed to evaluate the IPTS-68 by Planck-law thermometry have taken two forms; a struggle for accurate spectral-radiance ratios while using as a reference temperature the relatively secure thermodynamic values provided up to 725 K by the recent gas thermometry of Guildner and Edsinger, or an evaluation of the ratio of thermodynamic temperatures of fixed points both of which, though themselves thermodynamically uncertain, lie in the range of temperature that permits relatively accurate measurements.

Each of these two types of measurement is useful. The merit of the former is obvious; it meets the declining accuracy of gas thermometry and carries the thermodynamic temperature measurement process into the "natural" radiation range. The merit of the latter is that, once an adequate thermodynamic assignment can be made to its reference temperature, its temperature ratio is easily interpreted in thermodynamic terms.

1. Measurements by Coates and Andrews

Only recently has it become feasible to attempt seriously the use of Planck-law thermometry using a temperature as low as 725 K as a reference. One of the first such experiments was performed by Coates and Andrews.[65,66] The experiment was based upon the use of a photoelectric pyrometer,[67] in which radiation is collected by a mirror-based optical system, filtered, and detected by a photomultiplier tube. Ratios of the spectral radiances of the source at two temperatures and at several wavelengths can be observed in order to determine a thermodynamic value for the higher temperature.

The radiation source used by Coates and Andrews for their measurements between 440 and 630°C contained a graphite black body with a 3-mm platinum-disc aperture. Its temperature on the IPTS-68 was measured with a platinum resistance thermometer. A single wavelength, 812 ± 5 nm, was used with the NPL photon-counting pyrometer. The Kelvin thermodynamic temperatures corresponding to the T_{68} black body setting were determined with reference to Guildner and Edsinger's fitted value of the (t − t_{68}) difference at 440°C (−0.074°C).

Coates and Andrews determined (t − t_{68}) for many temperatures, and the work has been continued to the gold point by Andrews and Gu Chuanxin.[66a] Their results are shown in Figure 18 along with those determined in other experiments to be discussed presently.

2. Measurements by Jung

Jung of the PTB in Berlin also has measured Kelvin thermodynamic temperatures in the range 400 to 630°C with respect to Guildner and Edsinger's gas thermometry results. His experiment[68] differed from that of Coates and Andrews in several respects, as did his determinations of (t − t_{68}).

Jung employed a pyrometer containing a silicon photodiode radiation detector that exhibited a linear response over the range of his measurements. The operating wavelength was 974 ± 20 nm. Jung used two black body cavities with 3-mm apertures as radiation sources. Four platinum resistance thermometers were mounted on one black body to provide its IPTS-68 temperature. The other black body was immersed in an aluminum freezing-point cell to provide a fixed-temperature radiance standard.

The measurement sequence involved a periodic comparison between the radiance of the first black body, regulated at 456°C, and the second, regulated at the aluminum point, followed by a similar comparison with the first black body temperature changed to a new IPTS-68 temperature in the range 400 to 630°C. Thus the two measurements of the first

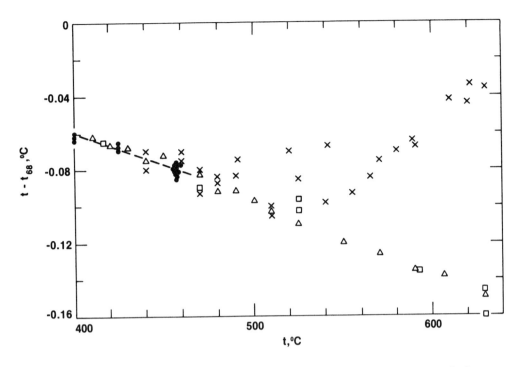

FIGURE 18. Differences between Kelvin thermodynamic temperatures and IPTS-68 temperatures in the range 400 to 630°C as determined by spectral radiation thermometry. The reference temperature in each measurement was derived from the calculated difference curve of Guildner and Edsinger[16] shown by the dashed line. The actual data of Guildner and Edsinger are shown (solid dots) to illustrate the variability in their measurements in this temperature range. The NPL photon-counting pyrometer results of Coates and Andrews[66] are shown by Xs. The triangles denote the results of Jung,[68] and the squares indicated the as-yet unpublished data of Bonhoure and Pello.[69]

black body were compared with each other, using an aluminum-point radiator as a continuing radiance standard.

Jung's measured differences are shown as the triangles in Figure 18. He estimated the uncertainty of his determinations of $(t - t_{68})$ at 630°C as ±0.02°C overall at the 99% confidence level.[70] (See also Chapter 4.) He also reported a preliminary thermodynamic measurement of the aluminum freezing-point temperature as 933.450 ± 0.022 K.

Bonhoure and Pello of the International Bureau of Weights and Measures have reported informally[69] yet other measurements of thermodynamic temperatures in the range 400 to 630°C. These were accomplished using two copper black bodies, from which the radiation at 1000 nm was selected by a double monochromator for detection by a cooled Ga-In-As-cathode photomultiplier. Their results, shown as squares, also are included in Figure 18.

3. Spectral Radiation Measurements Above 630°C

There is a troublesome problem associated with expressing the results of thermodynamic measurements above 630°C, in the thermocouple range of the IPTS-68. We have expressed such results in lower temperature ranges in two ways; as new, more accurate values for fixed-point temperatures, and as differences between the Kelvin thermodynamic scale and the relevant practical scale. The second of these methods gives the reader a ready correction that can be applied to practical-scale temperature values between the defining fixed points. For the EPT-76 and for the platinum resistance thermometer range of the IPTS-68, the levels of temperature nonuniqueness (see Chapter 4, Section II) scarcely are larger than typical thermometer imprecision levels; however, for the thermocouple thermometer range of the IPTS-68, the level of nonuniqueness (the variation in scale temperature values that can be

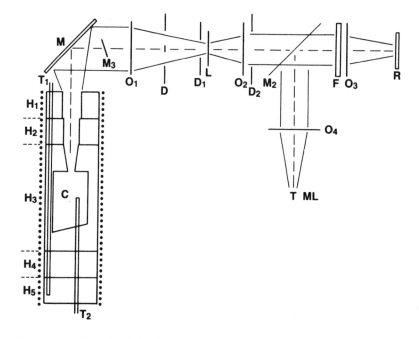

FIGURE 19. Spectral radiation thermometry apparatus of Bonhoure.[72] C, Cylindrical nickel radiation cavity, held within 0.1°C of the test temperature by zone heaters H_1 to H_5. The test temperature was measured on the IPTS-68 by two standard platinum-10 wt% rhodium vs. platinum thermocouple thermometers T_1 and T_2 that were protected by Al_2O_3 tubes. The cavity radiation was reflected from an aluminized quartz mirror M into the optical train containing lenses O and L, diaphragms D, an interference filter F operating at about 1000 nm, and a photomultiplier receiver R. The mirror M_2 was used to direct light from the mercury vapor lamp ML into the receiver. Mirror M_3 was used to check the alignment of the cavity.

obtained by the use of several approved thermometers) is as large as ±0.2°C.[71] (See also Chapter 4, Section V.)*

The magnitude of the IPTS-68 uncertainty above 630°C has caused relatively little difficulty to thermometrists who have evaluated the thermodynamic temperatures of the silver and gold points; these fixed points have been assigned definite values on the IPTS-68. However, experimenters measuring $(t_{th} - t_{68})$ above 630°C are faced with a dilemma: should they refer their measurements to the IPTS-68 by including in their apparatus a standard thermocouple thermometer, thus ensuring a large uncertainty (up to 0.2°C) in their results; or should they record their determinations of thermodynamic temperatures instead on platinum resistance thermometers, thus obtaining a more precise reference to a laboratory scale that has no formal standing?

We discuss this point more fully in the next section; for the moment, we only note that both options have been chosen, in recent[72] and in current[68] spectral radiation thermometry in the range above 630°C.

Bonhoure, a member of the International Bureau of Weights and Measures staff, has determined approximate differences between the Kelvin thermodynamic temperature scale and T_{68} in the range 630 to 1064°C in a long series of experiments. Figure 19 is a schematic drawing of his apparatus.

---

* The large uncertainty levels found by these and other experimenters are reflected in the text of the IPTS-68 as amended in 1975 (*Metrologia*, 12, 12, Section III.2, 1976);"... in general use, these precautions may not ensure an accuracy of better than ±0.2°C because of continually changing chemical and physical inhomogeneities in the wires in the regions of temperature gradients." But see also the work of Evans referred to in Chapter 6, Section IV.B.2. Evans and Wood provide the means for reducing the T_{68} uncertainty above 630°C to ±0.06°C. See Reference 83.

Table 16
DIFFERENCES BETWEEN THERMODYNAMIC AND PRACTICAL-SCALE TEMPERATURES BY SPECTRAL RADIATION THERMOMETRY OF HALL[63] AND BONHOURE[72]

t_{68} (°C)	$t_{th} - t_{48}$ (°C)[a]	$t_{th} - t_{68}$ (°C)[b]	$t_{68} - t_{48}$ (°C)[c]
419.58	0 (Ref temp)	—	+0.075
630.74	0 ± 0.2	0 (Ref temp)	+0.200
727	—	+0.35 ± 0.2	+0.47
779	+0.8 ± 0.4	—	+0.61
815	—	+0.5 ± 0.2	+0.71
910	—	+0.2 ± 0.2	+0.98
961	+0.4 ± 0.5	+0.1 ± 0.2	+1.12
1010	—	-0.2 ± 0.2	+1.27
1064	+0.7 ± 0.6	-0.05 ± 0.2	+1.43

Note: Figure 2 in Chapter 4 shows the differences $t_{68} - t_{48}$ graphically.

[a] Data from Reference 63.
[b] Data from Reference 72.
[c] Data from Reference 73.

Table 17
RATIOS OF HIGH-TEMPERATURE FIXED-POINT TEMPERATURES

Reference temperature (°C)	Measured temperature (°C)	Ref.
1064.43 (Au)	Cu 1084.890 ± 0.015	74
1064.43 (Au)	Cu 1084.87 ± 0.04	75
1064.43 (Au)	Cu 1084.87 ± 0.02	76
1064.43 (Au)	Cu 1084.88 ± 0.12	77
1084.88 (Cu)	Ag 962.05 ± 0.06	78
1064.43 (Au)	Ag 961.980 ± 0.015	74
1064.43 (Au)	Ag 962.05 ± 0.04	76
1064.43 (Au)	Ag 962.10 ± 0.18	79
1064.43 (Au)	Ag 962.06 ± 0.1	80
1064.43 (Au)	Ag 962.06 ± 0.20	72
1064.43 (Au)	Ag 962.06 ± 0.2	81
962.05 (Ag)	Cu/71.9 wt% Ag 779.97 ± 0.10	82

Bonhoure referred his measurements of temperature to the IPTS-68 temperature 630.74°C as determined by platinum resistance thermometry. This temperature was assumed to agree with the KTTS. Bonhoure's results are listed in Table 16; his overall estimated uncertainty, ±0.2°C, consists about equally of uncertainty in the IPTS-68 and imprecision in his radiation measurements. For comparison, values obtained by Hall and the differences ($t_{68} - t_{48}$) also are listed in Table 16.

We mentioned that radiation thermometrists can measure the ratio of the thermodynamic temperatures of fixed points above 630°C with substantially less uncertainty than they can achieve in determining ratios of the temperatures of such fixed points to those below 500°C. In Table 17, we summarize the results of a number of such recent determinations.

It is apparent from the values given in Table 17 that the relationships between the high-temperature scale fixed points are known within 0.1°C or less. There is as yet no connection to the thermodynamic temperature scale at that uncertainty level, however.

Table 18
KELVIN THERMODYNAMIC TEMPERATURES[a] OF FIXED POINTS ABOVE 0°C

Fixed point	t_{th} (°C)	t_{68} (°C)	Ref.
Water bp	99.975 ± 0.002	100.000	16
Water bp	99.971 ± 0.007		62
Tin fp	231.924 ± 0.003	231.9681	16
Zinc fp	419.514 ± 0.004	419.58	16
Sb fp[b]	630.59 ± 0.02	630.74	68
Sb fp	630.59		69
Sb fp	630.70 ± 0.05		66
Al fp[c]	660.300 ± 0.03	660.46 ± 0.2	68

[a] Uncertainties are stated at the 99% confidence level. If the original publication did not provide this information, we have estimated it.

[b] The antimony point is not a defining point of the IPTS-68, nor was the point actually realized in the experiments summarized in this table. Instead, all authors referred their difference measurements at this temperature to t_{68} = 630.74°C.

[c] The aluminum freezing point is not a defining fixed point of the IPTS-68. Its t_{68} value therfore must reflect the uncertainty of the scale in this range.

VII. KELVIN THERMODYNAMIC TEMPERATURES ABOVE 273 K

The Kelvin thermodynamic temperature scale above 273 K is not known nearly so well as it is below 273 K. We have presented in Table 5 the thermodynamic temperatures of the water boiling point and the tin and zinc freezing points as determined by Guildner and Edsinger. Their gas thermometry experiments covered the range 0 to 457°C. The thermodynamic scale from 0°C to slightly above the steam point has been realized also by Quinn and Martin, using the method of total radiation thermometry.

The pioneering work of Hall in applying the principles of spectral radiation thermometry to the measurement of thermodynamic temperatures below 500°C has been improved by Jung and others, although discrepancies in the resulting thermodynamic values remain to be resolved. Table 18 contains a summary of the thermodynamic temperatures of fixed points that so far have been determined. For comparison, the IPTS-68 values are given also.

It is of considerable interest to compare the temperatures as provided by the IPTS-68 with those measured or deduced for the thermodynamic scale. In Figure 20 we present a speculative view of the differences $t - t_{68}$ above 0°C. As the data of Table 18 show, thermodynamic temperatures are reasonably well known in the range of the IPTS-68 that is defined by the platinum resistance thermometer; the IPTS-68 itself is relatively well defined there, too. However, above 500°C, the uncertainty in the thermodynamic scale becomes substantial (witness the disagreement between the determinations of Coates et al.[66] and those of Jung and of Bonhoure shown in Figure 18*); above 630°C the IPTS-68 uncertainty reaches ±0.2°C.

The large uncertainties above 500°C notwithstanding, the measurements of Evans and Wood,[83] of Bonhoure,[72] of Crovini and Actis[48] (see Table 12), and of Bedford and Ma[82] all indicate that there is a thermodynamic inconsistency of 0.3 to 0.5°C in the range 630 to 1064°C.[84]

* Very recently, Coates et al. have recalculated their results; they now find good agreement with those of Jung (References 68 and 85).

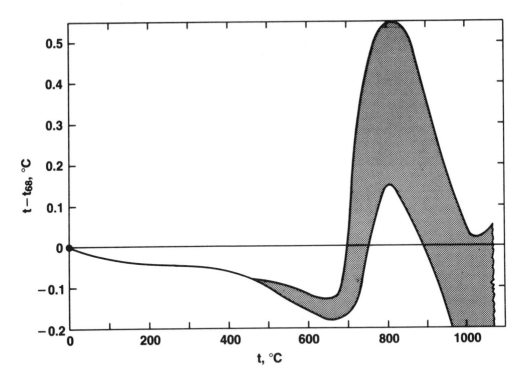

FIGURE 20. Approximate differences between the Kelvin thermodynamic temperature scale and the IPTS-68 above 0°C. The uncertainties indicated by the width of the curve above 500°C result in part from uncertainties in thermodynamic temperature determinations and in part from the nonuniqueness of the IPTS-68. Published data incorporated into this tentative curve are contained in References 16, 48, 60, 66, 68, 72, 82, and 83.

GENERAL REFERENCES

Schooley, J. F., Ed.-in-Chief, *Temperature, Its Measurement and Control in Science and Industry*, Vol. 5, Section 1, American Institute of Physics, New York, 1982.
Wensel, H. T., Temperature, in *Temperature, Its Measurement and Control in Science and Industry*, Vol. 1, Fairchild, C. O., Hardy, J. D., Sosman, R. B., and Wensel, H. T., Eds., Reinhold, New York, 1941, 3—24.
Corruccini, R. J., Principles of Thermometry, in *Treatise on Analytical Chemistry*, Vol. 8, Part 1, Kolthoff, I. M., Elving, P. J., and Sandell, E. B., Eds., Interscience, New York, 1968, chap. 87, sect. IIA.
Wolfe, H. D., Ed., *Temperature, Its Measurement and Control in Science and Industry*, Vol. 2, Section 2, Reinhold, New York, 1955, Papers 5, 6, and 7.
Quinn, T. J., *Temperature*, Academic Press, London, 1983, chap. 3.

REFERENCES

1. **Harker, J. A. and Chappius, P.**, A comparison of platinum and gas thermometers, including a determination of the boiling-point of sulphur on the nitrogen scale. An account of experiments made in the laboratory of the Bureau International des Poids et Mesures at Sèvres, *Philos. Trans. R. Soc. London*, 194A, 37, 1900.
2. **Callendar, H. L.**, On the practical measurement of temperature, *Philos. Trans. R. Soc. London*, 178A, 161, 1887.
3. **Beattie, J. A.**, The thermodynamic temperature of the ice point, in *Temperature, Its Measurement and Control in Science and Industry*, Vol. 1, Fairchild, C. O., Hardy, J. D., Sosman, R. B., and Wensel, H. T., Eds., Reinhold, New York, 1941, 74.
3a. **Beattie, J. A.**, Gas thermometry, in *Temperature, Its Measurement and Control in Science and Industry*, Vol. 2, Wolfe, H. D., Ed., Reinhold, New York, 1955, 63.

4. **Hoge, H. J. and Brickwedde, F. G.**, Establishment of a temperature scale for the calibration of thermometers between 14° and 83°K, *J. Res. Natl. Bur. Stand.*, 22, 351, 1939.
5. **Day, A. L. and Sosman, R. B.**, The nitrogen thermometer scale from 300° to 630°, with a direct determination of the boiling point of sulfur, *Am. J. Sci.*, 33, 517, 1912.
5a. **Day, A. L. and Sosman, R. B.**, High temperature gas thermometry, Publ. No. 157, Carnegie Inst. of Washington, 1911.
6. **Moser, H.**, High temperature gas thermometry, in *Temperature, Its Measurement and Control in Science and Industry*, Vol. 2, Wolfe, H. D., Ed., Reinhold, New York, 1955, 103.
7. **Barber, C. R.**, Helium gas thermometry at low temperatures, in *Temperature, Its Measurement and Control in Science and Industry*, Vol. 3, Part I, Herzfeld, C. M., Ed.-in-Chief, Reinhold, New York, 1962, 103.
8. **Birge, R. T.**, *Rev. Mod. Phys.*, 1, 1, 1929.
9. **Taylor, B. N., Parker, W. H., and Langenberg, D. N.**, *Rev. Mod. Phys.*, 41, 375, 1969.
10. **Cohen, R. and Taylor, B. N.**, *J. Phys. Chem. Ref. Data*, 2, 663, 1973.
11. **Moser, H.**, High-Temperature Gas Thermometry, in *Temperature, Its Measurement and Control in Science and Industry*, Vol. II, Wolfe, H. D., Ed., Reinhold, New York, 1955, 103.
11a. **Guildner, L. A. and Thomas, W.**, The measurement of thermodynamic temperature, in *Temperature, Its Measurement and Control in Science and Industry*, Vol. 5, Schooley, J. F., Ed.-in-Chief, American Institute of Physics, New York, 1982, 9.
12. **Pickering, S. F.**, Relations between temperatures, pressures, and densities of gases, U.S. National Bureau of Standards Circ. 279, December 3, 1925.
13. **Hirshfelder, J. O., Curtiss, C. F., and Bird, R. B.**, *Molecular Theory of Gases and Liquids*, 2nd ed., John Wiley & Sons, New York, 1964, chap. 3.
14. **Gray, D. E., Ed.**, *AIP Handbook*, 3rd ed., McGraw-Hill, 1972, sect. 4i.
15. **Berry, K. H.**, Measurements of thermodynamic temperature from 2.6 to 27.1 K, in *Temperature, Its Measurement and Control in Science and Industry*, Vol. 5, Schooley, J. F., Ed.-in-Chief, American Institute of Physics, New York, 1982, 21.
16. **Guildner, L. A. and Edsinger, R. E.**, Deviation of international practical temperatures from thermodynamic temperatures in the temperature range from 273.16 K to 730 K, *J. Res. Natl. Bur. Stand.*, 80A, 703, 1976.
17. **Dushman, S.**, *Scientific Foundations of Vacuum Technology*, 2nd ed., Lafferty, J. M., Ed., John Wiley & Sons, New York, 1962, sect. 1.12.
18. **McConville, G. T.**, Thermomolecular pressure corrections in helium vapor pressure thermometry: the effect of the tube surface, *Cryogenics*, 9, 122, 1969.
19. **Guildner, L. A. and Edsinger, R. E.**, Progress in NBS gas Thermometry above 500°C, in *Temperature, Its Measurement and Control in Science and Industry*, Vol. 5, Schooley, J. F., Ed.-in-Chief, American Institute of Physics, New York, 1982, 43.
20. **Beattie, J. A.**, The thermodynamic temperature of the ice point, in *Temperature, Its Measurement and Control in Science and Industry*, Vol. 1, Fairchild, C. O., Hardy, J. D., Sosman, R. B., and Wensel, H. T., Eds., Reinhold, New York, 1941, 74.
21. **Quinn, T. J. and Martin, J. E.**, Radiometric measurement of thermodynamic temperature between 327 and 365 K, in *Temperature, Its Measurement and Control in Science and Industry*, Vol. 5, Schooley, J. F., Ed.-in-Chief, American Institute of Physics, New York, 1982, 103.
22. **Guildner, L. A. and Edsinger, R. E.**, National Bureau of Standards Gas Thermometer. II. Measurement of capacitance to a grounded surface with a transformer ratio-arm bridge, *J. Res. Natl. Bur. Stand.*, 69C, 13, 1965.
23. **Guildner, L. A., Stimson, H. F., Edsinger, R. E., and Anderson, R. L.**, An accurate mercury manometer for the NBS gas thermometer, *Metrologia*, 6, 1, 1970.
24. **Yntema, Y. L. and Schneider, W. G.**, *J. Chem. Phys.*, 18, 641, 1950.
24a. **Gammon, B. E. and Douslin, D. R.**, *Proc. Fifth Symposium on Thermophysical Properties*, American Society of Mechanical Engineers, New York, 1970, 107.
25. The International Practical Temperature Scale of 1968, *Metrologia*, 5, 35—44, 1969 (Table 7).
26. **Berry, K. H.**, NPL-75: A low-temperature gas thermometry scale from 2.6 K to 27.1 K, *Metrologia*, 15, 89, 1979.
27. **Kistemaker, J. and Keesom, W. H.**, *Physica*, 12, 227, 1946.
28. **Keller, W. E.**, *Phys. Rev.*, 97, 1, 1955.
29. **Plumb, H. and Cataland, G.**, Acoustical thermometer and the NBS Provisional Temperature Scale 2-20 (1965), *Metrologia*, 2, 127, 1966.
30. **Clement, J. R.**, The 1958 He-4 Temperature Scale, in *Temperature, Its Measurement and Control in Science and Industry*, Vol. 3, Part 1, Herzfeld, C. M., Ed.-in-Chief, Reinhold, New York, 1962, 67.
30a. **Brickwedde, F. G., van Dijk, H., Durieux, M., Clement, J. R., and Logan, J. K.**, The 1958 He^4 scale of temperatures, *J. Res. Natl. Bur. Stand.*, 64A, 1, 1960.

31. **Keesom, W. H. and Walstra, W. K.**, *Physica*, 7, 985, 1940 and *Physica*, 13, 225, 1947.
31a. **Keller, W. E.**, *Phys. Rev.*, 97, 1, 1955.
32. **Kilpatrick, J. E., Keller, W. E., and Hammel, E. F.**, *Phys. Rev.*, 97, 9, 1955.
33. **Colclough, A. R.**, Low frequency acoustic thermometry in the range 4.2—20 K, with implications for the value of the gas constant, *Proc. R. Soc. London*, A365, 349, 1979.
33a. **Colclough, A. R.**, A low frequency acoustic thermometer for the range 2-20 K, in *Temperature, Its Measurement and Control in Science and Industry*, Vol. 4, Plumb, H. H., Ed.-in-Chief, Instrument Society of America, Pittsburgh, 1972, 365.
34. **Klein, H.-H., Klempt, G., and Storm, L.**, Measurement of the thermodynamic temperature of ^{4}He at various vapour pressures by a noise thermometer, *Metrologia*, 15, 143, 1979.
35. **Durieux, M., Astrov, D. N., Kemp, W. R. G., and Swenson, C. A.**, Derivation and development of the 1976 Provisional 0.5 K to 30 K Temperature Scale, *Metrologia*, 15, 57, 1979.
36. **Cetas, T. C. and Swenson, C. A.**, A paramagnetic salt temperature scale, 0.9 to 18 K, *Metrologia*, 8, 46, 1972.
37. **Durieux, M., Astrov, D. N., Kemp, W. R. G., and Swenson, C. A.**, The 1976 provisional 0.5 K to 30 K temperature scale, *Metrologia*, 15, 65, 1979.
38. **Gugan, D. and Michel, G. W.**, Dielectric constant gas thermometry from 4.2 to 27.1 K, *Metrologia*, 16, 149, 1980.
38a. **Gugan, D.**, Dielectric constant gas thermometry (DCGT): a new method of accurate thermodynamic thermometry, in *Temperature, Its Measurement and Control in Science and Industry*, Vol. 5, Schooley, J. F., Ed.-in-Chief, American Institute of Physics, New York, 1982, 49.
39. **Gugan, D.**, Surface-fitting of helium isotherms: application to the temperature scale 2.6—27.1 K, in *Temperature, Its Measurement and Control in Science and Industry*, Vol. 5, Schooley, J. F., Ed.-in-Chief, American Institute of Physics, New York, 1982, 55.
39a. **Gugan, D.**, The analysis of ^{4}He isotherms: density and dielectric virial coefficients, and the accuracy of NPL-75, *Metrologia*, 19, 147, 1984.
40. **Kemp, R. C., Besley, L. M., and Kemp, W. R. G.**, Constant volume gas thermometry from 13.8 to 83.8 K, in *Temperature, Its Measurement and Control in Science and Industry*, Vol. 5, Schooley, J. F., Ed.-in-Chief, American Institute of Physics, New York, 1982, 33.
41. **Kemp, R. C., Besley, L. M., and Kemp, W. R. G.**, Communication No. CCT/84-8 to the 15th Meeting of the CCT, June 1984, Bureau International des Poids et Mesures, Sèvres, France.
42. **Steur, P. P. M., van Dijk, J. E., Mars, J. P., ter Harmsel, H., and Durieux, M.**, Measurements with a gas thermometer between 4 and 100 K, in *Temperature, Its Measurement and Control in Science and Industry*, Vol. 5, Schooley, J. F., Ed.-in-Chief, American Institute of Physics, New York, 1982, 25.
43. **Steur, P. P. M.**, Determination of temperatures between 4 K and 100 K with a gas thermometer, Thesis, KOL Leiden, 1983.
44. **Johnson, J. B.**, Thermal agitation of electricity in conductors, *Phys. Rev.*, 32, 97, 1928.
45. **Nyquist, H.**, Thermal agitation of electronic charge in conductors, *Phys. Rev.*, 32, 110, 1928.
46. **Garrison, J. B. and Lawson, A. W.**, *Rev. Sci. Instrum.*, 20, 785, 1949.
47. **Pickup, C. P.**, A high-resolution noise thermometer for the temperature range 90—100 K, *Metrologia*, 11, 151, 1975.
47a. **Pickup, C. P.**, A high-accuracy noise thermometer for the range 100—150°C, in *Temperature, Its Measurement and Control in Science and Industry*, Vol. 5, Schooley, J. F., Ed.-in-Chief, American Institute of Physics, New York, 1982, 129.
48. **Crovini, L. and Actis, A.**, Noise thermometry in the range 630—962°C, *Metrologia*, 14, 69, 1978.
48a. **Crovini, L. and Actis, A.**, Noise thermometry and related experiments at IMGC, in *Temperature, Its Measurement and Control in Science and Industry*, Vol. 5, Schooley, J. F., Ed.-in-Chief, American Institute of Physics, New York, 1982, 133.
49. **Klempt, G.**, Errors in Johnson noise thermometry, in *Temperature, Its Measurement and Control in Science and Industry*, Vol. 5, Schooley, J. F., Ed.-in-Chief, American Institute of Physics, New York, 1982, 125.
50. **Storm, S.**, informal communication to W. R. G. Kemp et al.; quoted in **Kemp, R. C., Besley, L. M., and Kemp, W. R. G.**, Communication No. CCT/84-8 to the 15th meeting of the CCT, June 1984, Bureau International des Poids et Mesures, Sèvres, France.
51. **van der Ziel, A.**, *Noise: Sources, Characterization, Measurement*, Prentice-Hall, New York, 1970, 54.
52. **Kamper, R. A.**, Survey of noise thermometry, in *Temperature, Its Measurement and Control in Science and Industry*, Vol. 4, Plumb, H. H., Ed.-in-Chief, Instrument Society of America, Pittsburgh, 1972, 349.
53. **Soulen, R. J., Jr., and Van Vechten, D.**, Noise thermometry at NBS using a Josephson junction, in *Temperature, Its Measurement and Control in Science and Industry*, Vol. 5, Schooley, J. F., Ed.-in-Chief, American Institute of Physics, New York, 1982, 115.
54. **Berglund, P. M., Collan, H. K., Ehnholm, G. J., Gylling, R. G., and Lounasmaa, O. V.**, *J. Low Temp. Phys.*, 6, 357, 1972.
55. **Weyhmann, W.**, *Meth. Exp. Phys.*, 11, 485, 1974.

56. **Hudson, R. P., Marshak, H., Soulen, R. J., Jr., and Utton, D. B.,** *J. Low Temp. Phys.*, 20, 1, 1975.
57. **Marshak, H.,** Nuclear orientation thermometry, *J. Res. Natl. Bur. Stand.*, 88, 175, 1983.
58. **Marshak, H.,** Nuclear orientation thermometry from ~0.001 to ~1.2 K, in *Temperature, Its Measurement and Control in Science and Industry*, Vol. 5, Schooley, J. F., Ed.-in-Chief, American Institute of Physics, New York, 1982, 95.
59. **Richmond, J. C. and Nicodemus, F. E.,** Blackbodies, blackbody radiation, and temperature scales, in *Self-Study Manual on Optical Radiation Measurements*, Part I, NBS Tech. Note 910-8, 1985, chap. 12.
59a. **Halliday, D. and Resnick, R.,** *Physics for Students of Science and Engineering*, combined ed., John Wiley & Sons, 1962, chap. 47.
60. **Quinn, T. J. and Martin, J. E.,** Radiometric measurement of thermodynamic temperature between 327 K and 365 K, in *Temperature, Its Measurement and Control in Science and Industry*, Vol. 5, Schooley, J. F., Ed.-in-Chief, American Institute of Physics, New York, 1982, 103.
61. **Ginnings, D. C. and Reilly, M. L.,** Calorimetric measurement of thermodynamic temperatures above 0°C using total blackbody radiation, in *Temperature, Its Measurement and Control in Science and Industry*, Vol. 4, Plumb, H. H., Ed.-in-Chief, Instrument Society of America, Pittsburgh, 1972, 339.
62. **Quinn, T. J. and Martin, J. E.,** informal communication to the 15th Meeting of the Consultative Committee for Thermometry, June 1984, International Bureau of Weights and Measures.
62a. **Quinn, T. J. and Martin, J. E.,** Radiometric measurements of the Stefan-Boltzmann constant and thermodynamic temperature between $-40°C$ and $+100°C$, *Metrologia*, 20, 163, 1984.
63. **Hall, T. J.,** The radiation scale of temperature between 175°C and 1063°C, *Metrologia*, 1, 140, 1965.
64. **Heusinkveld, W. A.,** Determination of the differences between the thermodynamic and the Practical Temperature Scale in the range 630 to 1063°C from radiation measurements, *Metrologia*, 2, 61, 1966.
65. **Coates, P. B.,** The NPL photon-counting pyrometer, in *Temperature Measurement, 1975*, Billing, B. F. and Quinn, T. J., Eds., Conference Series No. 26, The Institute of Physics, London, 1975, 238.
66. **Coates, P. B. and Andrews, J. W.,** Measurement of thermodynamic temperature with the NPL photon-counting pyrometer, in *Temperature, Its Measurement and Control in Science and Industry*, Vol. 5, Schooley, J. F., Ed.-in-Chief, American Institute of Physics, New York, 1982, 109.
66a. **Andrews, J. W. and Gu Chuanxin,** The thermodynamic temperature of the gold and silver points, BIPM Document CCT/84-39.
67. **Quinn, T. J. and Chandler, R. R. D.,** The freezing point of platinum determined by the NPL photoelectric pyrometer, in *Temperature, Its Measurement and Control in Science and Industry*, Vol. 4, Plumb, H. H., Ed.-in-Chief, Instrument Society of America, Pittsburgh, 1972, 295.
68. **Jung, H. J.,** An optical measurement of the deviation of International Practical Temperatures T_{68} from thermodynamic temperatures in the range from 730 K to 930 K, *Metrologia*, 20, 67, 1984.
69. **Bonhoure, J. and Pello, R.,** informal communication to the 15th Meeting of the Consultative Committee for Thermometry, BIPM Document (CCT/84-21), International Bureau of Weights and Measures, June 1984.
70. **Ku, H. H.,** Statistical concepts in metrology, in *Precision Measurement and Calibration; Statistical Concepts and Procedures*, National Bureau of Standards Special Publ. 300, Vol. 1, 1969, 296.
71. **McLaren, E. H. and Murdock, E. G.,** New considerations on the preparation, properties, and limitations of the standard thermocouple for thermometry, in *Temperature, Its Measurement and Control in Science and Industry*, Vol. 4, Plumb, H. H., Ed.-in-Chief, Instrument Society of America, Pittsburgh, 1972, 1543.
72. **Bonhoure, J.,** Détermination radiométrique des températures thermodynamiques comprises entre 904 et 1338 K, *Metrologia*, 11, 141, 1975.
73. The International Practical Temperature Scale of 1968, *Metrologia*, 5, 44, 1969.
74. **Jones, T. P. and Tapping, J.,** A photoelectric pyrometer temperature scale below 1064.43°C and its use to measure the silver point, in *Temperature, Its Measurement and Control in Science and Industry*, Vol. 5, Schooley, J. F., Ed.-in-Chief, American Institute of Physics, New York, 1982, 169.
75. **Coates, P. B. and Andrews, J. W.,** *J. Phys. F*, 8, 277, 1978.
76. **Ricolfi, T. and Lanza, F.,** *High Temperatures — High Pressures*, Vol. 9, 1977, 483.
77. **Righini, F., Rosso, A., and Ruffino, G.,** *High Temperatures — High Pressures*, Vol. 4, 1972, 471.
78. **Ohtsuka, M. and Bedford, R. E.,** Measurement of the thermodynamic temperature interval between the freezing points of silver and copper, in *Temperature, Its Measurement and Control in Science and Industry*, Vol. 5, Schooley, J. F., Ed.-in-Chief, American Institute of Physics, New York, 1982, 175.
79. **Coslovi, L., Rosso, A., and Ruffino, G.,** *Metrologia*, 11, 85, 1975.
80. **Jung, H. J.,** Determination of the difference between the thermodynamic fixed-point temperatures of gold and silver by radiation thermometry, in *Temperature Measurement, 1975*, Billing, B. F. and Quinn, T. J., Eds., Conference Series No. 26, Institute of Physics, London, 1975, 278.
81. **Quinn, T. J., Chandler, T. R., and Chattle, M. V.,** *Metrologia*, 9, 44, 1973.

82. **Bedford, R. E. and Ma, C. K.**, Measurement of the melting temperature of the copper 71.9% silver eutectic alloy with a monochromatic optical pyrometer, in *Temperature, Its Measurement and Control in Science and Industry*, Vol. 5, Schooley, J. F., Ed.-in-Chief, American Institute of Physics, New York, 1982, 361.
83. **Evans, J. P. and Wood, S. D.**, An intercomparison of high temperature platinum resistance thermometers and standard thermocouples, *Metrologia*, 7, 108, 1971.
84. **Preston-Thomas, H., Bloembergen, P., and Quinn, T. J.**, Supplementary information for the IPTS-68 and the EPT-76, Bureau International des Poids et Mesures, 1983, 7.
85. **Coates, P. B., Andrews, J. W., and Chattle, M. V.**, *Metrologia*, 21, 31, 1985.

Chapter 6

MODERN THERMOMETERS

I. INTRODUCTION

There are so many different temperature-measuring devices available to present-day thermometrists that their very enumeration is difficult. Special thermometers have been developed for use in particular temperature ranges, for especially demanding environments, for unusual geometric or time-response situations, to achieve specific resolution or accuracy goals, or to provide measurements within well-defined cost criteria. In order to provide a coherent discussion of many of these types of thermometers within a limited space, we categorize modern thermometers according to the physical principles upon which their operation relies. Among the types of thermometers that can be noted are, of course, the gas thermometers that occupied our attention in the previous chapter, along with other thermal-expansion thermometers; radiation thermometers; resistance thermometers; thermocouple thermometers; vapor-pressure thermometers; paramagnetic thermometers; Johnson noise thermometers; and diode thermometers. Although we mention many different types of thermometers, our discussion will be more thorough, the more precise the thermometer.

We should note immediately that the word "thermometer" — defined as "an instrument by means of which one can measure the temperature of an object or system with respect to a particular scale of temperatures" — generally has a more complex meaning now than it did in Galileo's time. Whereas Galileo's air thermoscopes and Fahrenheit's mercury thermometers were relatively simple, self-contained devices, one finds that many modern thermometers are complicated systems of instruments in which the temperature-sensing component is located far from the instrument that interprets its signal. Indeed, it is not unusual, particularly in an industrial environment, to find whole groups of temperature sensors connected by electrical cables to elaborate switching devices which enable the operator to measure and record their temperatures by means of a single interpreting instrument. Often the whole process is performed automatically, with subsequent calculations being performed and controls adjusted as a result of the measurements. (See, for example, References 1 and 1a.)

Perhaps because of the complexity of modern thermometry equipment, the word "thermometer" is used nowadays with two different meanings: in some cases, it means "temperature sensor," as in the terms "platinum resistance thermometer" or "thermocouple thermometer"; in others, it means the entire system that is used for a particular temperature measurement. Since our own custom embraces the former definition, we suggest the use of the term "thermometer system" whenever one desires to emphasize the complexity of a particular temperature measurement. In any case, we shall attempt to distinguish clearly which instruments are to be included. By thus modifying the earlier definition, we also mean to emphasize the fact that the resolution, precision, and accuracy of a temperature measurement generally depend upon the properties of an interacting group of instruments, rather than simply upon the qualities of a single temperature sensor. A block diagram showing the relationships existing in a typical modern thermometer system is presented in Figure 1.

Other useful definitions include the word "transducer" (from the Latin, "leading across") which is defined as a device that receives energy in one form and converts it to another. A "temperature sensor" or "temperature transducer" experiences a change in some measurable parameter such as dimension, viscosity, electrical resistivity, or quantum level population as a result of exposure to a change in temperature.

A "signal conditioner" is employed whenever it is desired to amplify or attenuate the sensor output, to modify the signal impedance, or otherwise to vary the quality of the sensor output. A signal conditioner is used only in particular instances.

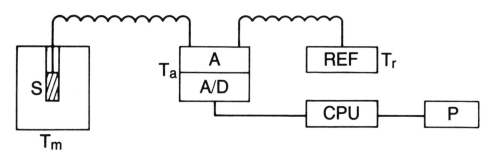

FIGURE 1. Block diagram illustrating the components of a modern thermometer system. The temperature sensor S, some device possessing a measurable property that reponds monotonically and with satisfactory sensitivity to changes in temperature, is placed in thermal contact with the medium whose temperature T_m is to be measured. An analog instrument A may be held at temperature T_a; its function is to provide an indication of the value of the temperature-dependent property of S, perhaps with reference to a measurement standard REF held at temperature T_r. Often, an analog-to-digital converter A/D is included in the system to provide digital information on the measurement to an electronic data processor CPU for recording on an automatic printer P.

An "analog" meter provides a measure of the temperature-sensitive property of the temperature sensor — a voltage, a resistance, a current, a pressure, a length, a frequency, or some other quantity. Often, the analog measurement is made with reference to a standard that must be shielded from mechanical shock, temperature variation, or some other disturbance.

A "transmitter" sends the measured signal to a recorder or to a remote indicator.

Many modern thermometer systems make use of "analog-to-digital converters" to provide the operator with numerical data that can be processed by an electronic calculator for automatic recording and perhaps for use in maintaining control of some operation.

Drift in the calibration, or other malfunction in any of the parts of a complex thermometer system will contribute to the imprecision and the systematic error of the indicated temperature; in order to maintain control on the accuracy of modern thermometers, the operator must continually monitor the level of performance of all of their components.

II. THERMAL EXPANSION THERMOMETERS

"Thermal expansion thermometers" could include a wide variety of present-day instruments, since many physical properties are influenced by the thermal expansion of the material involved. In this section, however, we restrict our discussion to include only gas-expansion thermometers, filled-system thermometers or pneumatic thermometers, bimetallic thermometers, and liquid-in-glass thermometers. Other types of thermometers, despite their response to the thermal expansion of their constituent materials, are discussed elsewhere.

We presented in Chapter 5 an extended discussion of gas thermometers as used in thermodynamics research. We add nothing further to that discussion here. Two types of thermometer/thermostat involving solid, liquid, and gas expansion, however, are used extensively in industry; although relatively imprecise, their ubiquity impels us to describe them briefly. We conclude the section with an up-to-date account of the liquid-in-glass thermometer.

A. Filled-System Thermometers

The use of pneumatic techniques for process control is very common in industry.[1,2] The chief advantages of pneumatic controls are their reliability and freedom from electrical interference. The main disadvantage is their relatively poor thermometric precision. There are several types of temperature sensor in which the working substance experiences an increase of pressure upon exposure to an increased temperature. This pressure increase can be utilized to operate either or both of a Bourdon-tube temperature display or a pressure-activated switch. This thermometer/thermostat is shown in schematic form in Figure 2.

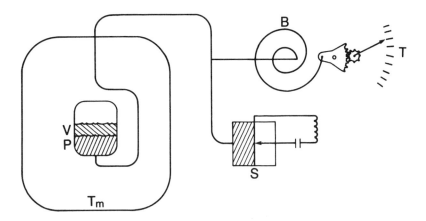

FIGURE 2. Schematic drawing of a pneumatic filled-system thermometer/thermostat. The temperature to be measured, T_m, surrounds the sensor. In this case, the sensor is chosen to be a pneumatic device that provides a sensitive indication of temperature through the changing vapor pressure of the liquid medium V. The temperature-dependent vapor pressure is transmitted by a fluid P that is relatively insensitive to changes in temperature. Through the action of a Bourdon tube that is mechanically linked to a dial, the vapor pressure measurement is converted to a temperature reading; a parallel path to a pressure-sensitive switch S provides the means for using the device as a thermostat.

The temperature-sensing bulb can be filled simply with a gas, for example N_2, so that it functions directly as a gas thermometer. This technique provides a bulb pressure that varies almost linearly with temperature, but it restricts the tolerable volume of the capillary, Bourdon tube, and switch in comparison with the bulb volume.

A greater temperature sensitivity can be achieved by use of an evaporative liquid in the bulb. Though nonlinear in its temperature-pressure relation, the vapor-pressure sensor offers much-improved temperature sensitivity over the gas thermometer and reduces the "dead space" effects arising from the volume of the capillary and indicator.

Yet another variation of the filled-system thermometer that functions also as a thermostat can be achieved by incorporating a thermally inert pressure-transmitting fluid (for example, a light oil) into the bulb along with the vapor-pressure substance. This technique accomplishes two goals; it minimizes the amount of vapor-pressure material in the system, and it minimizes the response of the thermometer to gradients in temperature along the connecting line to the switch and meter. It is this variation that is shown in Figure 2.

The accuracy of filled-system thermometers hardly ever is better than 1 to 3°C. They are not used for precise work in the form illustrated here.

B. Bimetallic Thermometers

Bimetallic thermometers, shown in schematic form in Figure 3, make up a rough-and-ready class of thermometer/thermostat that is used extensively in temperature measurement and control in the cooking, heating, and air conditioning industries.[2] The principle of operation of this thermometer is a simple one: two strips of unlike materials are bonded together; to the extent that their thermal expansion coefficients are different, the composite strip changes shape with changing temperature. If one end of the strip is held in a fixed position, then the motion of the other end can rotate an indicator needle on a dial or activate a switch. The principle of the bimetallic thermometer and a typical configuration for a bimetallic thermometer are illustrated in Figure 3.

Several alloys are used in the manufacture of bimetallic components. Invar, a nickel-steel alloy possessing a very low coefficient of thermal expansion, often is used as one strip. Brass or iron or nickel alloys often find use as the high-thermal-expansion comparison strip.

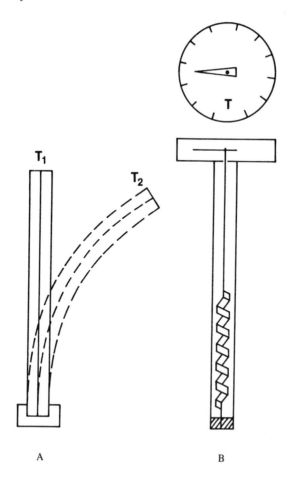

FIGURE 3. Schematic views of bimetallic thermometers. (A) The bonding of two strips with different thermal expansion characteristics produces bending as the temperature changes. (B) Forming the bimetallic strip into a helix provides a temperature-dependent rotation of an indicator needle.

The level of temperature uncertainty attainable with the use of bimetallic devices ranges from ±1% of the full-scale reading in carefully made models to several degrees Celsius in the rougher versions.

C. Liquid-in-Glass Thermometers

Liquid-in-glass thermometers deserve a favored place in the lore of thermometry, despite their limited temperature precision. From the earliest days of thermal science, they have provided an efficient answer to the question, "How hot is it?". There is no fumbling with wires, there are no "boxes" to manipulate, just a short stick that serves at once as a sensor and a visual indicator. (In some cases, such as the fever thermometer and the maximum/ minimum reading thermometer, the liquid-in-glass thermometer also serves the function of a recorder!)

At the present time, liquid-in-glass thermometers can be used in the temperature range 70 to 870 K (from −200 to almost 600°C). The lower limit is set by the freezing of liquids used in the low-temperature thermometers. The upper limit is necessitated by softening of the thermometer glass. When mercury is used as the thermometric fluid, the thermometer range can extend[3] from −35 (slightly above the mercury freezing point) to 595°C.

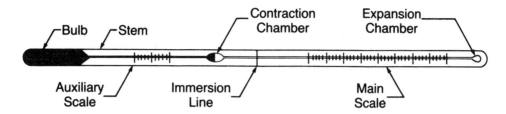

FIGURE 4. A typical liquid-in-glass thermometer of the partial immersion type. Adapted from Wise.[5]

There are many applications of liquid-in-glass thermometers. In numbers of thermometers in use, the leading application remains the measurement of household air temperatures, both indoors and out. Other widespread applications assist medical, biological, veterinary, air conditioning, automotive, chemical and other manufacturing personnel, and engineers employed in a wide range of processing industries in measuring and regulating temperature. Laboratory types of liquid-in-glass thermometers include clinical standard thermometers graduated from 90 to 112°F, calorimetric thermometers designed to measure temperature differences accurately, and thermometers recommended by organizations such as the American Society for Testing and Materials (ASTM). More than 100 varieties of laboratory thermometers are recommended for use by the ASTM; nine of these types are especially identified as suitable for reference standards.[4]

In this section, we discuss very briefly the use of liquid-in-glass thermometers as laboratory instruments. Even in this restricted category, there are hundreds of different configurations, each with its own characteristics of range, precision, and accuracy. References to more extensive discussions are provided for the interested reader.

The construction of a typical liquid-in-glass thermometer is shown in Figure 4. All such thermometers include a bulb, a reservoir in which most of the liquid is contained, and a stem, enclosing the capillary tube that extends from the bulb to the upper end of the stem. The bulb and the stem may be made of the same type of glass, particularly if the thermometer is intended for use at the highest temperatures, or they may be constructed of two different kinds of glass. Several special glasses have been compounded for use in liquid-in-glass thermometers.[6] They differ principally in their limiting temperatures of use, from normal-lead glass, limited to about 400°C, through borosilicate (460°C) to supremax, a glass that can withstand temperatures as high as 595°C.

The volume of the bulb in a mercury thermometer generally is more than 6000 times the volume contained in a 1°C length of capillary. If the liquid has a higher thermal expansion than mercury, then a relatively smaller bulb volume is used.

The stem of a laboratory liquid-in-glass thermometer should have an engraved scale and an engraved serial number. Also, it may include such visibility features as a colored backing material, a red line marking a significant temperature, or a special shape that magnifies the capillary image.

Mercury is the liquid used to fill most liquid-in-glass thermometers. The thermal expansion coefficient of this element, while temperature-dependent, is more nearly linear in temperature than is, say, the resistivity of pure platinum. Of course, it is the difference between the thermal expansion coefficient of the filling liquid and that of the glass comprising the bulb that determines the overall temperature linearity of the liquid-in-glass thermometer reading ℓ;

$$\Delta\ell/\Delta t = f[M, \beta(Hg, t), \beta(bulb, t)] \qquad (1)$$

in units of "scale degrees"/°C. In Equation 1, ℓ is the length of the mercury column; it is a function of M, the mechanical advantage of the bulb-capillary combination, and of the

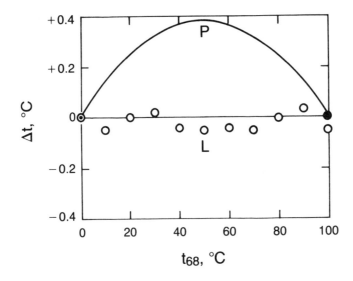

FIGURE 5. Deviation from linearity, in °C, of a typical liquid-in-glass thermometer in the range 0°C to 100°C compared with that of a platinum resistance thermometer. Both comparisons are made with respect to the IPTS-68, which approximates Kelvin thermodynamic temperatures in this range.

mercury and bulb volume thermal expansion coefficient $\beta = (1/v)(dv/dt)(°C)^{-1}$ which is approximately $3\alpha = (3/L)(dL/dt)(°C)^{-1}$. A typical value of α, the linear thermal expansion coefficient, for Hg is 4×10^{-5} $(°C)^{-1}$, while for borosilicate glass α is about 0.3×10^{-5} $(°C)^{-1}$.

The overall linearity of the mercury thermometer is surprisingly good; in fact, it is better than the linearity of the electrical resistivity of platinum, as is shown by the comparison in Figure 5. Note that $R_t(Pt) = R_0(1 + 3.985 \times 10^{-3} t - 5.86 \times 10^{-7} t^2)$, as given in Equation 10, Section IV.

Filling the glass with Hg-Tl alloys can extend the use of a thermometer down to $-56°C$. For even lower-temperature thermometry, toluene, pentane, and other organic filling liquids can be used. The practical lower limit for such thermometers is about 75 K ($-200°C$).

Since, unlike mercury, organic liquids "wet" glass, it is necessary to allow time for the capillary to drain after the temperature of an organic-filled thermometer is reduced. A wait of about 3 min for a change of 1 cm capillary length is considered adequate.

Often, a dye is added to organic filling liquids to promote easier reading. The manufacturer must ensure that the dye does not separate from the liquid, and particularly that it does not cloud the capillary.

The volume above the level of the filling liquid may be evacuated or it may be filled with an inert gas. Most modern liquid-in-glass thermometers are filled with dried nitrogen gas under pressure. This treatment retards the distillation of mercury, for example, that otherwise would limit the use of mercury thermometers to about 150°C.

Apart from its universal components — the bulb, stem, filling liquid, and gas or vacuum — a liquid-in-glass thermometer may include other features shown in Figure 4. Most gas-filled thermometers contain an expansion chamber at the upper end. Usually equivalent in volume to a 2-cm length of capillary, the expansion chamber reduces the buildup of gas pressure that otherwise would take place when the thermometer is heated near its upper limit. If the thermometer is designed for use only at temperatures higher than room temperature, it usually will contain two scales. The main scale will be offset from an auxiliary scale intended for use in measuring either the ice-point thermometer correction or the correction at another reference temperature. If a large gap in temperature separates the lower limit of the main scale from the reference point, a contraction chamber will shorten the length of capillary otherwise needed to separate the scales. Since the contraction chamber constitutes a secondary bulb, it should be maintained at the same temperature as the main bulb. An immersion line usually will be scribed above the contraction chamber to indicate the need for identical bulb and contraction-chamber temperatures.

The user of a properly made liquid-in-glass thermometer can employ it for thermometry over the whole range of its main scale. The careful manufacturer will engrave the scale on the thermometer stem, using well-defined but narrow graduations (line width no more than 20% of the distance between line centers). In addition, he will avoid capillaries with diameters less than about 0.1 mm; mercury will not flow easily in such capillaries, but instead will move by a series of jumps.[7]

Usually the scales on a laboratory thermometer are graduated in intervals of tenths, fifths, or halves of a scale unit. For any graduation interval up to 0.1 degree (either Celsius or Fahrenheit), the user may reasonably attempt to interpolate the position of the meniscus to the nearest 0.001 degree. For coarser graduation intervals of 0.2, 0.5, or 1.0 degree, readings to two decimal places should be achievable. Readings to one decimal place should suffice for any still coarser scale.

Techniques for measuring temperature accurately in practical situations can best be developed if one has an understanding of the methods involved in accurate calibrations of liquid-in-glass thermometers. These methods include proper "tempering" of the thermometer (that is, placing it so as to facilitate thermal equilibrium with the object to be measured), evaluation of necessary scale corrections, and accurate reading of the meniscus level.

Liquid-in-glass thermometers are calibrated in free-flowing liquid baths whenever possible, because the tempering conditions found there are particularly straightforward. A well-regulated, free-flowing, stirred liquid bath will provide a sizeable volume in which the temperature is constant within a few millidegrees. Moreover, the separation line between the regulated bath and the air above it is obvious, so that one need deal only with flowing liquid and still air — both fluids creating reasonably uniform temperatures. Note, however, that at the NBS, a slowly rising temperature (about one scale division per 3 to 10 min) is used in the calibration bath, because of the tendency of mercury to stick to the fine capillary as it recedes toward the bulb.

There are three obvious positions for placing liquid-in-glass thermometers in a medium whose temperature is to be measured. The same methods apply in calibration of the thermometers.[8] The three positions are designated by the depth to which the thermometer is immersed in the medium (usually a liquid bath). The three characteristics are called "complete," "total," or "partial" thermometer immersion. The technique designated "complete immersion" (rarely employed in modern thermometry), is shown in Figure 6 as position "c." Logically enough, it involves submerging the entire thermometer in the medium.

Reading the thermometer in this position demands the use of a bath that incorporates a clear fluid and a viewing window. For very precise work, the hydrostatic pressure on the thermometer during measurement must be the same as it was during calibration.

When using a thermometer at "total" immersion, one submerges most of the filling liquid of the thermometer, as shown in Figure 6 as position "t". Not only does a part of the thermometer protrude from the bath in the position of total immersion, but also about a 1-cm length of the thermometer liquid projects above the medium to facilitate reading the position of the meniscus.

Most common nowadays is the use of thermometers at "partial" immersion — either to an inscribed level as shown in Figure 4 or to a level specified by an engraved mark on the back of the thermometer. Figure 6 shows this immersion technique as position "p". The most precise readings of the thermometer in any position can be obtained by use of a cathetometer (i.e., a small telescope) as shown schematically in Figure 6.

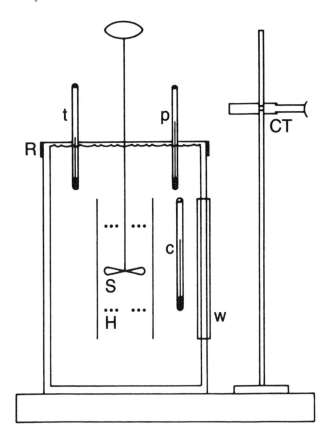

FIGURE 6. Positioning of liquid-in-glass thermometers in a fluid bath for calibration and measurement. H, S, Heater and stirrer to promote bath temperature uniformity; R, rotating mount for use in moving each of several thermometers in turn into position to be read by a cathetometer CT; C, thermometer in the "complete immersion" position; it must be read by viewing through a window W; P, thermometer in "partial immersion" position, T, thermometer in "total immersion" position.

Because the scale readings of a particular liquid-in-glass thermometer derive from the interplay of several different characteristics — liquid thermal expansion, glass thermal expansion, small variations in the capillary diameter, gas (if any) type and pressure, and thermal history of the bulb — it is necessary in general to calibrate such a thermometer throughout its working range. A "rule-of-thumb" at the NBS is to calibrate the liquid-in-glass thermometer at every 20 to 100 scale divisions, depending upon the thermometer range and scale interval, using as the reference thermometer a calibrated platinum resistance thermometer.[9]

The liquid-in-glass thermometer bulb expands inelastically whenever the thermometer is heated to high temperatures. This "stretching" effect arises from the increased pressure of the liquid and gas filling fluids acting in concert with an increased bulb-wall flexibility at elevated temperatures. Continued use at temperatures greater than 300°C can so distort the bulb as to cause the thermometer readings to become erratic. Most bulb-volume changes, however, correspond to less than 0.1°C and are reversible. The time needed for a bulb to recover from high-temperature stretching is variable, but it can approach 100 hr.

To compensate for small changes in bulb volume, the user should perform periodic recalibrations at the reference temperature of the thermometer (commonly the ice point).

Table 1
HYPOTHETICAL CORRECTIONS TO A PREVIOUSLY-CALIBRATED LIQUID-IN-GLASS THERMOMETER ACCORDING TO A NEW ICE-POINT MEASUREMENT

Scale reading (°C)	Original correction (°C)	New ice-point reading (°C)	New correction (°C)
+0.02 (Ice point)	−0.02	+0.01	−0.01
10.	−0.02		−0.01
20.	−0.03		−0.02
30.	0		+0.01
40.	−0.04		−0.03
50.	−0.01		0

All of the scale-reading corrections derived from the initial calibration then can be modified by the algebraic addition of a single number to reflect the new bulb volume. Table 1 shows a hypothetical set of calibration corrections adjusted in this way.

The use of a cathetometer is the best method for achieving accuracy in reading the meniscus level of liquid-in-glass thermometers. The method is sketched in Figure 6. Its main features are

- Optically magnifying the scale image at the level of the meniscus to facilitate the estimation of the meniscus position with respect to the engraved thermometer scale
- Minimizing parallax

A cathetometer mounted parallel to the test thermometer provides satisfactory magnification and minimizes parallax error. Skill at estimating the meniscus position within one tenth or one hundredth of the scale interval, as recommended above, can best be attained by practice.

We note here the NBS injunction against attempting to "build in" thermometer resolution by the selection of a very long thermometer or one with a very finely divided scale. In general, more reliable results will be achieved by using the cathetometer method described above to take full advantage of the resolution available in a more manageable thermometer.

When mercury thermometers receive rough treatment, perhaps in accidental dropping in the laboratory or in shipping, it is not uncommon to find the mercury column separated or to observe the presence of a bubble of gas in the bulb. These unhappy symptoms need not signal the end of the useful life of a thermometer — it is quite possible to restore the thermometer to a condition such that only a recalibration is needed to allow it to serve again with its original accuracy. The technique used at the NBS for this purpose is discussed in Reference 9. Basically, it consists of cooling the thermometer bulb by periodic immersion in powdered dry ice while continually manipulating the thermometer. As the cooling mercury contracts into the bulb, the operator can tap the thermometer gently on a thick rubber pad in order to reunite the mercury column. In the case that a gas bubble has formed in the bulb, the cooling process must be continued until all of the mercury has contracted into the bulb.

Both the precision and the accuracy of liquid-in-glass thermometry depend as strongly upon the capabilities of the user as they do upon the quality of the thermometer. A careful worker with good equipment can measure temperature reproducibly within 0.01°C; if the thermometer has been calibrated properly, the resulting temperatures may accurately represent the appropriate temperature scale at the same level. Because of the large existing variety of liquid-in-glass thermometers, however, and because of the strong influence of the user's technique, one cannot safely make general statements regarding either characteristic.

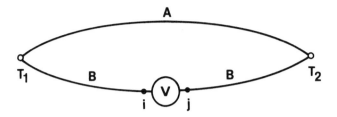

FIGURE 7. Simple thermocouple thermometer circuit. Dissimilar wires A and B are joined at temperatures T_1 and T_2. The original current-flow circuit of Seebeck has been modified by the insertion of a high-impedance potentiometer V (see Section IV.A) to emphasize the present-day thermometry technique.

III. THERMOCOUPLE THERMOMETRY

The thermocouple thermometer can be considered as the electrical analog of the liquid-in-glass thermometer — it provides an inexpensive though not highly accurate method for measuring temperature. The principles of thermoelectricity were discovered in the 19th century, nearly at the same time that high-impedance voltmeters and potentiometers were being developed. A "thermocouple thermometer system," of course, consists of both the thermocouple itself and the potentiometer or other device used for the measurement of its temperature-dependent electromotive force (emf).

The thermocouple thermometer enjoys widespread use in both scientific and industrial temperature measurement and control. In addition, there are many different types of thermocouples in service over a range of temperatures that reaches as low as 1 K and higher than 2000°C. For both of these reasons, there is an extensive literature on the use of thermocouple thermometers and on the sometimes peculiar problems that arise in their use. We provide references to some of that literature in this section as we briefly note the physical basis for thermocouple thermometry, offer information on several common types of thermocouple thermometers, and review recent progress in thermocouple thermometry research.

A. Seebeck and Peltier Effects

The relationship between temperature and the emf of a particular thermocouple cannot be predicted accurately by theory. By gaining an understanding of some of the physical effects that underlie thermoelectricity, however, one can better practice thermocouple thermometry.[10]

The existence of a thermoelectric current that is generated in a circuit composed of two dissimilar wires with different junction temperatures was discovered by Seebeck early in the 19th century.[11] He observed that current was induced to flow in a circuit composed of two dissimilar metal wires (Bi and Cu, and Bi and Sb) by making the two junction temperatures different. Measurements of the Seebeck effect can be made in terms of either the closed-circuit current or the open-circuit voltage as indicated in Figure 7. The latter technique commonly is used in practice to avoid the necessity of specifying the circuit impedance; a high-impedance potentiometer (see Section IV) provides nearly open-circuit conditions.

The simple circuit of Figure 7 forms the basis for all thermocouple thermometry. By choosing materials A and B for their properties in particular environments as well as for the magnitude of their combined Seebeck effect, one can construct an arsenal of thermocouple thermometers for use in a great variety of physical and chemical systems.

The Seebeck voltage E_{AB} for any given pair of materials connected as shown in Figure 7 can be written[10]

$$E_{AB} = \int_{T_1}^{T_2} S_{A,B}(T) \, dT \qquad (2)$$

The voltage is measured at the point V_{ij} in the figure.

Equation 2 shows that the thermal emf that is generated by a thermocouple depends upon the choice of the two temperatures T_1 and T_2. Thus in principle the thermocouple is a "difference thermometer". Often, the thermocouple is used to advantage to observe the difference in temperature between nearby objects. This is accomplished by establishing a thermal connection between one of the thermocouple junctions and the object at temperature T_1 and thermally connecting the other junction to the object whose temperature is T_2. The voltage E_{AB} is then a measure of the quantity $(T_2 - T_1)$ which is to be monitored. If it is desired to maintain $T_2 = T_1$, then the voltage V_{ij} can be used as the null-seeking input to a temperature-controlling circuit operating a heater on one of the objects.

The more general method for thermocouple thermometry involves the use of a fixed reference temperature for T_1, so that measured values of E_{AB} can be converted immediately to temperature values by the use of reference tables or graphs. The usual reference temperature for T_1 is 0°C; it is commonly attained by immersion of junction T_1 in a bath composed of melting ice. Figure 8 shows the emf-temperature relations for six of the more common combinations of wire pairs. In each case, the temperature of one of the junctions has been maintained at 0°C. The letter designations shown in Figure 8 were assigned by the Instrument Society of America in order to specify the emf-temperature relations for wire materials while avoiding reference to certain compositions that are protected by trademarks.

The quantity $S_{A,B}(T)$ in Equation 2 measures the sensitivity of the thermocouple pair. It is called the Seebeck coefficient or the thermoelectric power. Note that S is *not* a constant; it is both material and temperature dependent. The value of S can be determined from the slope of the corresponding emf-temperature curve as shown in Figure 8.

$$S_{A,B}(T) = dE_s/dT \qquad (3)$$

where $S_{A,B}(T)$, usually expressed in units of $\mu V/°C$, is defined as the voltage increment dE_s that is produced by a unit temperature difference $dT = T_2 - T_1$ between the junctions of a thermocouple composed of materials A and B at temperature $T = (T_1 + T_2)/2$. Values of the Seebeck coefficient for three common thermocouple types are shown in Figure 9.

In the manufacture and testing of thermocouple materials, it is advantageous to be able to evaluate the two materials of a particular combination individually. This is accomplished in most instances by measuring the thermal emf of a test wire in combination with a pure platinum wire. The National Bureau of Standards maintains a supply of a well-annealed standard platinum wire with the designation Pt67. By forming a thermocouple from a test wire "B" and a length of Pt67 wire, the operator can determine $E_{Pt67,B}(T)$ and $S_{Pt67,B}(T)$. From this information, the conformance of wire "B" to its nominal specifications can be ascertained regardless of the conformance of its usual partner.[13] The use of a platinum reference wire for calibration is shown in Figure 10.

If pairs of thermocouple wires A and B are connected in electrical series as shown in Figure 11, then the thermal emf $E_{A,B}(T)$ is multiplied by the number of wire pairs used. Such an arrangement is called a "thermopile." The thermopile provides an enhanced temperature sensitivity in cases where the voltage sensitivity of the measuring equipment is lacking or where the operator wishes to obtain an average temperature over an extended portion of the test object. The gain in measurement sensitivity is offset in part by increases in the electrical circuit noise and other measurement problems. The increased sensitivity of present-day potentiometers and potentiometric voltmeters has reduced somewhat the need for thermopiles.

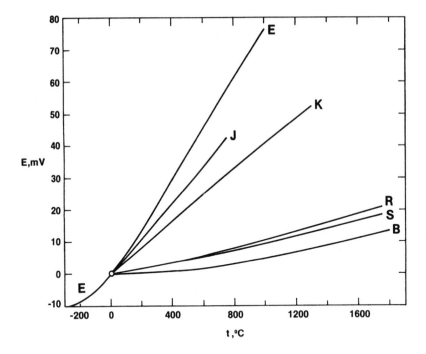

FIGURE 8. The emf-temperature relations for six standard thermocouple thermometers that are identified by ISA letter designations and appropriate compositions by weight as follows; Type E, (Ni + 10% Cr) vs. (Cu + 43% Ni); Type J, Fe vs. (Cu + 43% Ni); Type K, (Ni + 10% Cr) vs. (Ni + 2% Al + 2% Mn + 1% Si); Type R, (Pt + 13% Rh) vs. Pt; Type S, (Pt + 10% Rh) vs. Pt; and Type B, (Pt + 30% Rh) vs. (Pt + 6% Rh). The reference temperature in each case[12] is 0°C.

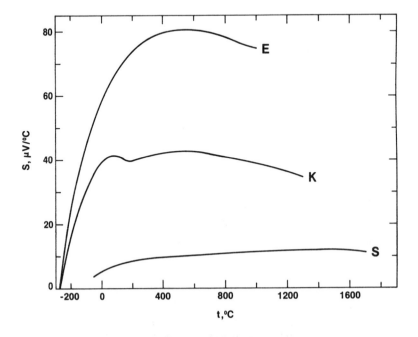

FIGURE 9. Temperature dependences of the Seebeck coefficients for three standard types of thermocouples. Note that no reference temperature is used in the determination of S. Instead, a uniform temperature difference is used. The data shown in this figure were taken from Reference 12.

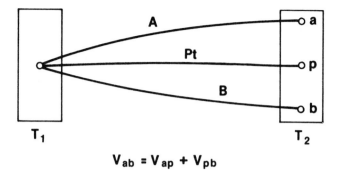

$$V_{ab} = V_{ap} + V_{pb}$$

FIGURE 10. Determining the thermal emfs of the materials of a thermocouple pair independently of each other by inclusion of a reference wire of Pt. Wire A or wire B may be tested against the reference wire by measuring V_{ap} or V_{pb}, respectively. The sum of those two voltages should equal V_{ab} as measured directly.

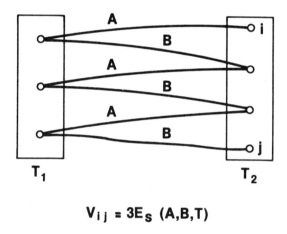

$$V_{ij} = 3E_s(A,B,T)$$

FIGURE 11. A "thermopile," or multiple thermocouple connected in electrical series.

If pairs of thermocouple wires A and B are connected in electrical parallel as shown in Figure 12, then they also provide a temperature-averaging capability. The parallel electrical connection introduces the possibility of loop currents within the individual thermocouple pairs, however, so that special care must be taken in this application; the calibration curves must be nearly identical for all the couples used, the Seebeck coefficients involved must be essentially constant over the temperature range to be spanned, and the electrical resistances of all the couples must be nearly the same. Otherwise, the measured temperature does not represent the average of the existing temperatures.

Shortly after Seebeck's discovery, Peltier[14] found a related phenomenon; passing a current through a thermocouple junction changes its temperature. The quantity of heat liberated or absorbed is

$$Q_p = \int_{T_1}^{T_2} \pi_{A,B}(T) \, I \, dT \tag{4}$$

where I is the current applied from A to B and $\pi_{A,B}(T)$ is the Peltier coefficient corresponding

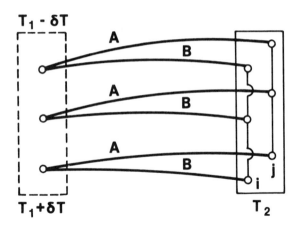

FIGURE 12. A temperature-averaging multiple thermocouple connected in electrical parallel.

to metals A and B at temperature T. In analyzing experimental results on the Peltier effect, one must account first for the effects of Joule heating of the circuit elements.

The Peltier effect is used extensively in thermoelectric heating and cooling. In ordinary thermocouple thermometry, the use of high-impedance voltage-measurement devices prevents distortion of the junction temperatures by the Peltier effect, since no current is allowed to flow in the thermocouple circuit.

B. The Laws of Thermocouple Circuits

We have seen that the thermocouple circuit of Figure 7 can be used to obtain reference curves for particular combinations of materials as shown in Figure 8 and as listed in References 10 and 12.

As a result of considerable experimentation with the simple circuit of Figure 7, a number of "Laws of Thermoelectric Circuits" have been established.[15,16] These laws can be stated as follows:

1. Law of Homogenous Metals

"A thermoelectric current cannot be sustained in a circuit composed of a single homogeneous metal, however varying in cross section, by the application of heat alone."

Applying this law to the circuit of Figure 7, we see that no voltage V_{ij} can appear if wires A and B are chemically and physically the same, no matter what values T_1 and T_2 might take. Note carefully, however, that kinked or work-hardened sections in a wire render it inhomogeneous and thus immune to the requirements of this particular law. Similarly, the absorption of impurities by a section of wire creates an inhomogeneous situation to which Law 1 does not apply. In either case, an unlucky positioning of the inhomogeneous section of wire within a region where the temperature varies will surely generate an undesirable thermoelectric emf — usually an emf of unpredictable sign and magnitude.

Law 1 provides that the position of the voltmeter in Figure 7 does not affect the emf V_{ij} so long as both wires are homogeneous. The voltmeter could be placed anywhere along wire A, wire B, or at either junction.

2. Law of Intermediate Metals at a Single Temperature

"The algebraic sum of the thermoelectric emfs in a circuit composed of any number of dissimilar metals is zero if all of the circuit is maintained at the same temperature."

Referring once again to Figure 7, Law 2 shows the significance of maintaining two different junction temperatures. If $T_1 = T_2$, then no emf can be detected at V_{ij}.

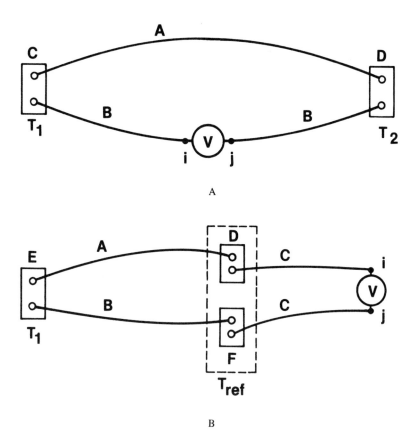

FIGURE 13. (A) Schematic illustration of the law of intermediate metals at a single temperature. The voltage V_{ij} is unaffected by the presence of materials C and D, whatever their compositions, so long as they serve only to connect wires A and B at uniform temperatures T_1 and T_2. (B) Combination of Laws 1 and 2. By Law 1, materials E, D, and F do not affect the circuit voltage V_{ij} so long as they are maintained in an isothermal condition. By Law 2 wires C do not affect the circuit voltage so long as they remain physically and chemically the same.

In Figure 8, all of the curves pass through the point (0,0) because the reference junction temperature in each case is 0°C.

In combination, Laws 1 and 2 are employed to great advantage in remote thermometry using thermocouples. One feature of such measurements often is the use of solder or brazing compound to form the thermocouple wire junctions. By Law 2, the insertion of solder or brazing metal has no effect upon the circuit emf, providing all of the junction remains uniform in temperature. A second advantage is that voltage-measurement connections to a thermocouple circuit can be made using a convenient signal wire C, providing that the connections A-C and B-C are maintained at the same temperature and providing also that the wires C are themselves homogeneous. Figure 13 illustrates these features.

In Figure 13b, we note that the circuit emf commonly is measured by means of connecting wires C (most often composed of copper in view of its thermoelectric uniformity) that are fastened to the thermocouple wires A and B within a uniform environment held at the reference temperature T_{ref}.

Figure 14 shows in schematic form a simple example of the modern use of thermocouple thermometers for laboratory measurements of the temperature profile in a capillary tube that connects a gas thermometer bulb GT to its cutoff valve (cf. Figure 5, Chapter 5). Wire A is connected in common to all of the measuring junctions and to a terminal strip of uniform

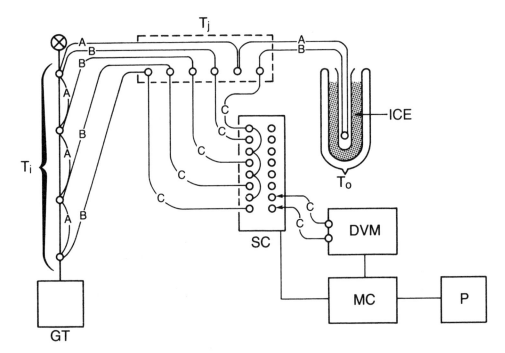

FIGURE 14. Schematic drawing of an actual multiple-thermocouple-thermometer circuit involving four measuring junctions between wires A and B. The thermometers are connected so that the reference junction at T_o is connected in turn to each of the measuring junctions that are arranged along the test specimen above GT. Signal wires of material C (copper) conduct the thermocouple emfs from a connecting point held at temperature T_j to a scanner SC and thence to a digital voltmeter DVM for processing by a microcomputer MC and printing by a printer P.

temperature T_j. The wires B form the second legs of each of four thermocouple thermometers. The reference junction is held at 0°C by an ice bath. The scanner SC connects the voltmeter DVM in turn to each of the measuring thermocouple thermometers, providing a measure of the capillary temperature profile referred to the ice point. In practice, the laboratory microcomputer can be programmed to record the digital voltages sequentially and to calculate the capillary temperatures from them.

Laws 1 and 2 dictate that a carefully prepared circuit of this type will provide values of emf that reflect only the properties of wires A and B operating between the temperatures along the capillary and T_0. By suitably loading the computer's memory with the appropriate thermocouple reference tables and measurement instructions, the operator readily can produce a profile printed explicitly in temperature values.

3. Law of Intermediate Metals at Different Temperatures

"The algebraic sum of the thermoelectric emfs in two circuits composed respectively of metals A-B and B-C is the same as the emf of a circuit composed of metals A-C, providing that the junction temperatures are the same."

This law is shown schematically in Figure 15. If all the wires are homogeneous, then $V_{ij} + V_{kl} = V_{mn}$. The major consequence of this law is that the emf-temperature relation that is characteristic of the thermocouple combination A-C can be obtained without measuring the emf directly. Many thermocouple thermometer reference tables are generated by algebraic summation, at each table temperature, of the results of separate measurements of A and C against B, where B is a reference wire such as platinum.

4. Law of Intermediate Temperatures

"If a given two-junction circuit produces an emf V_1 when its junction temperatures are

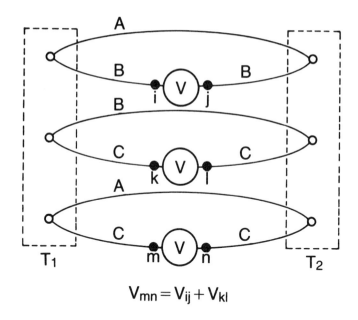

FIGURE 15. Schematic illustration of the Law of Intermediate Metals at Different Temperatures. If the respective junction A-B, B-C, and A-C are held at two uniform temperatures in the manner shown, then the voltage V_{mn} equals the sum of the voltages V_{ij} and V_{kl}.

Table 2
ADJUSTING TYPE E THERMOCOUPLE EMF-TEMPERATURE VALUES FOR A DIFFERENT REFERENCE TEMPERATURE

t(°C)	emf(t) (mV) ($t_r = 0°C$)	emf At t = 100°C(mV) ($t_r = 0°C$)	t(°C)	Adjusted emf(t)(mV) ($t_r = 100°C$)
−50	−2.787	+6.317	−50	−9.104
0	0		0	−6.317
75	+4.655		75	−1.662
150	+9.787		150	+3.470
350	+24.961		350	+18.644
700	+53.110		700	+46.793

T_1 and T_2 and produces an emf V_2 when its junction temperatures are T_2 and T_3, then the same circuit will produce an emf equal to $V_1 + V_2$ when its junction temperatures are T_1 and T_3.''

Correct application of this law is very important in case the operator wishes to use a particular thermocouple circuit with a different reference junction temperature than the one for which a set of emf-temperature values is available. Table 2 illustrates the proper technique for adjusting the existing table values in this circumstance.

C. Standard Thermocouple Thermometers and Their Calibration

We have mentioned already that the Instrument Society of America assigned a letter designation to each of several types of thermocouple. The purpose for this step was to allow the specification of an emf-temperature relation for each type without specifying its com-

Table 3
ISA STANDARD THERMOCOUPLES

ISA designation	Approximate composition[a] (positive leg listed first)	Useful temperature range (°C)
Noble metal types		
Type S	(Pt + 10% Rh) vs. Pt	−50 to 1767
Type R	(Pt + 13% Rh) vs. Pt	−50 to 1767
Type B	(Pt + 30% Rh) vs. (Pt + 6% Rh)	0 to 1820
Base metal types		
Type E	(Ni + 10% Cr) vs. (Cu + 43% Ni)	−270 to 1000
Type T	Cu vs. (Cu + 43% Ni)	−270 to 400
Type J	Fe vs. (Cu + 43% Ni)	−210 to 1200
Type K	(Ni + 10% Cr) vs. (Ni + 2% Al + 2% Mn + 1% Si)	−270 to 1372

[a] All compositions are given in weight percent.

position. Over the years a number of manufacturing concerns had developed particular thermocouple wire compositions that proved especially appropriate for use in thermometry. These include, for example,[17] Chromel™-constantan, a forerunner of ISA Type E thermocouples; copper-constantan (ISA Type T); iron-constantan (ISA Type J); and Chromel™-Alumel™ (ISA Type K). By specifying the emf-temperature relations by the letter designation rather than the compositions, the ISA could assure that those manufacturers or any others could deviate from the trademarked compositions so long as any manufactured thermocouple pairs could meet the published table values. Table 3 contains a list of seven types so identified by the ISA and currently accepted as American National Standards in the consensus temperature standard ANSI MC 96.1. Also included in the table are the useful ranges of the thermocouple thermometers as established[18] by the ASTM Committee E-20.

Representative samples of wire of each of these thermocouple types were studied extensively at the NBS in order to develop reference tables of emf vs. temperature over the range of use for which each type is suitable. The reference tables were published, along with Seebeck coefficient data and considerable supplementary information, in the NBS Monograph 125,[12] issued in 1974. Since that time, the tables have been reprinted, in whole or in part, in many other publications. Figures 8 and 9 show portions of these tables in graphical form.

The calibration of standard thermocouples consists of the determination of their emf values at a sufficient number of temperatures so that they may later be used to measure temperature on a particular scale with a stated accuracy. The process may include the following steps; annealing, test junction assembly, emf measurement, and construction of emf-temperature tables or equations or both. In discussing calibration procedures here, we generally follow the techniques practiced at the NBS.[19,20]

1. Annealing

Most base-metal thermocouples (ISA Types E, J, T, and K) are annealed during manufacture. This annealing is considered to be satisfactory for most thermometric purposes, so that the calibration process for base-metal thermocouples usually does not include an annealing step. For noble-metal thermocouples (Types S, R, and B), on the other hand, annealing has been demonstrated to be effective in promoting more uniform calibration results. For this reason, the NBS practice is to anneal all noble-metal thermocouples prior to calibration. The thermocouples are heated to about 1450°C in air by the passage of electric current through their length while they are suspended loosely between supports. After approximately 45 min at this temperature, they are annealed at 750°C for about 30 min, then slowly cooled to room temperature.

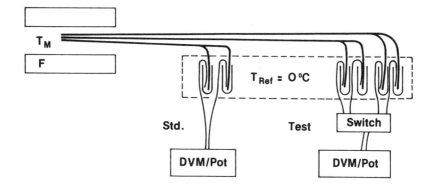

FIGURE 16. Schematic drawing of a comparison thermocouple calibration. All test thermocouples are placed in close thermal contact with the standard thermocouple inside the furnace F. Each thermocouple wire joins a copper emf-sensing wire at separate ice-point junctions. If two emf-measurement devices, usually high-impedance digital voltmeters (DVM) or potentiometers (Pot) are available, then simultaneous test and standard emf measurements can be obtained at each test temperature T_M.

2. Test Junction Assembly

If the thermocouple is to be calibrated by comparison with a standard thermocouple, then the test wire or wires usually are welded or brazed to the measuring junction of the reference thermocouple. By creating a single measuring-junction bead containing all of the thermocouples to be measured, the operator can minimize the temperature gradients among them. If the thermocouple is to be calibrated by immersion in temperature fixed-point cells, then it usually is handled as a separate unit.

3. emf Measurement

Two types of calibration — comparison and fixed-point — are in common use. The former type is illustrated schematically in Figure 16. In the comparison calibration, the variable-temperature junction, composed of one or more test wires and the reference-thermocouple junction in a single bead, is placed in a temperature-controlled environment. The reference junction of each thermocouple is placed separately in an ice-point cell. Then the temperature-controlled environment is set to one of the selected temperatures and stabilized so that the emf of each thermocouple may be measured. Depending upon the measurement precision desired, the emf may be monitored using a high-impedance digital voltmeter or a potentiometer.

Ideally, the test temperature would be maintained at a constant value T_M while the operator measured all the test thermocouples and the standard thermocouple. However, it is difficult to attain completely uniform temperatures in a reasonably short time in most furnaces. The use of two potentiometers or DVMs allows the simultaneous measurement of the standard thermocouple and one test thermocouple, so that a small temperature drift can be tolerated. This technique also permits many redundant measurements to be made, such as the exchange of the two potentiometers and the simultaneous measurement of each leg of the test thermocouple against a platinum standard.

It is possible also to perform the comparison calibration using a platinum resistance thermometer as the reference; if this technique is used, then the uncertainties of the test temperatures can be reduced.

If the fixed-point calibration method is to be employed, then each test thermocouple is placed in turn into fixed-point cells that provide reference temperatures throughout the range of interest. For the Type S thermocouple, for example, the NBS calibration includes measurements at the gold, silver, antimony, and zinc freezing-point temperatures.

4. Construction of Reference Tables

The calibration results may be utilized most easily in the form of a table of values of emf vs. temperature over the calibration range or as an equation that has been fitted to the calibration points. The NBS offers both these options. In the case of the fixed-point calibration from the zinc point to the gold point, the equation

$$e = a + bt + ct^2 + dt^3 \qquad (5)$$

is used; it offers an uncertainty level as small as 0.2°C in the range 630 to 1064°C.

If an extended range of temperature is to be covered, for example, from 0 to 2000°C, then the uncertainties of the table values depend upon the selection of the calibration points. If the points are not more than 200°C apart, then the temperature may be interpolated accurately within 2°C.

Another method of presenting calibration results is that of a difference table. In this case, the user is offered a table of differences between the emf-temperature values of a reference thermocouple and the test thermocouple. This technique is most often used for comparison calibrations.

The user of thermocouple wire often can avoid the expense and trouble of its initial calibration. Several manufacturers warrant that thermocouples constructed from their supplies of wire will conform within stated limits to published emf-temperature reference tables. Of course, this warranty does not apply once the thermocouple has seen substantial use.

D. Special Problems in Thermocouple Thermometry

The thermocouple thermometer possesses an apparent simplicity that often deceives its users. The sensor appears to be a tiny detector that evaluates the temperature exactly at the location of the measuring junction. In certain commercially available thermocouple thermometer systems, the reference junction is contained within a handy digital voltmeter, so that there is no messy ice bath to manipulate.

However, there are several problems that are peculiar to thermocouple thermometry. Each one arises from the nature of thermocouple measurements; that is, from the fact that the thermometric quantity is measured in terms of a small, steady voltage. Any spurious source of voltage in the thermocouple circuit directly contributes to the temperature measurement error.[21]

What are the primary sources of error in thermocouple thermometry? There are many. They include at least the following problems that have been discussed by various authors:

- Deviations from specifications in wire manufacture
- Existence or development in service of work-hardened, kinked, or inhomogeneous sections of wire in a portion of the thermocouple circuit that passes through a temperature gradient[22,23]
- Use of low-impedance measuring instrumentation, leading to "loop-current" errors that arise from the flow of substantial electrical currents within the thermocouple circuit
- Presence of substantial electrical leakage paths or accidental electrical grounds in the circuit[24]
- Presence of electromagnetic interference, either at the measuring junction or along improperly shielded extension wires
- Use of switching apparatus that introduces spurious, sometimes variable voltages
- Use of extension wires that do not match the emf-temperature relation of the thermocouple wires themselves or that introduce unwanted emfs in their connections to the circuit
- Calibration drift, especially in high-temperature service, arising from diffusion of impurities from materials surrounding the thermocouple[25]

Taken together, the problems noted above can render thermocouple thermometry worthless — that is to say, so much in error that the operator would be better off with only a visual estimate of the temperature of the test object. On the other hand, careful attention to the details of calibration, installation, and measurement[26] can reduce inaccuracies to levels less than 0.1°C.

E. Thermocouples for Special Applications in Thermometry

Kinzie discusses more than 200 types of thermocouples in his excellent source book on thermocouple thermometry. In addition, he lists about 500 references to published papers that describe, for the most part, efforts to measure temperature in difficult situations by the use of thermocouples.[27]

Burns and Hurst have presented other aspects of the use of thermocouples for special applications in thermometry, including cryogenics, high temperatures, nuclear environments, and corrosive systems.[28]

Special Technical Publication 470B of the American Society for Testing and Materials, an excellent desk reference for thermocouple thermometry to which we have referred often in this chapter, also offers information regarding the use of thermocouple thermometers in special applications.[29]

Finally, the series of symposia on temperature measurement and control provide a continuing source of new information on the applications of thermocouples to special situations in thermometry. The most recent of these symposia took place[30] in 1982.

In this section, we present brief sketches on the use of thermocouples for the measurement of low temperatures and of high temperatures.

1. Thermocouples For Use in Cryogenics

In the range below room temperature, the standard thermocouple types E, T, and K all may be used. Of the three, Type E has proven to be most useful in terms of lower wire thermal conductivity and higher Seebeck coefficient. A specialized alloy (gold + 0.07 at % Fe),[31] appears to be very satisfactory when used as the negative leg in combination with the (Ni + 10 wt% Cr) wire that constitutes the positive leg of the Types K and E thermocouples. Table 4 lists thermal emfs and Seebeck coefficients of these two thermocouple types.

One can see readily from Table 4 that the gold-iron thermocouple is more sensitive for thermometry below 40 K and that the Type E thermocouple is more sensitive above 100 K. Details of the measurement to be made might dictate the choice of either in the intervening range.

An interesting experiment was reported recently on the use of a superconductive thermocouple.[32] Armbrüster and his colleagues have prepared a Kondo-alloy wire composed of iron-doped gold (about 1 ppm Fe) that, despite the very low concentration of iron, still develops a Seebeck coefficient of approximately 0.75 μV/K at 1 K. By using a superconductive SQUID detector (see Chapter 5, Section III.C) in conjunction with a (NbTi superconductive wire) vs. (Au-Fe wire) thermocouple, these workers were able to measure temperatures as low as 0.02 K.

2. Thermocouples for High Temperatures

Of the standard base-metal thermocouples, Type K is regarded as the most versatile. Possessing the most favorable combination of high sensitivity, stability, oxidation resistance, and economy, the Type K is used in most thermometry applications involving temperatures in the range 900 to 1200°C. Despite these advantages, Type K thermocouples long have given evidence of substantial shortcomings, especially drifts away from calibration values of emf on long exposure to high temperatures, and short-term changes in thermal emf during heating in the range[33] 250 to 550°C.

Table 4
THERMAL EMFs AND SEEBECK COEFFICIENTS OF TYPE E AND (KP) VS. (GOLD + 0.07 AT % Fe) THERMOCOUPLES

T (K)	Type E		(KP) vs. (Au + 0.07 Fe)	
	E (T) (μV)	S (T), μV/K	E (T) (μV)	S (T) (μV/K)
0	0.00	−0.203	0.00	0.000
1	0.09	0.384	7.85	8.673
2	0.76	0.941	17.27	10.127
3	1.97	1.472	28.04	11.375
4	3.69	1.978	39.96	12.439
5	5.92	2.464	52.86	13.342
7	11.77	3.383	81.03	14.739
10	23.87	4.664	127.40	16.045
15	52.18	6.637	210.29	16.909
20	90.07	8.505	295.17	16.966
25	137.15	10.321	379.54	16.766
30	193.22	12.099	462.84	16.566
40	331.50	15.523	627.66	16.471
50	502.88	18.711	793.45	16.730
60	704.83	21.637	962.74	17.139
70	934.82	24.326	1136.32	17.575
80	1190.73	26.829	1314.22	18.002
100	1774.09	31.429	1682.46	18.810
120	2445.13	35.611	2065.91	19.513
140	3196.17	39.439	2462.15	20.094
160	4020.81	42.982	2869.12	20.592
180	4913.83	46.279	3285.35	21.019
200	5870.33	49.330	3709.45	21.383
220	6885.47	52.149	4140.36	21.698
240	7954.95	54.767	4576.81	21.930
260	9074.95	57.208	5017.33	22.129
273	9828.42	58.680	5305.96	22.267

Note: (KP): Positive leg of Type K thermocouple. Data from Tables 11.1 and 11.4 of Reference 29.

Over a period of many years, researchers have studied the nickel-based thermocouple systems in an effort to understand the basis for the stability limitations in Type K thermocouples and, if possible, to improve their performances.[34-37]

After considerable study and experimentation, it was concluded that any improvement in the nickel-based system must include substantial deviations from the published emf-temperature values for the Type K thermocouples. Once the necessity of this change was accepted, however, it was found possible to increase markedly the stability of nickel-based thermocouples by making small changes in the usual Type-K composition. New alloys that appeared to comprise an especially stable thermocouple were given the names nicrosil and nisil,[38] and reference tables were prepared listing the emf-temperature relations for the new system.[39] Table 5 contains selected values from Reference 39.

The nicrosil vs. nisil thermocouple system was developed in a collaborative effort involving manufacturers and scientists throughout the world. Some of the features of the development and the present status of the new thermocouple have been described in the literature cited above and in References 40 to 43. In recognition of the advantages of the nicrosil-nisil thermocouple for high-temperature thermometry, the Instrument Society of America recently has assigned the designation "Type N" to this thermometer. Reference tables based upon those in Reference 39 will be added to ASTM E-230, the thermocouple reference tables.[18]

Table 5
NICROSIL VS. NISIL THERMOCOUPLE

t (°C)	emf (µV)	S (µV/°C)	t (°C)	emf (µV)	S (µV/°C)
−270	−4345	0.339	550	18668	38.67
−260	−4335	1.575	600	20609	38.97
−250	−4313	2.926	650	22564	39.18
			700	24526	39.29
−200	−3990	9.934	750	26491	39.32
−150	−3336	16.04			
−100	−2406	20.93	800	28456	39.26
−50	−1268	24.33	850	30417	39.15
0	0	26.15	900	32370	38.99
			950	34314	38.79
50	1339	27.72	1000	36248	38.55
100	2774	29.63			
150	4301	31.42	1050	38169	38.29
200	5912	32.99	1100	40076	37.98
250	7596	34.32	1150	41966	37.61
			1200	43836	37.17
300	9340	35.43	1260	46048	36.55
350	11135	36.35			
400	12972	37.11			
450	14844	37.74			
500	16744	38.26			

Note: Data from Tables 7.3.3 and 7.3.5 of Reference 39.
Reference temperature: 0°C.

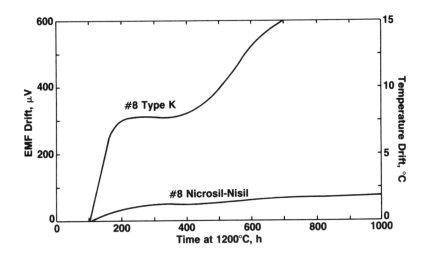

FIGURE 17. Long-term thermal emf drift in Type K and in nicrosil-nisil thermocouples upon exposure to air at 1200°C. Data from Reference 39, Figure 6.3.3.3 and Reference 43, Figure 4.

The superior stability of the nicrosil vs. nisil thermocouple to the Type K thermocouple has been demonstrated in many experiments. One of these is summarized in Figure 17, which shows data taken from References 39 and 43. The enhanced stability is ascribed to increases in the concentration of chromium and silicon (to Ni + 14.2 wt% Cr + 1.4 wt% Si) in the positive leg to change the mode of oxidation; and to increases in the concentration of silicon and magnesium in the negative leg (to Ni + 4.4 wt% Si + 0.1 wt% Mg) to

Table 6
HIGH-TEMPERATURE THERMOCOUPLES

Property	(W + 3% Re) vs. (W + 25% Re)	(Ir + 50% Rh) vs. Ir	(Pt + 20% Rh) vs. (Pt + 5% Rh)
Temp. limit in:			
Dry H_2	2760°C	NR	NR
Inert atm	2760°C	2050°C	1700°C
Oxidizing atm	NR	NR	1700°C
Vacuum	2760°C	2050°C	1700°C
Ave Seebeck coefficient	17.1 μV/°C	5.7 μV/°C	6.8 μV/°C
Stability with thermal cycling	Good	Fair	Good

Note: NR: Not recommended. Data from Reference 44.

provide an oxidizable diffusion barrier film and to suppress a magnetic transformation to temperatures below 0°C. The performance of the new thermocouple in nuclear reactor thermometry is improved because the concentrations of Mn, Co, and Cu have been markedly reduced. These elements, all present in Type K thermocouples, are particularly susceptible to neutron capture and thus prematurely degrade the sensor stability in that application.

Type B thermocouples can be used with moderate sensitivity at temperatures as high as 1700°C. In order to employ thermocouple thermometry at still higher temperatures generated in combustion, nuclear, and similar environments, several refractory-metal thermocouple materials have been developed. Prominent among these are tungsten-rhenium alloys, iridium-rhodium alloys, and platinum-rhodium alloys.[44-46] A summary of properties of selected thermocouples recommended for use at high temperatures is given in Table 6.

IV. RESISTANCE THERMOMETERS

The accurate measurement of temperature by means of resistance thermometers began with the work of Callendar at the end of the 19th century.[47] Siemens earlier had introduced the use of platinum wire as the resistance element because its advantages as a thermal element even then were very strong. These advantages included the following:

- Readily obtainable in rather pure form
- Resistance-temperature relation contains a dominant linear term
- Extremely high melting point
- Chemically and metallurgically stable

Callendar refined Siemens' construction methods to bring the thermometer to an enhanced level of stability. Using the platinum resistance thermometer (PRT) to estimate temperatures as high as the gold freezing point, Callendar developed resistance-temperature equations by calibration against a gas thermometer up to about 550°C.

In this section we discuss some of the features of resistance-measurement instruments, pure metal resistance thermometers, thermistor thermometers, and low-temperature resistance thermometers employing rhodium-iron alloys, doped-germanium sensors, and commercial compacted-carbon resistors.

A. Resistance Thermometer Measurement

We must not ignore the fact that the platinum resistance thermometer sensor only constitutes part of a thermometer system. As the resistance-measurement expert Mueller wrote,[48] "It is undoubtedly true that the term 'resistance thermometer' logically means the ensemble

Table 7
RANGES OF RESISTANCE AND SENSITIVITY REQUIRED FOR STANDARD PLATINUM RESISTANCE THERMOMETER MEASUREMENTS

	T ~ 14 K	T ~ 273 K (0°C)	T ~ 903 K (630°C)
R (PRT) (Ω)	0.033	25.5	84
dR/dT (μΩ/mK)	33.	100	80

of resistor, the means for connecting it to the resistance measuring equipment, the resistance measuring equipment itself and its accessories, such as battery, galvanometer and galvanometer reading device. There is very little occasion to use a single term to describe the ensemble, and the term 'resistance thermometer' has come to mean the resistor itself and the other things permanently fastened to it, such as protecting tube and connecting leads."

As we stated at the beginning of this chapter, we continue the usage of Mueller here; nevertheless we note that, strictly speaking, in order to make Callendar's platinum resistors into platinum resistance thermometers, it was necessary to make use of precise resistance-measurement (or, more commonly, resistance-comparison) instruments as well.

Use of standard platinum resistance thermometers requires the capability of measuring resistance in the range 0.02 Ω (for the low-temperature capsule thermometers at 13.8 K) to approximately 85 Ω. Measuring currents typically range from 1 to 4 mA as a compromise between practical voltage sensitivity and limited thermometer self-heating.[49] Table 7 shows the ranges of PRT resistances and sensitivities as dictated by the desire for millikelvin-level temperature resolution throughout the PRT range of the IPTS-68.

There are two basic methods for measuring resistance thermometers: the potentiometer and the bridge. In both cases, the thermometer resistances are compared with a reference resistor. These methods are illustrated in Figures 18 and 19.

1. Potentiometer Measurements

The potentiometer, first described by Poggendorf[50] in 1841, was developed with a galvanometer G to verify that two voltages were equal. The voltages were connected so as to oppose each other in the circuit loop that included the galvanometer. The galvanometer, composed of a multiple-turn coil of copper wire lightly suspended within the field of a permanent magnet, responds to even small currents by rotating about its supporting axis. Although the galvanometer still is used in some present-day potentiometric circuits, high-impedance voltmeters or DC amplifier "null detectors" are employed more frequently. The potentiometer principle, however, remains the same.

In the use of a potentiometer for resistance thermometer measurement, as shown in Figure 18, the thermometer resistance R_x is compared with a reference resistance R_s. The high-precision PRT is a low-impedance device (generally, R_x does not exceed 100 Ω). The resistances of the electrical leads, R(C), R(c), R(t), and R(T) therefore are not negligible. Use of a potentiometer for measurement, however, enables the user to obtain values of R_x despite the resistances of the voltage leads R(c) and R(t); this result is accomplished by reducing to zero the current in the galvanometer (or null detector) portion of the circuit.

In the measurement sequence, two test voltages are created; they are composed of the quantities $I_t R_x$ and $I_t R_s$. The thermometer current I_t is set by the operator to a value that is consistent with the self-heating properties and sensitivity of the thermometer resistor R_x and the reference resistor R_s. The reference resistor is placed in electrical series with R_x so that its voltage drop $I_t R_s$ involves the same value of current that occurs in the thermometer voltage $I_t R_x$; in the calculation of R_x, as we shall see, the current then cancels.

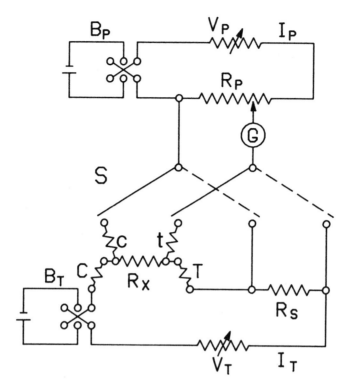

FIGURE 18. Potentiometric measurement of resistance thermometers. See text for details.

The reference resistor R_s is a carefully made fixed resistance in the range 0.1 to 10 Ω, usually contained in a temperature-regulated bath.

The variable potentiometer resistance R_p is determined by the setting of a linear set of decade resistors. The potentiometer current I_p is set to a value such that the potentiometer voltage $I_p R_p$ can be made equal in turn to the thermometer voltage and to the reference resistor voltage as switch S is used by the operator to place first the thermometer and then the reference resistance into the galvanometer circuit loop.

The battery voltages B_p and B_T are made reversible so that the effect of spurious thermal emfs in the leads can be eliminated. On reversing the potentiometer current and the thermometer current, the respective measuring voltages are reversed; however, since no connection in the galvanometer circuit loop is changed, any thermal emfs will remain constant and appear as a small constant imbalance in the galvanometer setting. This imbalance cancels when the two balance readings at reversed currents are averaged.

A determination of the thermometer resistance involves four balances. With the thermometer connected to the galvanometer circuit, R_p is adjusted until the galvanometer shows no change in its deflection upon reversal of both battery voltages; the resulting value of R_p may be designated R_p(thermometer). Two more balances with the reference resistance in the galvanometer circuit yields a potentiometer resistance value that we may call R_p(reference). Then

$$R_x = R_s[R_p(\text{th})/R_p(\text{ref})] \tag{6}$$

As we noted previously, the resistances of the thermometer leads need not influence the measurement. In order to obtain accurate values of the thermometer resistance, it is necessary that the potentiometer current remain constant within a small fraction, typically less than 1

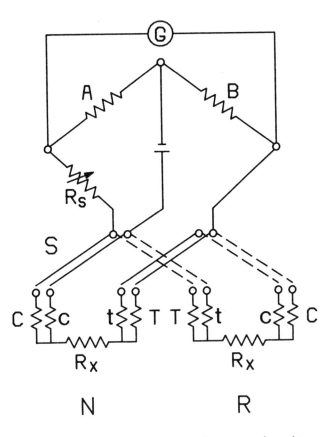

FIGURE 19. Measurement of resistance thermometers by resistance bridges. See text for details.

ppm, during the four balances. The reader should note that the measurements described do not involve the use of a standard cell, since it is a resistance ratio — not a voltage value — that is desired. The generally low voltages involved in platinum resistance thermometry usually dictate that the potentiometer resistor be a high-quality one, and that it be used near the maximum of its range.

The usual potentiometer is less convenient for thermometry than is a well-designed bridge. On the other hand, it has an advantage if a very low resistance must be measured, since the operator can install a low-resistance reference resistor and then adjust the measuring currents to suitable values. The resolution and accuracy of potentiometric measurements depend strongly upon the details of the measurement.

2. Bridge Measurements

The instrument used for many years at the National Bureau of Standards for platinum thermometer resistance measurements is the Mueller bridge. A variation of the Wheatstone bridge, first adapted to four-terminal platinum resistance thermometry measurements by Smith[51] in 1912, the Mueller bridge[52] provides resolution at the submicrovolt level and resistance accuracy at the level of 0.1 ppm or better, depending upon the detector used with it.

The Mueller bridge, shown in drastically simplified form in Figure 19, essentially provides a method for setting a high-quality linear set of decade resistances R_s equal to the resistance of the temperature-sensing element of a platinum resistance thermometer. This goal is accomplished in two steps: first, the thermometer leads, shown in the figure with letters C, c, t, and T, are connected to the bridge in the "normal" way, N, and the bridge resistance is adjusted to the value R_{s1}; this value is chosen to satisfy the requirement of a null reading in the galvanometer circuit G (recall that a high-impedance DC amplifier or voltmeter can be used in this circuit in place of an actual galvanometer). The second step involves reversing the thermometer leads as shown by the R position and performing a new balance, resulting in the bridge resistance value R_{s2}. The two balance equations are

$$R_{s1} + R_C = R_x + R_T \qquad (7)$$

$$R_{s2} + R_T = R_x + R_C \qquad (8)$$

where R_T and R_C are the resistances of the current leads of the thermometer. Note that R_C and R_T, or at least their differences, must remain constant during the period of measurement. The thermometer resistance then is the average of the two bridge resistances

$$R_x = (R_{s1} + R_{s2})/2 \qquad (9)$$

and is independent of the values of R_C and R_T.

The key to the success of the Mueller bridge in providing high precision measurements is the elimination of serious parasitic resistances, either by careful construction or by careful measurement. Some of the techniques used for this purpose are the provision of temperature control for the bridge resistors, the inclusion of a balancing resistor to equalize the fixed bridge resistors A and B, the commutation of A and B, the use of mercury-wetted contacts for critical connections, the use of special decade resistance elements, and the special placement of certain resistance elements. The circuitry of the Mueller bridge is described in some detail by Harris[53] and its calibration and use are discussed by Riddle et al.[54]

3. AC Measurements

The introduction of AC techniques, inductive voltage dividers, and integrated circuits has made possible the production of automatic-balancing, microprocessor-controlled bridges with very stable, linear-decading features. The AC technique for resistance thermometry generally removes the need for multiple balances in order to evaluate the thermometer resistance. On the other hand, the operative bridge parameter is impedance, not simply resistance, so that inductive and capacitive effects must be accounted for properly.

Several automatic resistance-thermometer bridges have been introduced during the past few years.[55-58] The bridge designed by Cutkosky for use in the platinum resistance thermometry laboratories of the National Bureau of Standards contains many elements in common with other bridges.[55] The bridge contains a five-stage transformer and five temperature-controlled, 10-Ω resistance standards. It is excited by 15 or 30 Hz square waves, with the result that high-frequency effects are minimized. A microprocessor controls the frequency generation, the selection of any of four test-thermometer input locations, the measuring current, and the balancing of the bridge. As originally desi241ed, the bridge measures resistances as large as 31.8 Ω. A later version will accept 100-Ω resistors. Its resolution is 1 $\mu\Omega$. For resistances greater than 10 Ω, the bridge is capable of 0.1 ppm accuracy. The bridge also is compatible with the IEEE 488 standard instrument control bus.

Figure 20 shows in block form a typical laboratory arrangement involving the use of microprocessor-controlled resistance bridges for the measurement and control of temperature. The laboratory that is depicted is one where high-temperature gas thermometry is studied in an attempt to realize the thermodynamic temperature scale (see Chapter 5, Section II.H.1). Under the overall control of a laboratory microcomputer MC that is operated through a terminal T, an automatic resistance bridge CB1 of the type designed by Cutkosky sequentially measures the resistance of four specially-prepared high-temperature resistance thermometers

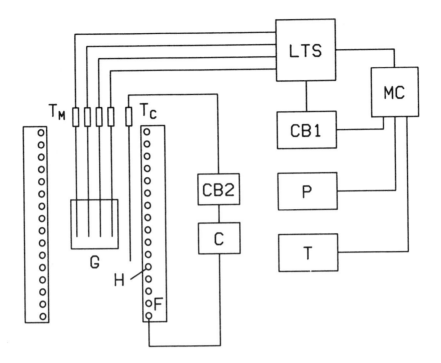

FIGURE 20. Block drawing showing the use of a microprocessor-based resistance bridge in the laboratory. See text for details.

T_M. The four thermometers are placed so as to provide information on the temperature of the gas thermometer bulb G that is located within a furnace F. Because other thermometers and reference resistors also are monitored by the bridge, a low-thermal scanner LTS (basically a multiple-unit four-pole switch) is used to select the particular thermometer that is to be measured. A printer P furnishes a record of the progress of the experiment as required. A second AC resistance bridge of an earlier design (CB2) monitors a control thermometer T_C; the bridge output provides the signal for a temperature controller C that activates the furnace heater coils H.

In considering the effectiveness of the various resistance-measurement instrumentation available, one should be aware of several facts:

- AC measurement of the resistance of a PRT may produce a different value than a DC measurement does, particularly because eddy-current effects may create different temperature gradients in the region of the sensor and, further, may generate a different level of self-heating in the sensor coil itself. In given instances, these effects may be negligible.
- The speed of measurements nearly always is faster with AC equipment, occasionally allowing better measurements to be made in situations where temperature drifts exist.
- The use of automatic measurements nearly always will allow better characterization of instrument precision because many more repetitions of a given measurement can be made. "Better data through automation" is not a foregone conclusion, however; untrained or careless personnel can obtain incorrect results using the best of equipment that is expertly prepared and programmed. The ultimate accuracy that automatic bridges can bring to resistance thermometry is not yet known. What *is* already clear, however, is the fact that the use of such bridges nearly eliminates the drudgery that has been a part of precise resistance thermometry for a century.

B. Pure Metal Resistance Thermometers

1. Standard Platinum Resistance Thermometers

Over an extended range of temperature, from about -75 to $500°C$, the resistance of a coil of very pure platinum wire may be represented closely by the equation[59]

$$R(t)/R(0) = 1 + 3.985 \times 10^{-3} t - 5.86 \times 10^{-7} t^2 \tag{10}$$

Use of Equation 10 allows one to calculate temperatures in the specified range that are accurate within approximately $\pm 0.1°C$. The nearly linear dependence upon temperature of the resistance of platinum, taken with its excellent stability, makes it the optimum choice for the active element of a resistance thermometer.

We briefly discussed in Chapter 4 the requirements and use of the SPRTs that are specified in the text of the IPTS-68. We have noted that these sensors, in conjunction with the best bridges, consistently will reproduce temperature measurements between 13.8 K and 630°C with accuracy levels — depending upon temperature — between 0.1 and 1.0 mK. At their best, the SPRTs are stable at about the 0.3 ppm level in resistance or 0.1 mK in temperature.

When Callendar took up the use of platinum resistors for the precise measurement of temperatures well above room temperature, the commercially available models were based upon a design by Siemens. The thermometers were composed of platinum wire wound upon a support made of clay and contained in an iron protective casing. The thermometer sensors were notoriously unstable in high-temperature use. Callendar modified their construction in his first experiments by winding a two-lead platinum resistor upon a blade-shaped support made of mica and protecting the resistor from its measurement environment with a porcelain or silica sheath. He also incorporated in the thermometer a second pair of shorted leads to facilitate compensation of their resistance by the use of bridge measurements. The resulting thermometers were found[60,61] to be reproducible within 0.01°C or so from 0 upwards to 600°C.*

Further progress in the design of standard platinum resistance thermometers, taking place at the National Bureau of Standards in the U.S. and at the National Physical Laboratory in England among other laboratories, emphasized the minimization of strain and thermal response time. Figure 21 shows some of the resulting designs.[62-65]

At the present time, the construction of the standard PRT is typically that shown in Figure 22 (cf. Figure 5, Chapter 4). Usually, the value of R(0°C) is 25.5 Ω, so that a resistance change of 1 Ω corresponds to 10°C. The resistor R is wound from about 60 cm of 0.1-mm-diameter high-purity platinum wire on an insulating former F. Mica still is the insulator in most common use in PRTs. The mineral has excellent insulation properties at ordinary temperatures, although it begins to fail[66,67] above 500°C. Other materials used as insulators include alumina, synthetic sapphire, and fused silica; each of these materials must be cleaned carefully, and all of them have presented problems in fabrication.

The form of the resistor coil is a compromise among the conflicting requirements of minimizing strain, minimizing thermal response time, and minimizing the fragility of the thermometer. The use of four leads, two connected to each end of the resistor coil, is standard. Generally, the winding is bifilar to minimize inductive effects. The choice of a simple helical winding places all of the resistor wire near to the sheath of the thermometer, minimizing its thermal response time; a coiled helix, on the other hand, requires fewer turns and thus less contact between the wire and the support. These and several other winding configurations still are used by various manufacturers.

The sensor leads L are nearly always made of platinum for at least a short distance above

---

* The resistor support used by Callendar is said to be a cross of mica, rather than a blade, in Reference 61. It is likely that the cross design was adopted by Callendar in later experiments. See also References 61a and 61b.

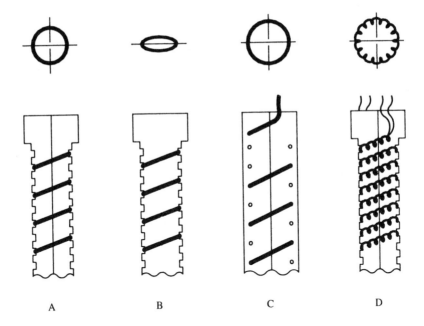

FIGURE 21. Principal designs used for high-precision platinum resistance thermometers. (A) Mica cross design of Callendar;[60] (B) calorimetric thermometer;[62] (C) "strain-free" design;[63] (D) double helix.[64]

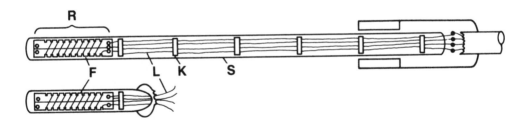

FIGURE 22. Typical construction of present-day long-stem and capsule PRTs.

the coil. Platinum generally is used over the whole length of each lead, but occasionally gold is chosen for the upper leads because of its ease in working. All the leads are made as nearly identical as possible to facilitate accurate measurement of the thermometer resistance. In long-stem PRTs, insulating spacers K prevent shorting of the leads within the sheath S. Most PRTs are hermetically sealed using tungsten wires that penetrate a soft glass seal. External leads connecting the thermometer to the measurement equipment generally are made of individually-insulated copper wires formed into small cables.[68]

Standard platinum resistance thermometers may be encased in borosilicate glass, metal, quartz, or fused silica sheaths. Dry air or a combination of oxygen with an inert gas usually is used to fill the sheath.

Recently, Berry of the National Research Council laboratories in Canada has studied the changes in resistance that can occur in standard PRTs as a result of various types of temperature cycling. For some time he has been concerned about the effects of oxidation of platinum upon its use in thermometers.[69] Because oxygen is needed in the thermometers to stabilize both the metal impurities in the platinum wire and the oxide insulation materials, most PRTs contain from 5 to 10 kPa partial pressure of oxygen at the time of construction. The presence of the oxygen, however, leads to the formation of one of two types of platinum oxides that Berry designates as "two-dimensional" and "three-dimensional." The two-dimensional growth appears to take place most rapidly at temperatures near 150°C; dissociation of this oxide occurs at about 450°C. Near 375°C, three-dimensional oxide growth begins; it may be accompanied by shifts in the R(0) value of the thermometer corresponding to 0.1 millidegree/hr. This growth continues at higher temperatures, in many cases accelerating. The three-dimensional oxide is dissociated[70] if the thermometer is held at temperatures as high as 630°C. The major source of error in thermometry lies in the measurement of the value of R(0) when the thermometer is in a different condition of oxidation than when the (higher) test-temperature resistance was measured. Berry recommends stabilizing the oxidation of the thermometer before using it; if temperatures in the difficult region from 350 to 400°C must be measured, then he suggests that a frequent measurement of R(0) be made. Use of an oxygen partial pressure as small as 1.3 kPa at 22°C is expected to minimize the formation of the three-dimensional oxide.

PRTs are mechanically fragile. Even slight bumping can result in a measurable increase in the thermometer resistance. Moreover, thermal shocks such as those caused by sudden immersion of a room-temperature thermometer in liquid nitrogen or by quenching of a thermometer from a temperature above 500°C can alter the calibration or, in extreme cases, break the glass thermometer sheath. While the best practice is to avoid mechanical or thermal shock, much of the resistance increase arising therefrom can be removed by annealing the thermometer for several hours at about 475°C, then allowing the thermometer to cool in still air.

The operator usually can determine the effect of depth of immersion upon the reading of a long-stem PRT. The simple process of changing the immersion depth often demonstrates the adequacy or inadequacy of a specific thermometer immersion; this technique fails, however, in the case where severe temperature gradients exist near the sensor coil. The user of a capsule thermometer, on the other hand, usually cannot modify its thermal tempering because the thermometer is enclosed within a cryostat. In that case, the adequacy of tempering must be evaluated on the basis of further experimentation.

It is interesting to consider that the resistor in a PRT generally does not experience the test temperature during a measurement. The usual measuring current is 1 to 4 mA; in response to this current, the thermometer resistor temperature rises above that of its surroundings (cf. Figure 4 of Chapter 3 and the discussion thereof). Typically, a 25-Ω thermometer coil might undergo a temperature rise of 0.3 to 1.2 mK/(mA)^2 when the PRT is immersed in an ice bath. Evaluation of the apparent temperature of the thermometer as a function of measuring current will permit the calculation of the zero-current temperature; the usual assumption is that the resistance (and thus the temperature) varies as the square of the measuring current. As is the case with the immersion depth, however, the operator usually determines the self-heating effect directly.

Less obvious corrections occasionally are necessary as a result of electromagnetic heating that can take place in the coil or owing to radiation being conducted to or from the coil through the walls of the sheath. The temporary removal of possible sources of high-frequency electromagnetic energy, or the installation of shielding, can reveal the existence of the former; "light piping" to or from the coil can be minimized by roughening the surface of the sheath to spoil its efficiency as a light pipe, or, if the radiation is entering from outside the system, by covering the thermometer installation with an opaque cloth.

2. High-Temperature Platinum Resistance Thermometers

Ever since the time of Callendar, thermometrists have been striving to make accurate temperature measurements up to the gold freezing point with PRTs.[60,63] In order to minimize strain and the introduction of impurities into the platinum resistance wire, Callendar wound the sensor coil on mica forms. This step accomplished its purpose, but at the same time it

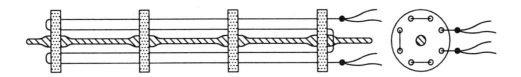

FIGURE 23. The NBS "bird cage" high-temperature platinum thermometer resistor. A central post of pure platinum supports several insulating discs made of sapphire, alumina, or silica. In turn, the discs provide the framework on which straight sections of relatively heavy (0.4 mm to 0.5 mm) Pt wire can be assembled into a series-wound resistance of about 0.25 ohms.

imposed an upper limit on the useful temperature range of the PRT. Since mica, a hydroxylated mineral insulator, begins to decompose to form H_2O in an irreversible fashion above 500°C, thermometers incorporating mica parts are susceptible to increased electrical leakage with high-temperature use.[67] The gradual degeneration of glass insulation through the mechanism of thermionic emission at temperatures above 800°C presents another limitation to the use of PRTs in high-temperature thermometry.

Despite these limitations, success has been achieved in the construction of laboratory prototypes of PRTs suitable for very precise work at temperatures as high as 1100°C. The very early work of Waidner and Burgess at the NBS,[63] in which the purest platinum then available enabled them to obtain reproducibility only at the level of about 1°C, was markedly improved by the higher purity of platinum used by Moser.[71]

In 1961, Barber and Blanke[65] suggested that the use of recrystallized alumina supports and sheath could extend the range of high-temperature thermometers. Their design incorporated four simple coils of 0.3-mm-diameter Pt wire, each one wound upon an alumina tube.[72]

During the Fourth Temperature Symposium, Evans and Burns reported data on the stability of the resistance of specially-designed PRTs at temperatures as high as 1100°C. In one design, discs of synthetic-sapphire insulator were spaced along a central Pt rod. Straight lengths of 0.4-mm Pt wire about 40 mm long were threaded through holes in the discs and welded in series at their ends, forming a "bird cage" resistor of about 0.25 Ω resistance at 0°C. This resistor configuration is shown in Figure 23. Other, more conventional designs also were employed. Several support materials were tried, including alumina, platinum, porcelain, and silica, besides the sapphire already mentioned. In some cases, their thermometers showed resistance drifts corresponding to temperature changes no larger than a few hundredths of a degree after several hundreds of hours of heating at 1000°C. They found that porcelain appeared to be unsatisfactory as a high temperature thermometer insulator but that synthetic sapphire, alumina, and platinum all could be employed successfully in thermometer construction.[73]

Building on the work of Evans and Burns, Curtis and Thomas[74] constructed several "bird cage" thermometers with R(0) values ranging from 0.18 to 0.25 Ω. They used synthetic sapphire and synthetic silica as insulation materials and connected the resistors so as to minimize the wire-to-wire voltages across the insulating discs. They also called attention to the need for minimizing the loss of heat from the thermometer by radiation along the sheath; this was accomplished by roughening the outer surface of the sheath.

To determine the effectiveness of the insulators at high temperatures, Curtis and Thomas connected the central platinum support rod to a fifth thermometer lead. Measuring the resistance between the resistor and the fifth lead, they found that the insulation resistance exceeded 2 MΩ even at temperatures greater than 1000°C.

Curtis and Thomas also observed the drift in R(0) as the thermometers were heated for extended periods of time to temperatures above 1050°C. Although R(0) generally dropped during the first 1000 hr, it usually rose again over the next several thousand hours. The overall R(0) resistance drift corresponded to as little as 0.002°C/100 hr at T > 1050°C, corroborating the excellent resistor stability found by Evans and Burns.

In a landmark paper that not only demonstrated conclusively the relative precision of high-temperature platinum resistance thermometers and Type S thermocouple thermometers, but also alerted the thermometric community to the need for a critical examination of the thermodynamic accuracy of the IPTS-68 in the range 630 to 1064°C, Evans and Wood carefully compared temperatures obtained using the two types of thermometer.[75,75a]

Their experiment was painstakingly prepared: 9 "bird cage" thermometers with R(0) ranging from 0.19 to 0.27 Ω were obtained commercially or constructed at the NBS; 8 platinum vs. (platinum + 10 wt% rhodium) thermocouple thermometers were prepared from several high-purity batches of wire; a group of fixed-point calibration cells was assembled, including the water triple point and tin, zinc, antimony, silver, and gold freezing points; a temperature-controlled furnace was fitted with a copper block for the direct comparison of as many as 4 thermometers at once; and sensitive electrical measurement equipment was assembled for the precise determination of the temperatures of both types of sensors.

The experimental procedure consisted of three fixed-point calibration sequences involving each thermometer, both resistance and thermocouple, interspersed by thermometer-comparison measurements. After the first calibration run, each thermometer was measured in the comparison furnace at 650, 750, 850, and 950°C. The thermometers were measured in batches of four, with positions and partners rotated according to a statistical design. A second comparison experiment was performed after the second calibration run; the furnace temperatures this time were 700, 800, 900, and 1000°C.

An annealing sequence was performed for each PRT after each temperature measurement, and the triple-point resistance was measured thereafter.

As the reader may imagine, an enormous quantity of data was generated in this experiment. The thermocouple thermometer comparison-block measurements were used to establish values of temperature on the IPTS-68 (t_{68}) according to the prescribed relation

$$E(t_{68}) = a + bt + ct^2 \qquad (11)$$

with the coefficients a, b, and c obtained by calibration at 630.74°C (as determined by the PRTs), and at the silver and gold freezing points.

The imprecision of the standard thermocouple thermometer in the Evans-Wood experiment was considerably reduced below the accepted value[76] for the 630 to 1064°C range, ±0.2°C. In this experiment, the 99%-confidence-level uncertainty (±3σ) for a single temperature determination in the range 630 to 1064°C was evaluated in the two comparison runs as ± 0.078 and ± 0.09°C. This improved precision was ascribed to the favorable conditions of the experiment — similar temperature gradients existed in the calibration and comparison measurements and undesired temperature changes and interfering electrical and mechanical disturbances could be minimized in the laboratory situation.

An artificial temperature scale for the platinum resistance thermometers was generated in the Evans-Wood experiment by use of another quadratic equation

$$W(\theta) = R(\theta)/R(0) = A + B\theta + C\theta^2 \qquad (12)$$

with the coefficients A, B, and C having been evaluated by calibration at the same temperatures used for the thermocouple thermometer calibrations, 630.74°C, the silver point, and the gold point. Note that Equation 12 is quite similar to Equation 10, which approximates the resistance-temperature relation for PRTs over most of their temperature range. However, the artificial temperature θ does not necessarily coincide with t_{68} except at the fixed-point temperatures.

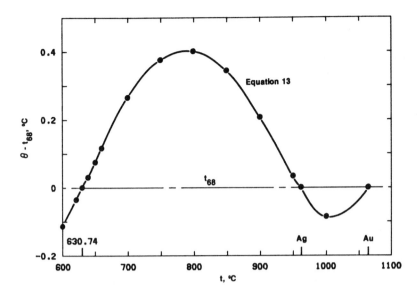

FIGURE 24. The differences between θ, an artificial platinum resistance thermometer temperature scale given by Equation 12, and the IPTS-68 (t_{68}) as derived from calibrated Type S thermocouple thermometers, in the range 630 to 1064°C. The t_{68} calibration points and the calculated differences based upon the use of Equation 13 are indicated in the figure. Data of Evans and Wood.[75]

Despite the improved precision of the Type S thermocouple thermometers found in this experiment, the PRTs performed substantially better. At the 99% confidence level, the imprecisions of a single PRT temperature measurement in the comparison runs were ±0.009 and ±0.011°C, demonstrating conclusively the superiority of the PRT in the 630 to 1064°C range.

In analyzing the two comparison runs, Evans and Wood found that the temperatures obtained by the use of Equation 12 for the nine PRTs (identified as θ in Equation 12) were markedly different from the IPTS-68 temperatures obtained using Equation 11 with the eight Type S thermocouple thermometers (t_{68}). These differences are shown in graphical form in Figure 24. The authors were able to express the differences in Figure 24 by an equation

$$F = \theta - t_{68} = (\theta - 1064.43)(\theta - 961.93)(\theta - 630.74)(A_0 + A_1\theta) \quad (13)$$

with $A_0 = -8.492168 \times 10^{-8}$ and $A_1 = 1.768905 \times 10^{-10}$. A more familiar form was given later by Evans:[75]

$$F = \theta - t_{68} = 54.84 \left(\frac{\theta}{1064.43} - 1\right)$$
$$\left(\frac{\theta}{961.93} - 1\right)\left(\frac{\theta}{630.74} - 1\right)\left(\frac{\theta}{480.81} - 1\right) \quad (13a)$$

(Note that the calibration points are unchanged in the second formulation; they still are 630.74, 961.93, and 1064.43°C. The last factor in Equation 13a is not related to a calibration temperature, but rather is dictated by the form of the equation.)

Evans and Wood estimated the uncertainty in temperature differences rendered by Equation 13 as ±0.054°C at the 99% confidence level. Coupled with the uncertainties connected with the use of an arbitrary PRT in the range 630 to 1064°C, they estimated that temperatures on the IPTS-68 could be approximated accurately within about ±0.07°C by the combined use of a high-temperature PRT calibrated as described above and the function F. This uncertainty is substantially less than the ±0.2°C uncertainty that accompanies the evaluation of the IPTS-68 by means of a single measurement with a Type S thermocouple thermometer.

Besides the clear evidence that PRTs are more precise than Type S thermocouple thermometers in the 630 to 1064°C range and the creation of a method for improving the precision with which the IPTS-68 can be realized there, the work of Evans and Wood raised an interesting question regarding the thermodynamic accuracy of the IPTS-68 in the thermocouple range. Referring again to Figure 24, one is moved to ask why the discrepancy between the temperatures defined by the two instruments should be so large as 0.4°C. It is true that Equation 12 is not based upon thermodynamic principles; yet a quadratic relation is known to provide a reasonable representation (within, say, 0.1°C) of the resistance-temperature behavior of PRTs for any range of calibration above about −100°C.

Subsequent research that *has* been based on thermodynamic thermometry indicates roughly the same discrepancy with respect to the IPTS-68 as did the PRT studies of Evans and Wood (cf. Figure 20 in Chapter 5).

Since the work of Evans and Wood, there have been several papers on the topic of high-temperature PRTs.[76-82] Chattle[77] prepared five 0.2-Ω "bird cage" thermometers and found them to be reproducible within about 0.01°C despite several hundreds of hours of heating at 1100°C, providing that a suitable annealing procedure was followed. Sawada and Mochizuki[78] constructed ten 25-Ω PRTs of the double-helix configuration that was introduced by Meyers (Reference 64 and Figure 21d). Lead wires of platinum and gold were used, along with insulators of silica and alumina and a filling gas of dried air. The stability of these thermometers was generally satisfactory. A drop of a few mΩ in R(0) during the first few hundreds of hours of heating to 1000°C usually slowed on further heating. On a long-term basis, the thermometers generally proved to be reproducible within about 10 millidegrees, so long as the proper annealing procedures were followed after prolonged heating. Anderson[79] was interested in the physical phenomena accompanying prolonged heating in platinum wire. Using a short test thermometer, he observed intergrain slippage and a so-called "bamboo" structure by means of electron microscopy. He found that even the incorporation of single-crystal samples of wire did not forestall a minimum level of irreproducibility in PRTs that were subjected to high temperatures for extended periods of time.

With the recent interest in revising or replacing the IPTS-68, work on high-temperature PRTs has focused increasingly upon the construction of thermometers with the "best" stability in the face of long-term exposure to temperatures above 1000°C. Evans has been the leader in the NBS effort, while thermometer construction projects have been initiated in the People's Republic of China, in Australia, in the Federal Republic of Germany, in France, and in Italy. Some of the results of these programs have been published.[80-85]

All of the successful thermometers have employed fused silica for the sheaths and for the internal supporting components. The overall dimensions have been kept approximately the same as those of standard PRTs in order to retain geometric compatibility with existing fixed-point cells.

Two different approaches have been used in recent thermometer construction to reduce the effect of high-temperature electrical leakage. One approach has been the well-tested use of sensors with R(0) values of 0.2 to 0.25 Ω. The Chinese thermometers, for example, incorporate 0.4-mm-diameter wire wound on a notched blade support like that shown in Figure 21b; in some cases the windings are elliptical in shape, in others, nearly circular. A second approach has been to introduce electrical guard circuits either into or around the thermometer to control leakage currents electrically. The NBS Automatic-Balancing Bridge[55] has been particularly useful in this latter effort, since it incorporates an active guard circuit in a fifth thermometer-lead connection.

Several sensor geometries have proved useful. Among them are

- Bifilar (noninductive) helix wound on crossed silica formers (NBS and Australia)[81]
- Bifilar helix wound on a single notched silica blade (PRC)[84]
- Compound bifilar helix ("door spring") wound on crossed silica formers (NBS and Japan)[81]
- Parallel wires threaded through insulating discs and joined in electrical series ("bird cage")(NBS)[85]

The outstanding characteristics of successful high-temperature PRTs up to the present time are their stability and the uniformity of their R-T relations. In particular thermometers, the R(0) values have been stable within an amount corresponding to 1 mK after cycling to 1100°C; typically, the drift in R(0) corresponds to about 1 mK/100 hr exposure to high temperatures.

The uniformity of the R-T relations of the new thermometers can best be appreciated in the following example: seven high-temperature thermometers prepared in different laboratories (three at the NBS and four at the NIM, Beijing) were calibrated by Evans according to the IPTS-68 procedure (measurements at the water triple point, at the tin freezing point and at the zinc freezing point), thus allowing him to evaluate the coefficients in Equation 10. Using these coefficients, he then could estimate the freezing-point temperatures of Al, Ag, and Au by extrapolation of Equation 10. This step was accomplished by determining the resistance ratio $W(T) = R(T)/R(0)$ at the aluminum freezing point, at the silver freezing point, and at the gold freezing point, using carefully-prepared freezing-point cells of the type discussed in Chapter 3. Finally, using Equation 10, he calculated $t'(Al)$, $t'(Ag)$, and $t'(Au)$ for each thermometer. In all, Evans made 14 determinations of each temperature in this manner. The calculated standard deviation of the set of measurements at $t'(Al)$ was 4 mK; at $t'(Ag)$, 10 mK; and at $t'(Au)$, 15 mK. The significance of these results[86,86a] is that groups of high-temperature platinum resistance thermometers appear to exhibit similar quadratic R-T relations at high temperatures just as they do at lower ones; therefore, the derivation of interpolation equations should be straightforward, given suitable fixed-point temperature assignments.

Not all thermometers prepared and tested in all laboratories have displayed such excellent levels of stability and uniformity. Nevertheless, the successful work that has taken place in several laboratories indicates that PRTs can be used to improve greatly the precision with which the IPTS-68 can be realized in the range 630 to 1064°C.

3. Resistance Temperature Detectors (RTDs)

Many types of thermometer sensors are based upon the temperature-varying resistances of pure metals. These range from the standards-type PRTs through the newer high-temperature PRTs just described to miniature wire- and film-type platinum, copper, and nickel sensors prepared for research applications or heavy-duty industrial use. In this section, we concentrate our attention on the last group, which are known generically as "resistance temperature detectors" (RTDs).

RTDs are not intended to provide temperature measurements of the best accuracy or precision. Instead, they are intended to survive the environments found in laboratory experiments and in industrial facilities, permitting routine thermometry to be accomplished at modest levels of accuracy and expense despite hostile conditions.

The configurations used in constructing RTDs are both numerous and varied. Figure 25 shows three fairly typical sensor designs, chosen variously for small size, for resistance to mechanical shock, and for fast response.

Both the 5th and the 6th symposia on *Temperature, Its Measurement and Control in*

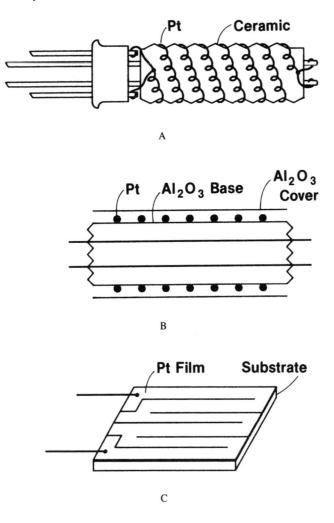

FIGURE 25. Typical RTD configurations. (A) Miniature sensor composed of a compound platinum wire helix wound on a ceramic form.[87] (B) Shock-resistant sensor; the platinum resistor is covered by an alumina coating.[88] (C) Fast-responding platinum film sensor deposited on an insulating substrate and coated with a thin insulation.[89]

Science and Industry featured numerous contributions on specialized RTDs, their properties, and their many uses.[87-106] Many other RTDs are available commercially; some are described in the references just given, and details on others can be obtained from sales literature.[107]

The stability and reliability of more than seventy 100-Ω RTDs was assessed by Carr in 1972.[94] The RTDs all were four-lead thermometers with metal sheaths. Their external diameters were either 0.47 or 0.63 cm. Table 8 shows the properties tested by Carr and typical results he obtained.

More recently, Mangum and Evans[98] performed a series of tests on some 60 platinum RTDs in the range 0 to 250°C. Their R(0) values varied between 50 and 2000 Ω, although most were 100 Ω; the number of leads varied from 2 to 4, and the overall lengths of the sensors ranged up to 13 cm. Mangum was interested principally in the stability of the sensors. Following 24-hr heating periods at 235°C, he found changes in R(t) from 0 to 100°C varying from as little as 1 mK to as much as 1 K. Approximately half the thermometers drifted in resistance by an amount corresponding to 0.015°C or more; one fourth of them changed by

Table 8
INDUSTRIAL 100-Ω RTD PROPERTIES[a]

Property	Result (no. tested)
Insulation resistance	
25°C	0.6 to 10^6 MΩ (79)
100°C	6 to 10^5 MΩ (21)
660°C	0.008 to 1 MΩ (10)
Thermal emf	Typically <10 μV (~0.03°C)
	Worst case 68 μV (~0.2°C)
Self-heating	0.06 to 0.11°C at 5 mA (14)
Calibration drift	8 of 11 survived 6000 hr at 660°C
	Worst case calibration drift 1.1°C (8)
Thermal cycling	4 survived 1000 cycles between 260 and 50°C;
	2 of these survived 1100 cycles between 760 and 80°C
	10 survived 1000 cycles between 650 and 260°C;
	5 of these survived 7000 more cycles; typical R(0) drift after 2000 cycles, +0.1°C
Time response	Average response time, 5.5 sec (15)
	Shortest, 2.2 sec
	Longest, 7.7 sec
Immersion depth	Negligible for 9 of 11 tested;
	2 thermometers exhibited temperature-sensitive lengths of ~30 cm

[a] See Reference 94.

as much as 0.05°C. Mangum attributed the changes in resistance to strain arising from the mismatch between the thermal expansion properties of the resistor and its support, as well as to the effects of moisture in the sensor.

In other work reported during the 6th *Temperature* symposium, McAllan[97] discussed the behavior of several types of industrial RTDs as a result of heating cycles at temperatures as high as 960°C. The R(0) values of 26 thermometers ranged from 10 to 100 Ω. He found that such thermometers, with care, can be used reproducibly in that range within 0.05°C.

Curtis[88] described the effects of thermal stress upon a variety of RTDs. His paper shows many typical construction techniques for industrial thermometers and offers suggestions as to the mechanisms that underlie temperature-cycling drifts in sensor resistance.

Connolly[100] was interested in the drifts in the resistances of industrial RTDs that might accompany their use in the temperature range up to 250°C. As a measure of this drift, he observed the change in R(0) of some 87 RTDs, mostly from 1 manufacturer, after 24 hr at 260°C. He found that more than 70% of the sensors tested drifted by no more than the equivalent of 0.0012°C in this test.

Actis and Crovini[101] compared a variety of fitting techniques for standard 100-Ω RTDs over the range −100 to +420°C. They found that the best sensors could reproduce temperature measurements in the range 0 to 420°C within ±0.01°C when the resistance was fitted by a modified quadratic equation of the type used in the IPTS-68.

Industrial RTDs typically range in R(0) from 25 to 2000 Ω, in length from 1 cm upwards, in diameter from 3 to 9 mm, in number of electrical leads from 2 to 4, and in application temperature from −220 to 850°C. Actual temperature accuracy limits range from ±0.002°C to larger values. Automatic reading, digital displays, and computer compatibility all are readily available. Many manufacturers offer control options with their thermometers.

There are industry-wide standards for RTDs in Germany (DIN 43760, issued by the Deutches Institut für Normen) and in England (BS 1904, issued by the British Standards Institute). The standards[95] promote uniformity in such properties as the R vs. T relations for specific values of R(0) and the minimum ratio for R(100)/R(0). In the U.S., the principal standard is RC21-4-1966, issued by the Scientific Apparatus Makers Association; other industry-wide RTD standards are in preparation within the E-20 Temperature Measurement Committee of the American Society for Testing and Materials.

Many of the manufacturers of RTD instruments acknowledge explicitly that the overall accuracy of the modern temperature-measurement system reflects combined systematic and random deviations from true values both in the sensors and in whatever bridge, potentiometer, power supply, reference resistor, or linearization circuitry may accompany them to produce the observed value of temperature. Often, the advertised accuracy of temperature readings for commerical equipment is restricted deliberately to levels consistent with the poorest expected overall performance of the equipment offered for sale. It should be emphasized, therefore, that the user nearly always can improve upon the performance of the commercial instruments by careful calibration and maintenance of the individual components. The use of fixed points or reference thermometers measured in the setting of one's own laboratory can help to evaluate the accuracy, stability, and other thermometric properties of a thermometry system.

C. Thermistor Thermometers

The thermistor was developed as an inexpensive substitute for platinum in industrial temperature measurement and control applications. Among the desired properties were economy of manufacture, small size, and good temperature sensitivity.

Many chemical compounds have been studied from this point of view since the 1930s.[108] Semiconducting materials were favored because of their strong temperature coefficient of resistivity. Some of the first thermistors were UO_2, $MgTiO_{3-x}$, CuO, Ag_2S, and FeO_x. The electrical resistivity was found to vary widely among these first materials and the hundreds of other compounds tested later. In addition, the temperature coefficients of resistance varied between large negative values and large positive ones. Many of the early applications of thermistors involved their use to compensate temperature-dependent changes in the resistance of copper-based electrical components. In some cases, the onset of a phase change from semiconductor to metal, accompanied by a marked decrease in resistivity with increasing temperature, raised intriguing possibilities for the use of these substances as thermal switches in electrical circuits.

The great variation existing among thermistors in chemical composition, in conduction phenomena, and in electrical resistance makes their overall classification somewhat difficult. In the use of thermistors for careful temperature measurements, however, there is less variation in these properties. Most of these thermistors are prepared from mixtures that are predominantly Mn and Ni oxides, with small amounts of other oxides or glass added to adjust their resistance values. Various techniques are used to form the mixtures into two lead glass-covered beads, discs, or other shapes 1 to 3 mm in diameter.[109]

Thermistor thermometers can be used over the range −150 to +600°C; however, if one restricts the temperature range to about −80 to +300°C, one can obtain thermistor temperature measurements with far better reproducibility. The tendency of early thermistor types to exhibit large drifts in resistance has been overcome by more careful preparation, selection, and handling techniques.[110,111]

A typical thermistor temperature sensor exhibits a negative temperature coefficient[112] such as is shown by the steeper curve in Figure 26. This curve contrasts markedly with the one for platinum, also shown in Figure 26; the approximately linear, positive slope of the platinum resistance-temperature curve (about +0.4%/°C) is small compared with the variable slope of the thermistor curve (as much as 4%/°C). The resistance-temperature relation for thermistors often is described by the Steinhart-Hart equation[113]

$$1/T = A + B(\log R) + C(\log R)^3 \qquad (14)$$

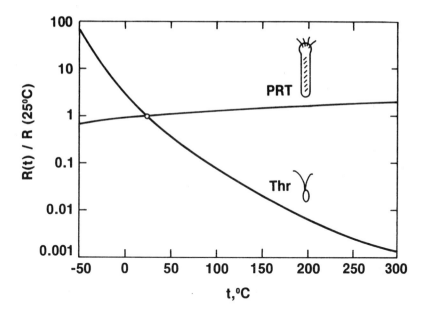

FIGURE 26. Comparison of the resistance-temperature relations of a typical thermistor thermometer (Thr) and a typical platinum resistance thermometer (PRT). Both curves are normalized to unity at 25°C.

in which the coefficients A, B, and C are to be determined by calibration of the thermistor thermometer against a standard thermometer.

In a 2-year study of the stability of thermistors, Wood et al. of the NBS[114] examined more than 400 bead and disc thermistors with R(25°C) values of 2000, 10,000, 15,000, and 30,000 Ω that were supplied by six manufacturers. Various batches were aged at 0, 30, and 60°C for periods of time up to 778 days, with measurements of resistance being taken at gradually increasing time intervals. The disc-type thermistors showed a general drift of resistance with aging time at each temperature; the drifts typically corresponded to -0.1°C/year. The bead-type thermistors, however, showed a remarkable stability — many samples exhibited resistance drifts that corresponded to temperature changes less than 0.001°C/year. Notwithstanding the fact that the sensor temperatures were nearly constant over that time, the stability levels demonstrated in the study showed that thermistor thermometry can be achieved with millidegree precision.

Because of the nature of their construction, the disc-type thermistors generally have provided better interchangeability (i.e., more uniform sample-to-sample R(T) and dR/dT values) than have the bead-type thermistors. In an effort to combine the uniformity of electrical properties typical of the disc-type thermistors with the stability shown in the NBS study by the bead-type thermistors, LaMers et al.[115] prepared special glass-coated disc-type thermistors and compared their stability and aging properties at temperatures up to 250°C with epoxy-coated discs otherwise prepared in the same way. The glass-coated samples showed substantially better stability levels.

Noting the fact that many users employ thermistors for temperature measurements of moderate precision in the 0 to 70°C range (mainly for biological and medical applications), Mangum and Thornton of the NBS[116,116a] characterized the phase-equilibrium temperatures of melting-point cells of gallium (mp = 29.772°C), and Mangum[117] similarly examined reproducibilities of triple-point cells of succinonitrile (tp = 58.0805°C) for use as reference temperatures in that range. Mangum calibrated a set of five bead-type thermistors according to Equation 14, using measurements at the triple point of water, the melting point of gallium,

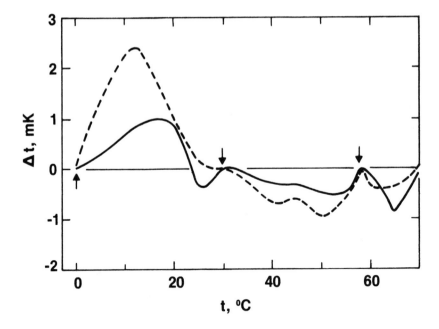

FIGURE 27. Differences between the temperatures of two thermistor thermometers as calculated from the three-constant Steinhart-Hart equation (Equation 14) and the t_{68} temperature of the comparison bath as derived from measurements using a standard PRT. Data from Table 4 of Reference 117.

and the triple point of succinonitrile to evaluate the coefficients A, B, and C. The R(T) values were measured using a 10-µA constant-current supply, a reference resistor, and a 6-digit digital voltmeter.

Comparing R(t) values for the same thermistors at t = 12, 20, 25, 30, 35, 40, 45, 50, 55, 60, 65, and 70°C (determined using a standard PRT and a 400-Hz resistance bridge), Mangum fitted these data with the equation

$$1/T = A + B(\log R) + C(\log R)^2 + D(\log R)^3 \qquad (15)$$

The equation fitted all the calibration data within less than 1 mK.

Reducing the number of calibration points to 3 (i.e., using Equation 14) over the same range noticeably degraded the accuracy with which R(t) in the range 0 to 70°C could be calculated; the deviation of the 3-point Steinhart-Hart equation (Equation 14) for 2 of the 5 thermistors is shown in Figure 27. Nevertheless, the experiment showed that, given a restricted range of temperature, use of a 3-point calibration can enable one to practice thermistor thermometry that is accurate within ±0.005°C. For selected thermistors, the 3-point calibration error may even be reduced to ±0.001°C.

In summary, we see that thermistor thermometry can provide an inexpensive solution to temperature measurement problems in the range −80 to +300°C. Thermistors are very small and relatively inexpensive. Because of their generally high R values the leads resistance usually is negligible; therefore a relatively inexpensive measuring system may be used (viz., a constant-current supply of perhaps 10-µA capacity, a reference resistor, and a high-quality digital voltmeter). Under conditions of restricted temperature cycling and range, their stability and accuracy reach the level of a few millikelvins.

D. Rhodium-Iron Resistance Thermometers

During the past 15 years, there has appeared a new type of resistance thermometer that

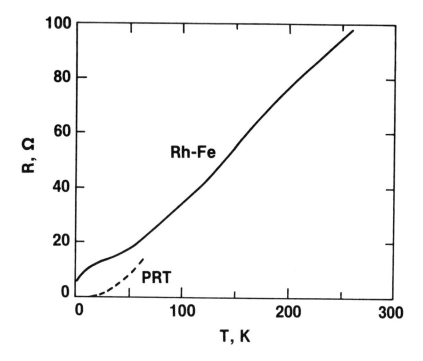

FIGURE 28. Resistances at low temperatures of thermometers with R(273 K) = 100 Ω. Upper curve, Rh + 0.5 at % Fe. Lower curve, Pt.

has markedly improved thermometry in the field of cryogenics. The thermometer was first described by Rusby[118] of the National Physical Laboratory in England. Its development was sparked by the observation[119,120] that the resistivity of iron-doped rhodium retains a strong temperature dependence at temperatures well below 1 K, and thus could extend considerably the useful range provided by platinum resistance thermometers. An alloy composed of 0.5 at % Fe and 99.5 at % Rh proved both sensitive and stable.[121,122] Thermometers made from this alloy now are available commercially.

Figure 28 shows the difference in the low-temperature resistances of a Rh-Fe thermometer and a platinum thermometer, each with R(273 K) = 100 Ω. The Rh-Fe thermometer resistance at 15 K is approximately 11 Ω, a value that is easily measured with an imprecision of only a few ppm. On the other hand, at 15 K the platinum thermometer resistance is about 0.12 Ω; measuring its resistance within a few ppm is difficult. Moreover, the standard capsule PRT resistance at 273 K is not 100 Ω, but only 25 Ω. Temperatures below 20 K can be measured by means of platinum resistance thermometers,[123] but doing so is a demanding task.

In the next section, we discuss briefly the use of doped germanium resistance thermometers in cryogenics. Let us anticipate that discussion by commenting that any given germanium thermometer has neither the range nor (in general) the stability that has been demonstrated by the Rh-Fe thermometer.

The upper curve in Figure 28 shows that the resistance of a Rh-Fe thermometer is not a linear function of temperature. Rusby has recommended that the R-T curve be fitted by a power series

$$T = \sum_{k=0}^{n} a_k (AR + B)^k \tag{16}$$

(cf. Equation 2, in Reference 122) where the coefficients A and B are chosen so that the quantity (AR + B) varies between +1 and −1. Over the range 0.5 to 27 K, Rusby found that a least-squares fitting of Equation 16 at tenth order with about 20 calibration points resulted in a standard deviation of the fit, σ, defined by the equation

$$\sigma^2 = \sum_{i=1}^{m} [T_i(\text{calc}) - T_i(\text{meas})]^2/(m - n - 1) \quad (17)$$

of about 0.2 mK. In Equation 17, m is the number of calibration points and n the order of the fit (the number of terms in the fitting equation). The use of fewer calibration points, or the lack of a calibration point in a particular range, increased the imprecision of the fit correspondingly.[122]

The stability of the Rh-Fe thermometers evidently is excellent. As noted in Equation 16, it is the thermometer resistance, not its resistance ratio, that is utilized in preserving a calibration. This technique places a strong emphasis upon the stability of the R-T relation. Both Rusby[122] and Besley[124] have evaluated the stability of several Rh-Fe thermometers. In examining the stability of three thermometers used in the gas thermometry studies of Berry,[125] Rusby found that, over a period of several years that entailed continued use of the thermometers, no shift was detected in R vs. T beyond the 0.2 mK average level of the measurements. Besley performed a systematic study of 9 Rh-Fe thermometers by cooling them some 30 times from room temperature to either 90 or 6 K. The stability of his cryostat temperature was approximately 0.05 mK; the limit of his resistance precision was about 10^{-5} Ω. Only one of the nine thermometers tested showed a detectable shift in its characteristic resistance — the change at 90 K corresponded to 1.0 mK, and that at 6 K to 0.07 mK. It is true that relatively few Rh-Fe thermometers have been manufactured thus far; nevertheless, the stability of those tested is remarkable.

The Rh-Fe thermometer currently is manufactured in a geometry that is very similar to that of the capsule platinum resistance thermometer — a four-lead coil wound on an insulating former and encapsulated in a ^{4}He atmosphere within a metal sheath of diameter 5 mm and length 3 cm [R(273 K) = 47 Ω] or 5 cm [R(273 K) = 100 Ω]. The thermometer sensitivity is a fairly strong function of temperature below 100 K, ranging for a thermometer of R(273 K) = 100 Ω from 0.15 Ω/K to perhaps four times that value. The thermometer retains considerable sensitivity to temperatures of about 0.1 K. Thermometer self-heating appears to be most noticeable in the range 1 to 2 K, where the helium filling gas condenses, and below 0.5 K. By using 0.1 mA measuring current up to 2 K, however, one can substantially reduce the self-heating there.[122] Typically, measuring current values of 0.1 to 0.3 mA are used below 20 K.

The excellent stability, extended range, and relatively large signal voltage provided by the Rh-Fe thermometer make it a logical choice for the most precise thermometry in the range from about 27 to 0.5 K. Even below the latter temperature, thermometry using Rh-Fe has been accomplished.[120,126] Because the R-T relation is not linear, there appears to be a need for several calibration points in order to define accurately a particular temperature scale; however, given a calibrated thermometer, the user apparently can rely on its indications within a fraction of a millikelvin.

E. Semiconducting Thermometers in Cryogenics

In previous sections, we have discussed the use of high-purity platinum thermometers and rhodium-iron thermometers for precise thermometry in cryogenics. There are many other types of resistance thermometers that are used at low temperatures when some factor other than the highest possible precision is paramount. In this section, we call attention to the existence of these thermometers without extensive discussion.

There have been several summaries of thermometry methods in cryogenics; resistance thermometers generally have not been singled out for special treatment.[127-133] In the references noted, however, one can find information on doped germanium and other semiconductor thermometers, on compacted-carbon and carbon-glass thermometers.

The use of doped germanium resistors in thermometry dates from the investigations of Estermann et al.[134,134a] and Gerritson[135] just after World War II. A good discussion of the physics of the temperature dependence of semiconducting compounds is given by Friedberg in the 3rd *Temperature* symposium.[136] Kunzler et al. have provided a good introduction to the construction of these small (typically no larger than about 5 mm diameter and 10 mm long), four-lead thermometers.[137]

The low-temperature resistance of a germanium thermometer varies inversely with temperature, but in a complicated way. Furthermore, no two germanium thermometers follow the same R-T relation. Typically, a 1-decade decrease in temperature is accompanied by a 2-decade increase in the thermometer resistance; thus it is not easy to use a single thermometer over a range of temperature that spans more than about 1 decade. On the other hand, it is possible to achieve microkelvin temperature sensitivity with the use of a germanium thermometer.

Until the advent of the Rh-Fe thermometer, the germanium thermometer was used to maintain national scales of temperature below 30 K.[138] With their help, different laboratories could compare their scales at the millikelvin level.

In recent years, however, the stability of germanium thermometers for measurements at the millikelvin level has come into question.[139-141] For the most part, extensive testing has indicated that most of the thermometers are stable within a fraction of a millikelvin. However, shifts in low-temperature resistance corresponding to temperature changes as large as 20 mK accompanied repeated cycling of some thermometers to room temperature. No superiority of either n- or p-type material could be established.

The small size and generally high quality of the germanium thermometers makes them attractive for all applications but those most demanding of reproducibility.

Swinehart[142] has outlined a new technique for producing germanium thermometers in a planar geometry by impurity diffusion. He has found that this method produces thermometers with more uniform R-T characteristics than those shown by thermometers cut from boules; he has also noted that the R-T curve is less steep than for bulk resistors.

The forward voltage of a semiconductor diode biased by a constant current has been used for some years to measure temperatures that range from above 300 K to below 1 K. The temperature dependence of the diode voltage typically shows two distinct regions that are characterized by quite different sensitivities. The high-temperature intrinsic-conductivity[143] region, in which the temperature dependence is approximately linear, gives way in the range 100 to 25 K to the extrinsic conductivity regime that provides a more nearly exponential low-temperature dependence. Both regions can be utilized in thermometry.

Good discussions of the techniques involved in diode thermometry have been provided in several papers.[144-146] It is not always appreciated by users of diode thermometers that the accuracy and the reproducibility of the temperatures derived therefrom depend as critically upon the care with which the biasing current is controlled as upon the accuracy of the voltage measurement. In terms of our earlier remarks about the use of the word "thermometer" where we should use the word "sensor," the reader is advised to use special care in transforming a "diode sensor" into a "diode thermometer." With the use of very careful techniques, diode thermometry can provide temperatures that are accurate within a few millikelvins.[145]

Ohte et al.,[147] building upon the discussion of silicon transistor thermometers by Verster,[148] have suggested that temperature reproducibility within $\pm 0.1°C$ in the range -50 to $+200°C$ can be obtained with only a 1-point calibration.

One of the first empirical thermometers used in cryogenics was the carbon radio resistor.[149] Possessing the attractive properties of high sensitivity and low cost, they have continued to provide rough and ready thermometry since their discovery. The temperature reproducibility of the carbon resistor thermometer after cycling to room temperature is not particularly good — Lindenfeld[150] suggests that thermometers that are handled carefully while at room temperature may reproduce 4 K within ±0.004 K, while Anderson[151] would limit the effectiveness of this procedure to about ±0.01 K at 1 K. An equally important characteristic is the stability at a fixed low temperature. Drifts in resistance corresponding to temperature shifts of more than 0.002 K have been detected over a period of hours in carbon resistors.[152]

Clement and Quinnell[149] suggested that the R-T relation of the carbon thermometer could be fitted satisfactorily by the equation

$$\log R + [K/(\log R)] = A + BT \tag{18}$$

Other formulations also are used for the resistors typical of various manufacturers.[133]

The advent of metal film resistors for use in electrical circuits has interrupted the manufacture of carbon-composition resistors. Rubin, however, points out a number of sources of both the better-known and some newer resistors.[133]

Resistance thermometers made by impregnating porous glass with carbon-containing material have been available for some time.[153] These units appear to be stable at fixed temperatures and to provide greater temperature sensitivity than the carbon composition thermometers, but their reproducibility on cycling from room temperature seems not as good as that of the germanium thermometers. On the other hand, in the presence of high magnetic fields, which strongly modify the resistances of Rh-Fe and germanium thermometers, carbon-glass thermometers show only small, isotropic variations in resistance.[154] A recent study of carbon-glass thermometers by Ricketson and Grinter is directed toward the development of more precise fitting of their calibration curves.[155] Another recent paper describes similar thermometers made in China.[156]

V. RADIATION THERMOMETERS

Radiation thermometry is based upon the detection of energy that is emitted by every substance in consequence of its temperature. The character of that radiated energy is by no means the same for all substances, however.

We learned in Chapter 2 that the emission of radiant energy is governed by the laws of quantum mechanics — a substance radiates by virtue of its relaxation from one discrete energy level to another of lower energy. In order to evaluate the temperature of a given substance by analyzing its radiation, then, we must determine the makeup of its energy levels and which of these might be occupied at a given temperature.

Radiation usually is characterized by its frequency or its wavelength rather than its energy, simply because radiation detectors usually are characterized by their wavelength or frequency responses. Figure 29 shows a portion of the electromagnetic spectrum, expressed in terms of wavelength in meters. Even though the range that is depicted covers 16 decades, the figure does not portray all of the available spectrum.

From the point of view of thermometry, however, the range shown in Figure 29 is sufficient. All of the systems of practical interest in temperature measurement radiate most strongly within the range shown. The spectrum of energy levels in molecular gases reflects the energies involved in atom-atom vibration and in molecular rotation, which occur typically in the infrared. As a gas is heated or otherwise excited to the point that any molecular bonds are broken and some or all of its electrons dissociate from its nuclei, its characteristic radiation lies in the ultraviolet and X-ray ranges; such energetic situations exist normally in

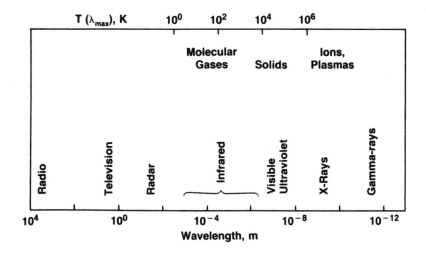

FIGURE 29. Schematic description of the electromagnetic spectrum. Typical wavelengths associated with excitations in various substances are indicated. The Wien-law temperatures corresponding to particular wavelengths are shown at the top of the figure.

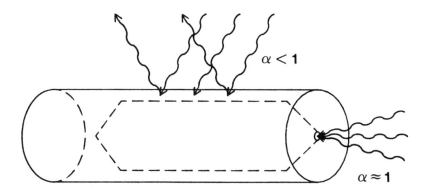

FIGURE 30. Construction of a black body from nonblack materials.

the sun and in other stars and can be created in the laboratory in electric discharges or in nuclear fission or fusion experiments.

A. Radiation Properties of a Black Body

A broad range of characteristic wavelengths can be found in solids as a result of the availability of vibrational modes of excitation that arise in solid lattices.[157] The presence of conduction electrons in metals provides an even greater range of interaction frequencies between metallic solids and electromagnetic radiation. In studying the radiative properties of matter, scientists a century ago found that blocks of different materials emit quite different amounts of radiant energy, even when they are held at the same temperature. Following the observation that the centermost surfaces of a group of hot objects appear substantially brighter, they prepared "cavity radiators" such as the one that is shown in Figure 30.

In this figure is illustrated an essential difference between the *outside* surface of an opaque material and the *inside* surface. If the material is an ordinary metal, more than half of any visible or infrared radiation that strikes the outer surface will be reflected; on the other hand, virtually all the radiation entering the cavity through its tiny port will be absorbed, generally after many internal reflections. The quantity α, the radiative absorptance, provides a measure

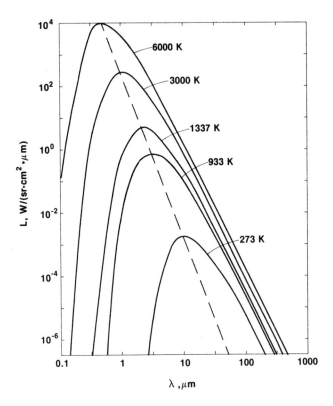

FIGURE 31. Spectral radiances of a black body at five different temperatures. The dashed line shows the wavelength dependence of the radiance maxima as predicted by the Wien distribution law.

of the absorptive power of various materials (see below). Such a cavity can be described as "black" in the sense of being a perfect absorber of radiation.

It was found that, if they are held at the same temperature, cavity radiators that are carefully prepared from different opaque materials appear equally bright when viewed through the entrance aperture. One can suppose that the radiation field inside the cavity comes to full equilibrium with the interior walls of the cavity, whereas on the outer surface it does not.[158]

Well-made cavity radiators have come to be known as "black bodies." When a black body is held at a particular temperature, its spectral radiance, that is, its radiant energy as a function of wavelength, is always the same. In Figure 31, we show the spectral radiance curves for a black body at each of five temperatures. There are several striking features to this figure. One notices first the enormous variation in peak radiance with the temperature of the black body; at 6000 K, roughly the temperature of the sun's surface, a black body radiates energy at about 10 million times the maximum rate that is found at 273 K. Any device that is to detect radiation from sources at both temperatures must be extremely versatile in its response. The second remarkable feature of the curves in Figure 31 is that the spectral radiance curves, plotted in this fashion, all possess the same shape; a single "pattern" can be used to trace all of them. Finally, one notes a surprising regularity in the wavelengths at which the maxima in the spectral radiance curves occur; this regularity exists because of a special relation between the peak-radiance wavelength λ_{max} and the black body temperature:

$$\lambda_{max} T = 2898 \; \mu m \cdot K \qquad (19)$$

This relation is derived from the Wien displacement law.[159]

As we noted in Chapter 2, Planck succeeded in representing the black body spectral-radiance curves by means of an equation[160,161]

$$L(\lambda,T) = \frac{c_1}{\lambda^5[e^{-(c_2/\lambda T)} - 1]} \quad (20)$$

The Planck equation can be expressed in several ways. As given above, the spectral radiance is commonly expressed in units of watts/[steradian · meter^2(area) · meter (wavelength)]. The first and second radiation constants, c_1 and c_2 respectively, have the values

$$c_1 = 1.191\ 0621 \pm 0.000\ 0064 \times 10^{-16}\ \text{W} \cdot \text{m}^2/\text{sr}$$

$$c_2 = 1.438\ 786 \pm 0.000\ 045 \times 10^{-2}\ \text{m} \cdot \text{K}$$

Although originally suggested on empirical grounds, we recall that the equation soon was given a theoretical basis by Planck by adopting two assumptions. One assumption is that the radiation from a black body is emitted by atomic oscillators whose energy levels are quantized with respect to a characteristic frequency according to the relation

$$E = (n + 1/2)h\nu \quad (21)$$

where E is the energy of the nth quantum state, n is an integral index of the quantum state, h is a natural constant, and ν is the frequency. The second assumption made by Planck is that the black body radiation itself is emitted in quantized amounts that correspond to a change by the atom from a higher-energy quantum state to a lower one.

The success of Planck's ideas, as we know, marked the beginning of the quantum theory of matter. Planck's constant, h, which relates the energy of the atomic radiators given by Equation 21 to their radiance in Equation 20, appears in the radiance equation through the constants c_1 and c_2, equal respectively to $2c^2h$ and to hc/k. In SI units, Planck's constant has the value[161]

$$h = 6.626\ 176 \pm 0.000\ 036 \times 10^{-34}\ \text{J} \cdot \text{sec}$$

Planck's success also implies that an accurate measurement of the spectral radiance of a black body will allow the accurate evaluation of its temperature by the use of Equation 20.

Planck's equation can be integrated over all wavelengths to produce the Stefan-Boltzmann equation

$$P = \sigma T^4 \quad (22)$$

which shows that the total radiance from a black body is determined by its thermodynamic temperature. The Stefan-Boltzmann constant, σ, is

$$\sigma = 5.670\ 32 \pm 0.000\ 71 \times 10^{-8}\ \text{W}/(\text{m}^2 \cdot \text{K}^4)$$

Note that its uncertainty (~125 ppm) is rather large in comparison with those of other natural constants.[161,161a]

Differentiating Planck's equation allows evaluation of the wavelength at which the spectral radiance is a maximum. This result provides an expression similar to that derived earlier on classical grounds by Wien (Equation 19).

We see that the existence of Planck's equation permits the evaluation of the thermodynamic temperature of a black body by several methods that appear quite feasible. It should be obvious, too, that one can evaluate ratios of thermodynamic temperatures by the use of either the Planck law or the Stefan-Boltzmann law; one needs only a detector that responds to different levels of radiation in a well-known fashion. We have presented already (in Chapter 4, Section II.B and Chapter 5, Section VI) brief discussions of radiation thermometry experiments involving the determination of black body radiances.

B. Radiation Properties of Real Materials

If radiation of wavelength λ is projected onto a slab of material, the general result is that a fraction of the radiation is reflected, another fraction is absorbed, and the remainder is transmitted. Indicating the reflectance by the symbol (ρ), the absorptance by (α), and the transmittance by (τ), we see that

$$(\rho) + (\alpha) + (\tau) = 1 \tag{23}$$

In general, the ability of a material to reflect radiation depends upon the wavelength of the radiation and upon the material's index of refraction. In addition, it depends strongly upon the condition of its surface; a highly polished surface reflects radiation more effectively than does a rough one. A material tends to emit radiation better in a wavelength region where it absorbs strongly.

The intrinsic ability of a material to absorb radiation is measured by the Lambert law[162] $I = (I_0 - I_\rho) \exp(-\alpha_L d)$, where $(I_0 - I_\rho)$ is the intensity of the radiation within the material (i.e., after any reflection has already taken place). The coefficient α_L clearly is different from the quantity (α) that appears in Equation 23.

The radiation-absorbing properties of liquids and solids are quite different from those of gases, as we mentioned above; gases absorb primarily those wavelengths that correspond to characteristic vibrational and rotational excitations. Among solids, too, there is great variation in absorptive power. Metals tend both to reflect and to absorb strongly throughout the infrared and visible ranges of the spectrum, while nonmetals often transmit radiation in extended wavelength regions.[163]

According to the law of Kirchhoff,[164] the directional spectral absorptance $\alpha_\lambda(\theta,\phi)$ of a substance is everywhere equal to its directional spectral emittance

$$\alpha_\lambda(\theta,\phi) = \epsilon_\lambda(\theta,\phi) \tag{24}$$

providing only that thermal equilibrium exists. Here θ and ϕ define the polar coordinates of the radiation direction.

Using Equation 24, the definition of a black body is simply $\alpha = \epsilon = 1$. It follows necessarily from Equation 23 that $\rho = \tau = 0$ for a black body.

The radiative properties of any real material, $L_{\lambda T}$, can be evaluated by comparison with a black body radiator, producing values of the spectral emittance $\epsilon_{\lambda T} = L_{\lambda T}/L_{\lambda BT}$, where $L_{\lambda BT}$ is the spectral radiance of a black body at wavelength λ and temperature T as given by Equation 20.

As it happens, there is no real substance for which $\epsilon = 1$. Only by preparing a material in the form of a cavity radiator can a practical black body be created. (Even for a cavity radiator, the emissivity only approaches unity in proportion to the ratio of its aperture area to the total area of the interior cavity.) If the emissivity of a material ϵ_λ is constant over all wavelengths, then it is said to be a "gray body." For most substances, the emissivity depends strongly on both wavelength and temperature, as well as on other characteristics of the radiation. Typical emissivities for a variety of solid materials are given in Table 9.

Table 9
NORMAL TOTAL EMISSIVITIES OF SOLID SUBSTANCES[165]

Metal	T(K)	ϵ_n	Wavelength (μm)
Metal			
Aluminum, polished	800	0.05	
Aluminum, oxidized	800	0.33	
Brass, polished	600	0.03	
Brass, oxidized	600	0.6	
Copper, polished	300	0.02	
Copper, black oxidized	300	0.78	
Iron, polished	600	0.12	
Iron, cast rough, oxidized	600	0.95	
Steel, polished	300	0.09	
Steel, rough oxidized	300	0.81	
Tungsten, filament	300	0.03	
Nonmetal			
Alumina	800	0.65	
Brick, white refractory	1300	0.29	
Brick, rough red	300	0.93	
Carbon, lampblack	300	0.95	
Ice	273	0.97	
Paint, oil	373	0.94	
Paint, flat black	373	0.98	
Water, ocean	300	0.96	
Tungsten, ribbon	2000		
		0.437	0.25
		0.474	0.30
		0.475	0.40
		0.462	0.50
		0.448	0.60
		0.436	0.70
		0.382	1.0
		0.288	1.5
		0.227	2.0

We already have mentioned that gases generally do not exhibit radiation patterns that approach black body radiation. Their energy level schemes include a finite number of lines and bands that represent molecular rotation and vibration and also atomic excitation.[166] We treat the determination of their temperatures separately.

C. General Features of Radiation Thermometers

Radiation thermometers that are used to determine the temperature of a test substance by analyzing the radiation emitted by it can be called "passive" devices. Passive radiation thermometers in general are comprised of windows, lenses, mirrors, light guides, and prisms, whose function it is to focus the radiation on the various parts of the instrument as required; filters, sectored discs, and "gray" wedges to adjust the amount of radiation admitted to the instrument; reference sources to provide *in situ* calibration of the instrument; and detectors to determine the amount of radiation or to compare the radiation from two sources.

Radiation thermometers in which a probe beam is used to determine the relative populations of the energy levels of gaseous test substances can be called "active" devices. These thermometers generally incorporate some or all of the features of passive instruments in addition to the probe beam, which often is one or more laser sources.

Bearing in mind our discussion of the radiation properties of solids, one must realize that each of the components that must pass the radiation from the test body necessarily will

Table 10
PROPERTIES OF RADIATION DETECTORS

Detector	M (W)	Lc(μm)
CdS photoconductor	10^{-16}	0.8
PbS photoconductor	10^{-10}	3.0
PbTe photoconductor (90 K)	10^{-10}	6.
InSb photoconductor	10^{-9}	8.
S20 photomultiplier tube	10^{-15}	0.7
S4 photomultiplier tube	10^{-15}	0.6
Si photodiode	10^{-13}	1.1
Thermal detector	10^{-10}	—
Human eye	10^{-14}	0.7

Note: M = Minimum detectable power level in watts; Lc = longest useful wavelength in micrometers.

weaken and distort that radiation through the mechanisms of reflection and absorption. Indeed, merely passing the radiation through air will alter its properties.[167]

Reference sources include cavity radiators and electrical lamps of various geometries.

Detectors include the human eye, photoconductors, photomultiplier tubes, phototransistors, photodiodes, and thermal detectors. Table 10 shows the properties of some of these.[168]

A great many types of radiation thermometers are available, ranging from one-of-a-kind research units to relatively simple and inexpensive commercial devices. Radiation thermometers can be categorized by their purposes as well as by their components. In the following discussion we present brief notes on many of the available devices.

1. Radiation Thermometers for Thermometry Research

In research on the Kelvin thermodynamic temperature scale, one can find in use both total radiation thermometers and spectral radiance thermometers. In both cases, the source of the radiation is a cavity radiator designed to emit radiation that is nearly "black" (ε = 1). We described both types of thermometer in Chapter 5, Section VI; we only summarize that discussion here to emphasize the types of components in use.

The best present example of a total radiation thermometer is the instrument of Quinn and Martin,[169] which is based upon an earlier development by Ginnings and Reilly[170] (see Figure 17, Chapter 5). In this device, the cavity radiator emissivity is estimated as 0.99992. Its temperature is measured on the IPTS-68 by means of eight platinum resistance thermometers mounted in various positions on its walls.

By first measuring the thermal radiation emitted by the cavity at T = 273.16 K, the principal defining temperature of the KTTS, and then measuring the radiation emitted by the cavity at test temperatures ranging from 240 to 390 K, Quinn and Martin could obtain the ratio of the two temperatures by use of Equation 22:

$$T = 273.16 \, [P'(T)/P'(273.16)]^{1/4} \tag{25}$$

where P' is a quantity that is proportional to the total power radiated from the cavity. Use of the ratio technique avoids the necessity of accurately measuring the total hemispherical radiation. It also avoids the uncertainty inherent in the Stefan-Boltzmann constant.

The detector in the Quinn-Martin experiment is itself a cavity that is operated as a calorimeter. Initially held at 2 K by weak thermal contact with a pumped ^{4}He bath, the cavity is warmed slightly by the radiation from the test cavity upon the opening of a shutter. After equilibrium conditions have been achieved, the shutter is closed and the same temperature rise is achieved by substituting a carefully measured rate of electrical heating of the detector cavity. The electrical power then is used as P' in Equation 25 to calculate T.

Their discussion of measurement uncertainties indicates that the instrument of Quinn and Martin is accurate within ±9 mK (99% confidence) at 365 K, about 20 ppm.

We also have described (in Chapter 5, Section VI.B) the spectral radiation thermometers built by Jung[171] and by Bonhoure[172] in which the spectral radiance of a cavity radiator at a wavelength λ and test temperature T is compared with the same quantity at a reference temperature T_0, using Equation 20:[173,174]

$$\frac{L(\lambda,T)}{L(\lambda,T_0)} = \frac{\epsilon(\lambda,T)}{\epsilon(\lambda,T_0)} \frac{[e^{-(c_2/\lambda T_0)} - 1]}{[e^{-(c_2/\lambda T)} - 1]} \qquad (26)$$

In the experiment of Jung, a set of four calibrated platinum resistance thermometers was used to adjust the temperature of a cavity radiator to a known value of T_{68}. A second cavity radiator was immersed in an Al freezing-point cell. The radiation from this cavity served as a "standard lamp" for automatic calibration of the detector system.

The radiation thermometer contained a multicavity interference filter that provided a very stable instrument wavelength at approximately 974 nm. Two detectors were used; both were commercial, highly linear silicon photovoltaic units. A rotating mirror alternately directed radiation from the two black body cavities into the filter-detector system.

Using as the reference temperature Guildner and Edsinger's value[175] of T_{68} that corresponds to 729.15 K, Jung measured the thermodynamic temperature corresponding to T_{68} in the range 650 to 900 K. His estimate of the total uncertainty of his temperature measurement is ±0.02 K at 99% confidence. This low uncertainty level was achieved in part by the use of the second reference black body that was held at the aluminum point, but it also reflects the very careful techniques and instrument calibrations accomplished by Jung.

An interesting attempt to utilize for precision thermometry the unique capability of optical fibers to carry radiation with negligible loss has begun at the NBS.[175a] The technique involves the deposition of an opaque coating of refractory metal, such as iridium, onto the end of a single-crystal sapphire fiber as long as $1/2$ m. According to preliminary measurements by Dils, Reilly, and Geist, such a device can provide radiation that is nearly black in character. Use of an appropriate filter and a linear photodiode detector then can permit the determination of temperature ratios from the Planck law in the usual way. The advantages of this technique over more familiar ones discussed previously are its small size and freedom from certain aperture corrections.

2. Commercial Radiation Thermometers

Commercial radiation thermometers range from very simple devices that project the radiation from a hot target onto a blackened thermopile to sophisticated instruments that automatically compare the single- or two-wavelength radiation level from the target with that emitted by a self-contained reference lamp. The principles behind the use of these various techniques are discussed by Smith, Jones, and Chasmar,[176] Worthing and Halliday,[177] and Campbell.[178]

Figure 32 shows in schematic form a simple radiometer. Thermal radiation enters the instrument through a lens L and is focused by a mirror M onto a blackened thermopile T. The voltage output of the heated thermopile then is a measure of the source temperature. The response time of this type of radiometer generally ranges from a fraction of a second to several seconds. The accuracy is no better than 1% of full scale.

The instrument can be modified by incorporation of a filter to select a spectral range that avoids the absorption bands of carbon dioxide and water vapor, and by the use of a photovoltaic detector. Such improvements can reduce measurement errors by as much as a factor of three.[179]

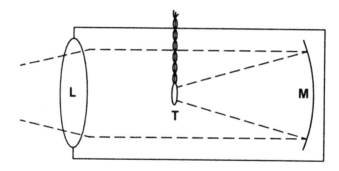

FIGURE 32. Schematic view of a simple radiometer incorporating an entrance lens L, a focusing mirror M, and a thermopile detector T.

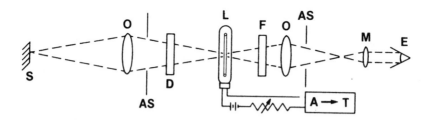

FIGURE 33. The disappearing-filament optical pyrometer. The components are: S, source; O, objective lenses; AS, aperture stops; D, optional gray-wedge or sectored disc; L, reference lamp; F, filter; M, microscope objective; E, observer's eye. The ammeter current A generally is calibrated to read in terms of the temperature T.

In Figure 33 we show schematically the "disappearing filament optical pyrometer." In this type of radiation thermometer, the radiance of a calibrated lamp L is compared visually with that of the source S. The image of the source is focused on the filament of the reference lamp and then the lamp current is varied until the filament no longer can be distinguished from the source. The filter used with this type of instrument commonly admits radiation of fairly wide bandwidth; the observer's eye-response also enters directly into the instrument sensitivity.[179]

The accuracy of the disappearing-filament radiation thermometer depends upon several factors, particularly the stability of the reference lamp, the careful selection and placement of the apertures and filters, and the emissivity of the object whose temperature is being measured. At best, the temperature precision in experienced hands has been found to be about ±0.3°C in the range 800 to 2200°C.[179] Commercial instruments generally are rated at no better than ±5°C accuracy under black body conditions (source emissivity equal to 1); for nonblack sources, the temperature error must be calculated from a relation of the type

$$1/T_s = (1/T_b) + (\lambda/c_2) \ln \epsilon \qquad (27)$$

where T_b is the actual temperature of the source (expressed in kelvins), T_s is the indicated temperature (also in kelvins), λ the wavelength, c_2 the second radiation constant, and ϵ the source emissivity. For example, the temperature of a source whose emissivity is 0.3 might be measured at a wavelength of 0.65 μm as 1500°C; its actual temperature, however, is some 190°C higher, an error of more than 10%.

In two-color pyrometry, applying the same principles mentioned above, the correct temperature of a nonblack object can be evaluated by use of the relation

Table 11
CHARACTERISTICS OF RADIATION DETECTORS[184]

Material	D(cm Hz$^{1/2}$/W)	τ# (μsec)	Δλ* (μm)
Si	4 × 10^{12}	~1	0.2—1
PbS	4 × 10^{10}	150—500	1—3
InAs	10^9	~2	1—3.6
InSb	4 × 10^7	<1	1—7
Thermistor bolometer	10^8	1000—1600	1.5—>12
Thermopile	10^9	~4000	1.5—>12

Note: # = Time constant of the detector, governing the maximum chopping frequency; * = wavelength range over which the figure of merit equals or exceeds the value stated.

$$1/T_s = (1/T_b) + \frac{1}{c_2}\left[\frac{\lambda_1 \lambda_2}{\lambda_2 - \lambda_1}\right] \ln\left[\frac{\epsilon_{\lambda_1}}{\epsilon_{\lambda_2}}\right] \tag{28}$$

Note that the correction term in Equation 28 disappears if the radiation source is "gray" (emissivity constant for all wavelengths), showing the power of two-color pyrometry for gray or near-gray targets.

The problem of nonblack emissivity in radiation thermometry has received considerable attention in recent years. Studies have been made on cavity radiators as well as on industrial targets. Ono[180] has presented evidence that an increase in the specularity of certain cavities can enhance their emissivities, as can the use of a conical bottom of angle 120°. The emissivities of metal tube cavities with apertures cut into the side were found to vary from 0.92 to 0.999, depending upon wavelength and size and placement of the aperture.[181] Beynon[182] gives an example of the use of radiation thermometers in an environment where the emissivity of the target, while considerably less than unity, is reasonably well known. A method for measuring the target emissivity during the time of temperature measurement has been outlined by Iuchi and Kusaka.[183]

By replacing the human eye with a different detector, one can extend the range and improve the precision of a radiation thermometer. There are many instruments that have been modified in this way. A comparison of the radiation from the source and the reference lamp generally is achieved in newer instruments by automatic alternation of the two beams. A wide range of filters, optical elements, and detectors has been incorporated to improve accuracy in temperature measurement in a variety of environments that in no way resemble laboratory conditions.

Warnke[184] has analyzed several detectors from the point of view of a characteristic figure of merit that is a function of wavelength and of beam chopping frequency:

$$D^2(\lambda, f) = \frac{A\Delta f}{N_\lambda^2} \left(\frac{cm^2 Hz}{W^2}\right) \tag{29}$$

where D is the figure of merit, A is the detector area in cm^2, Δf is the electrical bandwidth of the instrument in Hz, and N_λ, in watts, is the so-called "spectral noise equivalent power," that value of monochromatic incident radiation of wavelength λ that is required to produce an rms signal-to-noise ratio of unity. In Table 11, we list some of his findings.

Radiation thermometers that incorporate an imaging feature have been in use for some time. Such "thermal imagers" have obvious utility in finding defects in electronic components, "hot spots" in the walls of industrial ovens, or the locations of warm objects in an otherwise cool environment. The first examples of imaging thermometers included a television camera that was adapted to direct the image of a scene upon an array of nearly 500,000 silicon diodes, with the individual voltages being displayed by the vertical and horizontal sweep circuits of the camera approximately 30 times each second;[185] an infrared camera that could record the thermally discriminated image of a scene on a photographic film or present it visually;[186] a microscope equipped with an image-scanning mirror and a then newly developed InSb detector only 0.012 cm in size, which is capable of recording the scanned image once each second on the screen of an oscilloscope;[187] and a similar system for macroscopic scenes in which the detector data were analyzed and recorded by a dedicated computer.[188] The available stability, temperature range, temperature resolution, and temperature accuracy were very limited in these first devices; temperature sensitivities of 0.1 to 2.5°C and a temperature accuracy of ±5 K were reported. A summary of thermal imaging techniques was presented in 1975 by Agerskans.[189]

Wallace and Cade[190] have written a book that discusses thoroughly the use of infrared imaging in the diagnosis of surface tumors in humans. The basis for this type of thermometry is the physiological fact that malignant tissue generally exhibits a higher metabolic rate than normal tissue; thus surface tumors can be observed in thermography provided that the temperature sensitivity is adequate. As a rule, the presence of near-surface malignancy raises the skin temperature over the affected area by 1 to 3°C. The authors note that the emissivity of the human skin is very nearly unity for thermal radiation, so that the human acts approximately as a black body radiator. Typical temperature resolution in clinical thermography is somewhat better than 0.2°C. This type of measurement has been extended to the detection of deeper body tumors by the observation of microwave radiation at wavelengths of 1 and 3 cm. Edrich and Jobe[191] have shown that such radiation can be imaged in a manner that is similar to infrared techniques; the longer wavelength, however, is generated deeper in the body, thus revealing the presence of "hot" tissue that might escape ordinary thermal imaging.

A study of the utility of infrared imaging for energy conservation was performed recently by staff members of the NBS. This work was described by Hurley and Kreider.[192] Using a commercial thermography unit, the NBS group surveyed typical industrial installations for the presence of heat leaks in energy-intensive equipment. The unit employed a cooled detector of InSb, operating in the wavelength range 2 to 5.6 μm. A mechanical scanner swept 100 to 500 horizontal lines over the detector at a frequency as high as 25 Hz. The operating temperature sensitivity depended upon the range of temperatures selected in a particular sequence. The temperature profile of a particular object under study generally was established by including a contact thermometer on the surface being examined; this technique provided an *in situ* calibration.

There are several interesting radiation thermometers with application to measurements near room temperature and in the presence of electromagnetic radiation. Typically, these thermometers are used in conjunction with the medical application of microwave or other high-frequency radiation to human tissue. A clear indication of the temperature rise that accompanies the tissue irradiation is essential to the proper control of the radiation intensity; the use of ordinary metallic or semiconductive sensors in this application would be fruitless, since the high-frequency radiation would heat the metal components preferentially. One of these is the so-called "fluoroptic thermometer."[193] This thermometer makes use of the temperature-dependent ratio of the intensity of two spectral lines to provide values of temperature in the range −65 to 240°C. The two lines are among many that are emitted in fluorescence from a rare-earth phosphor located at the end of an optical fiber in response to ultraviolet exciting radiation that is transmitted from a lamp contained in the instrument control unit. The fluorescence radiation is filtered, the two desired lines are detected and amplified, and the resulting line intensities are analyzed by a microprocessor in the control unit. The temperature precision of the thermometer is about 0.1°C, and its accuracy varies from 0.5 to 1°C, depending upon range. Its advantages over commoner methods of thermometry include the immunity to electromagnetic radiation already mentioned, small size of the sensor, and relatively fast thermal response (a few seconds).

An experimental type of sensor that appears to be immune to microwave interference depends upon the change with temperature of the time constant for fluorescent decay of the red "R" lines of ruby.[194] The time constant in the ruby sample that was tested (0.05% Cr in Al_2O_3) varied approximately linearly from about 3.49 msec at 35°C to about 3.38 msec at 46°C; the magnitude of the time constant depends upon the concentration of the Cr impurity. By exciting the ruby sample with a pulse of blue-green radiation from a tungsten lamp and passing the returning fluorescence radiation through a red filter, Sholes and Small obtained a signal that could be time-averaged automatically to produce a measure of the time constant. Individual measurements agreed with the overall data curve within ±0.3°C.

A radiation thermometry technique that is in use currently with biomedical microwave installations depends upon the shift of the infrared absorption edge of GaAs with temperature.[195] In this technique, two tiny (0.25 mm diameter) plastic optical fibers transport narrow-band radiation of 0.9 nm wavelength to and from a thin GaAs sensor. The fraction of the light that is transmitted by the sensor decreases markedly as the sensor temperature is increased from 20 to 55°C. The light source itself is a GaAs light-emitting diode; the detector is a silicon photodiode. In use, the temperature dependence of the photodetector signal can be calibrated automatically by a microprocessor-controlled system that refers all temperature measurements to three calibrated platinum resistance thermometers contained in a temperature-controlled calibration well. The sensor reflects temperature accurately within ±0.1 to ±0.3°C, depending upon the time interval since the previous calibration cycle. Because of the small size of the sensors, as many as four of them can be used simultaneously within a 1.2-mm-diameter hypodermic needle in order to obtain temperature profiles during microwave irradiation of tissue.

Other microwave-compatible thermometers have been described by Cetas and Connor[196] and by Wickersheim and Alves.[197]

D. Temperature Measurement in Gases, Flames, Plasmas, and Stars

The measurement of temperature in hot gases, in flames, in plasmas, and in stars has a long history. At the "cooler" end of this range of temperatures, contact methods can be used for thermometry; such measurements[198] have been made up to about 3000°C. However, the growing uncertainties associated with thermometry at ever higher temperatures and the lack of dependable materials at extended temperatures have led increasingly to the use of radiation methods in these systems. Of course, the distances involved in astrophysical thermometry still restrict those measurements to radiation methods.

The study of stellar temperatures generally encompasses measurements of the continuous spectrum, observations of line and band spectra, and determinations based upon kinetic theory.[199] Each of these types of measurement can yield values of temperature in particular cases — often, however, the results have been found to be incompatible, even when the expected variation of the stellar temperature with depth have been incorporated into the analysis.

The spectral distribution of radiation from the surface of a star can be treated using the Planck equation and Kirchhoff's relation. This technique is complicated by the wavelength-dependent emissivity of the stellar surface and by the absorptive properties of the intervening earth's atmosphere. Surface temperatures of stars obtained in this way range from a few thousands of kelvins to a few tens of thousands.[200]

Temperature information can be obtained from measurements of both line and band spectra of stars. Usually these spectra are seen as absorption phenomena in a continuous radiation pattern from the star, although in the study of solar flares, for example, emission spectra extend into the X-ray range.[201] The study of stellar temperatures by means of the line and band spectra involves the evaluation of the temperature dependences of excitation and ionization. Excitation temperatures may be derived from the Boltzmann equation

$$N_u/N_\ell = (g_u/g_\ell) \exp(-h\nu/kT) \tag{30}$$

where the N are the relative numbers of atoms in an upper and a lower quantum level, the g are the respective statistical weights of the two levels, $h\nu$ is the energy separation between the two levels, and kT is a measure of the thermal energy of the system. The determination of the ratio of the N values from L_u/L_ℓ, the ratio of the spectral line intensities, usually is not straightforward. At very low atom densities, the ratio of the line intensities may be directly proportional to the atomic ratio. However, the presence of radiation damping, collision damping, thermal Doppler effect, or hyperfine broadening may distort the proportionality.[199]

Ionization temperatures are derivable from the use of the Saha equation

$$\frac{N_{r+1}}{N_r} = \frac{2}{N_e} \frac{u_{r+1}}{u_r} \frac{(2\pi mkT_e)^{3/2}}{h^3} \exp(-\chi_r/kT_e) \tag{31}$$

where N_r and N_{r+1} are the relative numbers of ions with ionization states r and r + 1; N_e is the number density of electrons; u_r and u_{r+1} are the partition functions of the states r and r + 1; m is the mass of the electron; χ_r is the ionization energy of the rth state of ionization; and T_e is the electronic temperature.[202]

If thermodynamic equilibrium cannot be verified for the system under consideration, then the "electronic temperature" has only limited significance. Stellar ionization temperatures typically range somewhat above temperatures that are determined from continuous-radiation measurements.

Kinetic temperatures of stars are determined from studies of stellar line profiles to produce thermal velocity values v for the radiating atoms. The temperature then can be obtained from the relation

$$v^2 = 1.664 \times 10^8 \, T/M \, (cm/sec)^2 \tag{32}$$

where M is the atomic weight of the radiating species. The existence of turbulence in the stellar envelope often will distort the results of this type of analysis, indicating extremely high temperatures in certain cases.[199]

High-energy-density discharges have been studied in the laboratory over a period of many years. They provide a controllable environment for the physical study of plasmas, which are interesting in their own right, and for analyzing the effects of space reentry and other heavily energetic environments on engineering materials. At peak rates of energy deposition, laboratory plasmas exhibit many of the properties of stars, and similar analytical techniques can be employed in their study.[203-203b]

In recent years, experimental research devoted to the creation of energy sources based upon thermonuclear fusion has brought forth a number of studies of temperature measurement in plasmas. These have focused on the evaluation of electron temperatures as determined from cyclotron or X-ray emission.[204-206] The temperatures that have been determined in these measurements are quite high, of the order of 10^7 K; expressed in energy units, typical values are 1 to 7 keV for electron and ion temperatures.

Increasingly, fusion researchers have employed lasers in plasma thermometry,[207] with corresponding gains in spatial discrimination and versatility. In fact, one can make the general statement that measurement of the temperatures of hot gaseous systems has been revolutionized by the introduction of laser-based measurements and dedicated electronic computers. These "active" radiation techniques offer the capability for extremely rapid, nonintrusive thermometry in many systems that formerly were measured by heavily protected contact thermometers, such as thermocouples. Bechtel, Dasch, and Teets have reviewed the use of lasers in combustion measurements.[208] Their summary contains a good description of several laser-based thermometry techniques, including absorption spectroscopy, laser-induced fluorescence, spontaneous Raman spectroscopy, coherent Raman spectroscopy, optical refraction, and tomography.

One of the best-developed of the laser-based methods for thermometry in combustion processes is known as Coherent Anti-Stokes Raman Spectroscopy (CARS). It can be applied to the rapid and localized measurement of temperature in many of the gases that commonly occur in combustion — N_2, CO, H_2, H_2O, and CO_2. Pulsed CARS measurements can be used to limit the time interval of a temperature determination to about 10 nsec; spatial volumes as small as a few cubic millimeters can be sampled. Extensive discussions of both the theory and the application to thermometry of the CARS technique can be found in the recent literature.[209,210]

The advantages of the CARS technique over other laser techniques include its higher signal levels, its generation of a narrow signal beam that is relatively accessible and that permits filtering of unwanted background radiation, and its very high spectral resolution in favorable cases.

Typically a CARS measurement will consist of three high-intensity laser beams, two of frequency ν_1, and a third of frequency ν_2. These three beams interact coherently through the third-order nonlinear susceptibility of Raman-active target molecules. When the frequency difference $\nu_1 - \nu_2$ coincides with a Raman-active molecular mode, an enhanced anti-Stokes signal $\nu_3 = 2\nu_1 - \nu_2$ is generated. If the frequency ν_2 is varied in proper fashion, or if ν_2 covers a broad spectral region, a large set of anti-Stokes lines can be observed. The signal intensities I_3 are given approximately by the equation

$$I_3 \approx I_1^2 I_2 \left[b + \sum_i n_i \sigma_i g(\nu_1 - \nu_2, \nu_i) \right]^2 \tag{33}$$

Here I_1 and I_2 are the input laser intensities, b is a background term, n_i is the population difference between two levels that are connected by a Raman transition, σ_i is the Raman cross section for a transition of frequency $\nu_1 - \nu_2$, and $g(\nu_1 - \nu_2)$ is the complex Raman lineshape function. In practice, the selected gas is chosen so that the population difference n_i varies strongly with temperature in the range of interest; then the relative intensities of the anti-Stokes signal lines provide a temperature-dependent profile that is characteristic of a particular gas. However, one should note that this profile also is modified by the pressure of the gas, by its concentration, and by the nature and concentrations of other gases that are present.

The complete analysis of CARS spectra is a complicated task, but it provides a wealth of information. Besides yielding multiple values of temperature, the analysis can show the extent to which thermal equilibrium is present in the test system. The accuracy of temperatures determined from these spectra depends strongly upon a full understanding of the optical properties of the target molecules. In an effort to achieve more accurate CARS thermometry, several groups of researchers are attempting to improve the level of understanding of the optical properties that are relevant in particular substances.[211]

The temperature distribution in flames can be determined by line-of-sight observations across the flow field by laser tomography.[212] This technique can be considered as the successor to the Abel inversion method[213] that involved a "layered" analysis of parallel radiation measurements through the temperature-varying region of an axially symmetric flame. Using several different analytical techniques, Semerjian and his colleagues[214] have obtained flame-temperature profiles that are accurate within ±1.7%.

The temperature of flames and other hot, gaseous systems also can be determined by the method of laser-excited atomic fluorescence, as we mentioned above. Zizak et al. recently have summarized the characteristics of a dozen variations on this method,[215] although they have not provided information on their respective temperature accuracies.

VI. OTHER HIGH-PRECISION THERMOMETERS

There are many other methods for the precise measurement of temperatures besides those discussed at length so far in this chapter. A few of these should be mentioned here, if only briefly, for the sake of completeness.

A. Vapor-Pressure Thermometry

The traditional boiling-point standards of temperature involve the determination of the pressure that accompanies the equilibrium between the liquid and vapor phases of a pure substance. However, because the saturation vapor pressure of a pure substance is a monotonic and sensitive function of temperature, vapor-pressure thermometry can provide quite accurate measurements of temperature over extended ranges. The principle underlying this method is illustrated in Figure 2 of Chapter 3; here, one sees the saturation vapor-pressure line that marks the region of coexistence of liquid water and its vapor phase. The vapor pressure of pure water varies from 611 Pa at 0°C to 101,325 Pa at 100°C; in contrast to this factor-of-160 change, the same temperature variation is accompanied by no more than a 40% increase in the pressure exerted by an ideal gas at constant volume, or in the resistance of a PRT.

"Thermometers" used in precise determinations of temperature by vapor-pressure measurements usually are not commercially made devices, however, but instead are constructed in the laboratory as part of a larger research apparatus. In general, one undertakes vapor-pressure thermometry by constructing an apparatus of the type depicted in Figure 7 of Chapter 3, including the following components: a supply of the working substance in highly pure form; a sensitive, accurate, and wide-range pressure-measuring device (a U-tube manometer or a well-calibrated Bourdon or piston gage); and a cell that can be connected to the pressure-measurement and supply systems by a suitable length of capillary tubing. Some type of thermostat commonly is used to achieve and maintain steady temperatures throughout the desired range.

Besides ensuring the purity of the working substance and the accuracy of the pressure indicator, one must determine that any discrepancies between the actual saturation vapor pressures realized in the cell and the pressure values provided by the indicator are minimized. Such discrepancies can arise from "cold spots" in the sensing tube (locations along the sensing tube that are colder than the cell itself), from thermomolecular flow in the sensing tube, and from hydrostatic pressure heads in the system (see Chapter 5, Section II.E and F). In addition, of course, one must ensure the presence of thermal equilibrium between the gas-liquid interface and any thermometers to be calibrated. Finally, one must apply the correct relation between vapor pressure and temperature in order to achieve temperature accuracy with this technique.

We already have touched upon the use of the isotopes of helium for vapor-pressure thermometry in cryogenics (see References 35, 37, and 38 in Chapter 4). The thermodynamic vapor-pressure/temperature relation for a liquid with a monatomic vapor can be written

$$\ln P = -L_0/RT + (5/2)\ln T + i_0 + \epsilon(T) - (1/RT)\int_0^T S_\ell dT + (1/RT)\int_0^P V_\ell dP \quad (34)$$

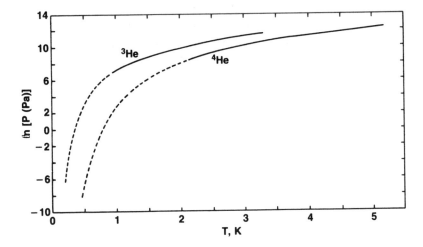

FIGURE 34. Vapor pressure/temperature relations for ^{4}He and ^{3}He. Solid curves; experimental measurements[216] against the NPL-75 gas thermometer scale. Dashed curves; calculated values using Equation 34.

In Equation 34 L_0 is the heat of vaporization at T = 0 K; R is the gas constant; i_0 is the chemical constant, which can be evaluated from the expression $i_0 = \ln[(2\pi m)^{3/2} k^{5/2}/h^3]$ where m is the mass of a single atom; $\epsilon(T)$ is a correction term that accounts for the nonideality of the vapor through the relation $\epsilon(T) = \ln(PV/RT) - (2B/V) - (3C/2V^2)$, where B and C are the second and third virial coefficients, respectively; and $S_\ell(T)$ and V_ℓ are respectively the molar entropy and the molar volume of the liquid, which are to be integrated along the saturation curve up to the indicated temperature and pressure limits.

With the exception of L_0, all of the terms in Equation 34 can be calculated for both isotopes of helium with some confidence. In principle, therefore, measurement of the vapor pressure at one well-known temperature would suffice if one intended to provide a calculated reference table for either ^{4}He or ^{3}He. In fact, considering the uncertainties involved in relating both measured and calculated vapor pressures to the NPL-75 gas-thermometer temperature scale, Durieux and his colleagues[216-218] have recommended a combination of the two techniques. Their results, which supplant the earlier T_{58} and T_{62} vapor-pressure relations, are shown in graphical form in Figure 34.

As finally recommended for use in providing EPT-76 temperatures that are accurate within ±1 mK, the vapor-pressure equations are quite different in form from Equation 34. For ease of computation, the forms shown in Table 12 have been employed. The reader should be advised that the achievement of accurate helium-vapor-pressure temperatures is substantially easier at temperatures above 1 K than it is below (see, for example, Figure 4 in Reference 217). Both the magnitude of necessary thermomolecular corrections and the very low vapor pressures involved conspire there to bedevil the thermometrist's efforts.

The vapor-pressure/temperature relation of no other substance — save, perhaps, water — has received such careful attention as has been lavished upon the data for the helium isotopes. However, equations and selected values of vapor pressure for many substances are given in the literature.[219,220]

B. Nuclear Quadrupole Resonance Thermometry

The variation with temperature of the nuclear quadrupole resonance (NQR) frequency of ^{35}Cl in KClO$_3$ is the basis for very precise thermometry from 50 K to above 400 K. The temperature dependence of the NQR frequency is given by

Table 12
VAPOR-PRESSURE/TEMPERATURE EQUATIONS FOR ^{4}He AND ^{3}He AS GIVEN BY DURIEUX AND RUSBY[218]

^{4}He from 0.5 K to 2.1768 K

$$\ln(P) = \sum_{k=-1}^{n} a_k T^k$$

Coefficients

a_{-1}	-7.41816	K
a_0	5.42128	
a_1	9.903203	K^{-1}
a_2	-9.617905	K^{-2}
a_3	6.804602	K^{-3}
a_4	-3.0154606	K^{-4}
a_5	0.7461357	K^{-5}
a_6	-0.0791791	K^{-6}

^{4}He from 2.1768 K to 5.1953 K

$$\ln(P) = \sum_{k=-1}^{n} a_k (T/T_c)^k + b[1 - (T/T_c)]^{1.9}$$

Coefficients

a_{-1}	-30.93285
a_0	392.47361
a_1	$-2,328.04587$
a_2	$8,111.30347$
a_3	$-17,809.80901$
a_4	$25,766.52747$
a_5	$-24,601.4$
a_6	$14,944.65142$
a_7	$-5,240.36518$
a_8	807.93168
b	14.53333

^{3}He from 0.2 K to T_c

$$\ln(P) = \sum_{k=-1}^{n} a_k T^k + b \ln(T)$$

Coefficients

a_{-1}	-2.50943	K
a_0	9.70876	
a_1	-0.304433	K^{-1}
a_2	0.210429	K^{-2}
a_3	-0.0545145	K^{-3}
a_4	0.0056067	K^{-4}
b	2.25484	

Note: All pressures are to be expressed in pascals, all temperatures in kelvins. Because of the form of the equations, failure to use all digits given above will result in significant computational errors. The present recommended value of the ^{4}He lambda-point temperature is 2.1768. The present recommended value of the ^{4}He critical-point temperature is 5.1953 K. The ^{3}He critical-point temperature T_c is given variously as 3.3158 to 3.3162 K.

$$\nu(T) = \nu_0 \left\{ 1 - \sum_i \frac{3A_i h}{2\omega_i} \left[\frac{1}{2} + \frac{1}{\exp(h\omega_i/2\pi kT) - 1} \right] \right\} \quad (35)$$

In Equation 35, ν_o is the frequency at 0 K, A_i is the reciprocal of the moment of inertia of the ith lattice mode whose frequency is ω_i, h is Planck's constant, and k is Boltzmann's constant.[221,222]

The NQR thermometer offers an interesting feature, somewhat analogous to a vapor pressure curve; because the resonant frequency is a unique function of temperature for a given system, the only need for calibration of such a thermometer is the original determination of this function. Subsequent thermometers utilizing the same system should follow the identical frequency-temperature relation.

Following work by Vanier[223], Utton[224] developed a ^{35}Cl/KClO$_3$ NQR thermometer whose temperature uncertainty did not exceed ±0.001 K in the range 50 to 300 K. The uncertainties rose to ±0.01 K as Utton's measurements approached 20 K, owing to the decreased sensitivity of the function described by Equation 35.

An automated commercial version of the ^{35}Cl/KClO$_3$ NQR thermometer was described recently by Ohte and Iwaoka.[225] In tests conducted at the NBS, this device provided temperatures on the IPTS-68 that were accurate within ±1 mK over the range 90 to 398 K. The data obtained in these tests agreed with those of Utton within the same uncertainty.

Thus, the NQR thermometer provides a real example of the continuing goal of thermometer manufacturers; to find a thermometer that follows a universal temperature reference relation without extensive factory treatment or calibration.

C. Quartz Resonance Thermometers

The piezoelectric properties of quartz have been known for a long time. The mechanical resonance oscillation of a suitably cut quartz crystal can provide an extremely stable time reference that is used to this day in moderately priced timepieces.

Many modes of oscillation can be obtained in quartz. For use in keeping time, a temperature-independent mode is desirable; by selecting a temperature-dependent mode, however, various researchers have been able to construct thermometers with reproducible frequency-temperature outputs.[226,227]

A commercial version of the quartz thermometer employs probe crystals that are cut so as to exhibit, as nearly as possible, only a linear frequency-temperature response.[228] A second, temperature-independent crystal serves as a stable reference oscillator. In use, the instrument presents a digital display of temperature that is derived from a beat frequency which itself arises from mixing of the probe and reference oscillator signals.

The commercial instrument is sensitive to temperature variations of magnitude 0.000 01 to 0.01°C, depending upon the sampling time that is chosen. Its range is −80 to 240°C. The thermometer accuracy is limited by a hysteresis effect that is thought to arise from thermal stresses in the probe-crystal electrical contacts. This hysteresis can amount to 0.02°C when the probe is cycled over its full range. Limited temperature excursions permit more accurate measurements; the manufacturer suggests that, over a 10°C cycle, the hysteresis should not exceed 0.001°C.

D. Paramagnetic Thermometers

We have mentioned the use of paramagnetic thermometry in connection with the development of the EPT-76 (Chapter 4, Section IV.E). Although this thermometry technique is restricted to the temperature range below 100 K, it deserves at least a brief discussion, because it offers good temperature precision with only modest effort.

The use of paramagnetism for thermometry is rooted in the discovery by Curie that, for many materials, the magnetic susceptibility χ (defined as the ratio of the magnetization to the applied magnetic field in the low-field limit) varies inversely with temperature:

$$\chi = C/T \tag{36}$$

Langevin offered an explanation of Equation 36 by supposing that the individual magnetic moments in magnetically active materials tend to align more and more easily with an applied magnetic field as thermal disturbances are reduced, thus leading to increased susceptibilities at lower temperatures. Weiss built upon Langevin's theory by postulating an extra interaction between the magnetic moments, leading to ferromagnetism below a particular temperature θ:

$$\chi = C/(T - \theta) \tag{37}$$

Following the introduction of the ideas of quantum mechanics, Van Vleck developed a fairly complete theory of magnetic susceptibilities,[229] deriving the Curie constant C for specific magnetic materials in terms of the quantum-mechanical descriptions of their lowest energy levels.

Although it is possible to use the results of Van Vleck and others to calculate from first principles the magnetic-susceptibility/temperature relations for many substances, thermometrists commonly do not measure the paramagnetic susceptibility quantitatively. Instead, the thermometric probe sample is placed within a concentric pair of cylindrical copper coils and a sensitive bridge is used to observe the changing value of the mutual inductance of the coil-pair through the relation

$$M = M_\infty + K'\chi$$
$$= M_\infty + K'C/(T - \theta) \tag{38}$$

where M_∞ is the coil-pair mutual inductance at infinite temperature and K' is a constant that depends primarily upon the proportion of the coil volume that is occupied by the sample.[230] sample.[230]

An alternative to the mutual-inductance-bridge measurement of the relative susceptibility is its measurement by means of an ultrasensitive magnetometer employing a superconductive quantum interference detector (SQUID)[231] (see Chapter 5, Section III). In this technique, a small paramagnetic sample is placed within an induction coil and the whole is incorporated into a superconducting tube in which magnetic flux has been trapped. The changing sample susceptibility affects the trapped field; this disturbance is reflected in a change in the induction-coil current, which is connected to a signal loop in the SQUID. The extremely high sensitivity of the SQUID permits the use of tiny (∼ 1 mg) paramagnetic samples to achieve temperature resolution equivalent to that obtained using gram-sized samples with an ordinary mutual inductance bridge.

An enormous amount of study[232-234] has been given to the magnetic properties of various paramagnetic substances, which include crystalline salts containing ions with partially-filled 3d electron shells (Ti, V, Cr, Mn, Fe, Co, Ni, and Cu), the 4d series (Zr to Ag), the 5d series (Hf to Au), the 4f series (the rare earths), and the 5f series (the actinides).

The principles involved in paramagnetic susceptibility thermometry are the same, regardless of the techniques to be used: The operator prepares a probe sample of a material for which the form of the susceptibility equation is known; utilizing either a mutual inductance bridge or a SQUID magnetometer, that person then performs a calibration by measuring the instrument output at enough known temperatures to evaluate the coefficients in the appropriate equation; then the paramagnetic susceptibility thermometer is brought into thermal equilibrium with the test object and the temperature is determined from the instrument reading at that temperature.

Some of the most precise paramagnetic thermometry known has been done by Cetas and Swenson,[235] by Cetas,[236] and by Rusby and Swenson.[216] In the range 0.5 to 83 K, these workers have achieved temperature reproducibilities within ±1 mK (and much better in particular instances).

E. Industrial Johnson Noise Thermometry

We discussed in Chapter 5, Section III, the determination of thermodynamic temperatures by the measurement of the Johnson noise in a carefully prepared laboratory resistor. The square of the Johnson noise voltage is given by

$$V^2 = \int 4Rkt d\nu \tag{39}$$

where R is the sensor resistance and the integration is performed over the sensitive frequency range of the detector.

Noise thermometry also can be used to determine the temperature of a resistor that resides in an industrial environment such as a nuclear reactor, a steel mill, or a traditional oil- or coal-fired power plant. In these cases, problems such as electrical interference, mechanical shock, and chemical corrosion introduce relatively large uncertainties into thermometry of all types, so that special approaches must be considered.

The major problem that appears in industrial noise thermometry is that of electrical noise generated in the long cables that often must be used to connect the sensing resistor to the measuring instruments.[237] This problem was attacked by Brixy et al. by the use of a resistor-comparison measurement technique coupled with two-channel correlation detection.[238] These methods were tested under a wide range of industrial conditions and found to yield temperatures in range −170 to 1000°C that proved to be accurate within ±0.5%. In one such instance, the sensor resistance changed during in-pile nuclear reactor measurements by some 8%; however, repeated measurements of the sensor resistance sufficed to maintain the stated temperature accuracy.

Changing sensor resistance is specifically excluded as a source of systematic error in the Johnson noise power thermometer (JNPT) developed by Borkowski and Blalock.[239] In this approach, the outputs of a voltage-sensitive preamplifier and a current-sensitive preamplifier are further amplified in an alternating sequence, then fed to integrating and multiplying circuits. The simplified preamplifier equations and the resulting noise power output equations are given by

$$\overline{V_n^2} = 4kTR \, \Delta f$$

$$\overline{I_n^2} = 4kT \, \Delta f / R$$

$$\sqrt{\overline{V_n^2}\, \overline{I_n^2}} = 4kT \, \Delta f \tag{40}$$

As shown in Equation 40, the JNPT output is not affected by changes in the sensor resistance, commonly a 50 to 300 Ω refractory-metal wire. The accuracy of the technique has been assessed by using the JNPT to measure the temperature of a well-controlled platinum resistance thermometer, for which the temperature also can be measured by conventional thermometry. Under favorable conditions, the JNPT is accurate within ±0.5%. Uncertainties arising from cable noise, however, can raise this inaccuracy figure to several percent.[240]

At the present time, efforts are under way to reduce long-cable uncertainties in JNPT thermometry.[241]

GENERAL REFERENCES

Betts, D. S., *Refrigeration and Thermometry Below One Kelvin,* Sussex University Press, London, 1976.
Sachse, H., *Semiconducting Temperature Sensors and Their Applications,* John Wiley & Sons, New York, 1975.
Hall, J. A., *The Measurement of Temperature,* Chapman and Hall, London, 1966.
Corruccini, R. J., Principles of Thermometry, in *Treatise on Analytical Chemistry,* Kolthoff, I. M., Elving, P. J., and Sandell, E. B., Eds., Vol. 8, Part I, Interscience, 1968, chap. 87, sect. IIB.
Quinn, T. J., *Temperature,* Academic Press, London, 1983.
Kinzie, P. A., *Thermocouple Temperature Measurement,* John Wiley & Sons, New York, 1973.
Committee E-20 on Temperature Measurement, American Society for Testing and Materials, *Manual on the Use of Thermocouples in Temperature Measurement,* ASTM Spec. Tech. Publ. 470B, American Society for Testing and Materials, Philadelphia, 1981.
Pollock, D. D., *Thermoelectricity, Theory, Thermometry, Tool,* ASTM Spec. Tech. Publ. 852, American Society for Testing and Materials, Philadelphia, 1985.
Temperature Handbook and Buyer's Guide, Measurements and Data Corp., Pittsburgh.
Benedict, R. P., *Fundamentals of Temperature, Pressure and Flow Measurements,* 2nd ed., John Wiley & Sons, New York, 1977, chap. 5 to 8.
Schooley, J. F., Ed.-in-Chief, *Temperature, Its Measurement and Control in Science and Industry,* Vol. 5, American Institute of Physics, New York, 1982, sect. IV to VII and X.
DeWitt, D. P. and Nutter, G. D., Eds., *Theory and Practice of Radiation Thermometry,* to be published as an NBS Monograph, 1985.
Richmond, J. C. and Nicodemus, F. E., Blackbodies, blackbody radiation, and temperature scales, in *Self-Study Manual on Optical Radiation Measurements, Part I, Concepts,* Nicodemus, F. E., Ed., NBS Tech. Note 910-8, 1985, chap. 12.

REFERENCES

1. **Moore, R. L.,** *Basic Instrumentation Lecture Notes and Study Guide,* Vol. 1, 3rd ed., Instrument Society of America, Research Triangle Park, N.C., 1983.
1a. **Hurley, C. W. and Schooley, J. F.,** *Calibration of Temperature Measurement Systems Installed in Buildings,* U.S. National Bureau of Standards Bldg. Sci. Ser. 153, Sect. 3, January 1984.
2. **Kerlin, T. W. and Shepard, R. L.,** *Industrial Temperature Measurement,* Instrument Society of America, Research Triangle Park, N.C., 1982.
3. **Busse, J.,** Liquid-in-glass thermometers, in *Temperature, Its Measurement and Control in Science and Industry,* Vol. 1, Fairchild, C. O., Hardy, J. D., Sosman, R. B., and Wensel, H. T., Eds., Reinhold, New York, 1941, 228.
4. Standard E1-80, Standard Specification for ASTM Thermometers, *Annual Book of ASTM Standards,* American Society for Testing and Materials, Philadelphia.
5. **Wise, J. A.,** *Liquid-in-Glass Thermometry,* U.S. National Bureau of Standards Monogr. 150, January 1976.
6. **Thompson, R. D.,** Recent developments in liquid-in-glass thermometry, in *Temperature, Its Measurement and Control in Science and Industry,* Vol. 3, Part 1, Herzfeld, C. M., Ed.-in-Chief, Reinhold, New York, 1962, 201.
7. **Hall, J. A. and Leaver, V. M.,** Some experiments in mercury thermometry, in *Temperature, Its Measurement and Control in Science and Industry,* Vol. 3, Part 1, Herzfeld, C. M., Ed.-in-Chief, Reinhold, New York, 1962, 231.
8. Standard E 77-80, *Annual Book of ASTM Standards,* American Society for Testing and Materials, Philadelphia.
9. **Wise, J. A. and Soulen, R. J., Jr.,** *Thermometer Calibration: A Model for State Calibration Laboratories.* NBS Monograph, 174, in press. Will supersede NBS Monograph 150, *Liquid-in-Glass Thermometry.* (See Ref. 5.)
10. Committee E-20 on Temperature Measurement, American Society for Testing and Materials, *Manual on the Use of Thermocouples in Temperature Measurement,* ASTM Spec. Tech. Publ. 470B, American Society for Testing and Materials, Philadelphia, 1981, chap. 2.

11. **Seebeck, T. J.,** *Proc. Royal Acad. Sci.,* Berlin 1822, p 265.
12. **Powell, R. L., Hall, W. J., Hyink, C. H., Jr., Sparks, L. L., Burns, G. W., Scroger, M. G., and Plumb, H. H.,** *Thermocouple Reference Tables Based on the IPTS-68,* U.S. National Bureau of Standards Monogr. 125, March 1974.
13. ASTM Standard E207-78, Standard method of thermal emf test of single thermoelement materials by comparison with a secondary standard of similar emf-temperature properties, *ASTM Annual Book of Standards.*
14. **Peltier, J. C. A.,** *Annales de Chemie et de Physique,* v56 (2nd series) 1834, 371.
15. ASTM Committee E-20, *Manual on the Use of Thermocouples in Temperature Measurement,* ASTM Spec. Tech. Publ. 470B, American Society for Testing and Materials, Philadelphia, 1981, 13.
16. **Reed, R. P.,** Thermoelectric thermometry: a functional model, in *Temperature, Its Measurement and Control in Science and Industry,* Vol. 5, Schooley, J. F., Ed.-in-Chief, American Institute of Physics, New York, 1982, 915.
17. Chromel and Alumel are trademarks of the Hoskins Manufacturing Co.
18. ASTM Standard E230-83, Temperature-emf tables for thermcouples, *ASTM Annual Book of Standards.*
18a. **Powell, R. L., Hall, W. J., Hyink, C. H., Jr., Sparks, L. L., Burns, G. W., Scroger, M. G., and Plumb, H. H.,** *Thermocouple Reference Tables Based on the IPTS-68,* U.S. National Bureau of Standards Monogr. 125, March 1974, 2.
19. ASTM Standard E 220, Calibration of thermocouples by comparison techniques, *ASTM Annual Book of Standards.*
20. Committee E-20 on Temperature Measurement, American Society for Testing and Materials, *Manual on the Use of Thermocouples in Temperature Measurement,* ASTM Spec. Tech. Publ. 470B, American Society for Testing and Materials, Philadelphia, 1981, chap. 8.
21. **Benedict, R. P.,** *Fundamentals of Temperature, Pressure, and Flow Measurements,* 2nd ed., John Wiley & Sons, New York, 1977, sect. 7.6, 7.7, and 7.9.
22. **Mossman, C. A., Horton, J. L., and Anderson, R. L.,** Testing of thermocouples for inhomogenieties: a review of theory, with examples, in *Temperature, Its Measurement and Control in Science and Industry,* Vol. 5, Schooley, J. F., Ed.-in-Chief, American Institute of Physics, New York, 1982, 923.
23. **Reed, R. P.,** Validation diagnostics for defective thermocouple circuits, in *Temperature, Its Measurement and Control in Science and Industry,* Vol. 5, Schooley, J. F., Ed.-in-Chief, American Institute of Physics, New York, 1982, 931.
24. **Anderson, R. L. and Ludwig, R. L.,** Failure of sheathed thermocouples due to thermal cycling, in *Temperature, Its Measurement and Control in Science and Industry,* Vol. 5, Schooley, J. F., Ed.-in-Chief, American Institute of Physics, New York, 1982, 939.
25. **Anderson, R. L., Lyons, J. D., Kollie, T. G., Christie, W. H., and Eby, R.,** Decalibration of sheathed thermocouples, in *Temperature, Its Measurement and Control in Science and Industry,* Vol. 5, Schooley, J. F., Ed.-in-Chief, American Institute of Physics, New York, 1982, 977.
26. Committee E-20 on Temperature Measurement, American Society for Testing and Materials, *Manual on the Use of Thermocouples in Temperature Measurement,* ASTM Spec. Tech. Publ. 470B, American Society for Testing and Materials, Philadelphia, 1981, chap. 12.
27. **Kinzie, P. A.,** *Thermocouple Temperature Measurement,* John Wiley & Sons, New York, 1973.
28. **Burns, G. W. and Hurst, W. S.,** Thermocouple thermometry, in *Temperature Measurement, 1975,* Billing, B. F. and Quinn, T. J., Eds., Conference Series No. 26, The Institute of Physics, London, 1975, 144.
29. Committee E-20 on Temperature Measurement, American Society for Testing and Materials, *Manual on the Use of Thermocouples in Temperature Measurement,* ASTM Spec. Tech. Publ. 470B, American Society for Testing and Materials, Philadelphia, 1981, chap. 3 to 5 and 11.
30. **Schooley, J. F., Ed-in-Chief,** *Temperature, Its Measurement and Control in Science and Industry,* Vol. 5, American Institute of Physics, New York, 1982, sect. VI.
31. **Sparks, L. L. and Powell, R. L.,** Low temperature thermocouples: KP, "normal" silver, and copper versus Au-0.02 at % Fe and Au-0.07 at % Fe, *J. Res. Natl. Bur. Stand.,* 76A, 263—283, 1972.
32. **Armbrüster, H., Kirk, W. P., and Cheshire, D. P.,** Very low temperature thermocouple devices: development and application techniques for temperature measurements, in *Temperature, Its Measurement and Control in Science and Industry,* Vol. 5, Schooley, J. F., Ed-in-Chief, American Institute of Physics, New York, 1982, 1025.
33. **Burley, N. A., Burns, G. W., and Powell, R. L.,** Nicrosil and nisil: their development and standardization, in *Temperature Measurement, 1975,* Billing, B. F. and Quinn, T. J., Eds., Conference Series No. 26, The Institute of Physics, London, 1975, 162.
34. **Dahl, A. I.,** The stability of base-metal thermocouples in air from 800 to 2200°F, in *Temperature, Its Measurement and Control in Science and Industry,* Vol. 1, Fairchild, C. O., Hardy, J. D., Sosman, R. B., and Wensel, H. T., Eds., Reinhold, New York, 1941, 1238.

35. **Potts, Jr., J. F. and McElroy, D. L.,** The effects of cold working, heat treatment, and oxidation on the thermal emf of nickel-base thermoelements, in *Temperature, Its Measurement and Control in Science and Industry,* Vol. 3, Part 2, Herzfeld, C. M., Ed.-in-Chief, Reinhold, New York, 1962, 243.
36. **Starr, C. D. and Wang, T. P.,** Effect of oxidation on stability of thermocouples, *Trans. ASTM,* 63, 1185, 1963.
37. **Burley, N. A. and Ackland, R. G.,** *J. Aust. Inst. Metals,* 12, 23, 1967.
38. **Burley, N. A.,** Nicrosil and nisil: Highly stable nickel-base alloys for thermocouples, in *Temperature, Its Measurement and Control in Science and Industry,* Vol. 4, Plumb, H. H., Ed.-in-Chief, Instrument Society of America, Pittsburgh, 1972, 1677.
39. **Burley, N. A., Powell, R. L., Burns, G. W., and Scroger, M. G.,** *The Nicrosil versus Nisil Thermocouple: Properties and Thermoelectric Reference Data,* U.S. National Bureau of Standards Monogr. 161, 1978.
40. **Burns, G. W.,** The nicrosil versus nisil thermocouple: recent developments and present status, in *Temperature, Its Measurement and Control in Science and Industry,* Vol. 5, Schooley, J. F., Ed.-in-Chief, American Institute of Physics, New York, 1982, 1121.
41. **Burley, N. A., Cocking, J. L., Burns, G. W., and Scroger, M. G.,** The nicrosil versus nisil thermocouple: the influence of magnesium on the thermoelectric stability and oxidation resistance of the alloys, in *Temperature, Its Measurement and Control in Science and Industry,* Vol. 5, Schooley, J. F., Ed.-in-Chief, American Institute of Physics, New York, 1982, 1129.
42. **Wang, T. P. and Starr, C. D.,** Oxidation resistance and stability of nicrosil-nisil in air and in reducing atmospheres, in *Temperature, Its Measurement and Control in Science and Industry,* Vol. 5, Schooley, J. F., Ed.-in-Chief, American Institute of Physics, New York, 1982, 1147.
43. **Burley, N. A., Hess, R. M., Howie, C. F., and Coleman, J. A.,** The nicrosil versus nisil thermocouple: a critical comparison with the ANSI standard letter-designated base-metal thermocouples, in *Temperature, Its Measurement and Control in Science and Industry,* Vol. 5, Schooley, J. F., Ed.-in-Chief, American Institute of Physics, New York, 1982, 1159.
44. **Committee E-20 on Temperature Measurement, American Society for Testing and Materials,** *Manual on the Use of Thermocouples in Temperature Measurement,* ASTM Spec. Tech. Publ. 470B, American Society for Testing and Materials, Philadelphia, 1981, chap. 3.
45. **Kinzie, P. A.,** *Thermocouple Temperature Measurement,* John Wiley & Sons, New York, 1973, chap. 5.
46. **Burns, G. W. and Hurst, W. S.,** Thermocouple thermometry, in *Temperature Measurement, 1975,* Billing, B. F. and Quinn, T. J., Eds., Conference Series No. 26, The Institute of Physics, London, 1975, sect. 2.2.2.
47. **Callendar, H. L.,** On the practical measurement of temperature, *Philos. Trans. R. Soc. London (A),* 178, 161, 1887.
48. **Mueller, E. F.,** Precision resistance thermometry in *Temperature, Its Measurement and Control in Science and Industry,* Vol. 1, Fairchild, C. O., Hardy, J. D., Sosman, R. B., and Wensel, H. T., Ed., Reinhold, New York, 1941, 162.
49. **Riddle, J. L., Furukawa, G. T., and Plumb, H. H.,** *Platinum Resistance Thermometry,* U.S. National Bureau of Standards Monogr. 126, 1973, sect. 4.4.
50. **Poggendorf,** *Ann. Phys. Chemie,* Vol. 54, 1841.
51. **Smith, F. E.,** On bridge methods for resistance measurements of high precision in platinum thermometry, *Phil. Mag.,* 24, 541, 1912.
52. **Mueller, E. F.,** Wheatstone bridges and some accessory apparatus for resistance thermometry, *Nat. Bur. Stand. Bull.,* 13, 547, 1916.
53. **Harris, F. K.,** *Electrical Measurements,* John Wiley & Sons, Inc., New York, 1952, chap. 7.
54. **Riddle, J. L., Furukawa, G. T., and Plumb, H. H.,** *Platinum Resistance Thermometry,* U.S. National Bureau of Standards Monogr. 126, 1973, App. H.
55. **Cutkosky, R. D.,** Automatic resistance thermometer bridges for new and special applications, in *Temperature, Its Measurement and Control in Science and Industry,* Vol. 5, Schooley, J. F., Ed.-in-Chief, American Institute of Physics, New York, 1982, 711.
56. **Kirby, C. G. M.,** An automatic resistance thermometer bridge, in *Temperature, Its Measurement and Control in Science and Industry,* Vol. 5, Schooley, J. F., Ed.-in-Chief, American Institute of Physics, New York, 1982, 715.
57. **Brown, N. L., Fougere, A. J., McLeod, J. W., and Robbins, R. J.,** An automatic resistance thermometer bridge, in *Temperature, Its Measurement and Control in Science and Industry,* Vol. 5, Schooley, J. F., Ed.-in-Chief, American Institute of Physics, New York, 1982, 719.
58. **Wolfendale, P. C. F., Yewen, J. D., and Daykin, C. I.,** A new range of high precision resistance bridges for resistance thermometry, in *Temperature, Its Measurement and Control in Science and Industry,* Vol. 5, Schooley, J. F., Ed.-in-Chief, American Institute of Physics, New York, 1982, 729.
59. **Stimson, H. L.,** *J. Res. Natl. Bur. Stand.,* 42, 209, 1949.
60. **Callendar, H. L.,** On the construction of platinum thermometers, *Phil. Mag.,* 5th Series, 32, 104, 1891.

61. **Heycock, C. T. and Neville, F. H.**, On the determination of high temperatures by means of platinum resistance thermometers, *J. Chem. Soc.*, 67, 160, 1895.
61a. **Beattie, J. A., Jacobus, D. D., and Gaines, J. M., Jr.**, An experimental study of the absolute temperature scale. I. The construction of several types of platinum resistance thermometers, *Proc. Am. Acad. Arts Sci.*, 66, 167, 1930—31.
61b. **Mueller, E. W.**, Resistance thermometry, in *Temperature, Its Measurement and Control in Science and Industry*, Vol. 1, Fairchild, C. O., Hardy, J. D., Sosman, R. B., and Wensel, H. T., Eds., Reinhold, New York, 1941, 162.
62. **Dickinson, H. C. and Mueller, E. F.**, Calorimetric resistance thermometers and the transition temperature of sodium sulphate, *Bull. B. S.*, 3, 641, 1907.
63. **Waidner, C. W. and Burgess, G. K.**, Platinum resistance thermometry at high temperatures, *Bull. B. S.*, 6, 149, 1909—10.
64. **Meyers, C. H.**, Coiled-filament resistance thermometers, *J. Res. Natl. Bur. Stand.*, 9, 807, 1932.
65. **Barber, C. R. and Blanke, W. W.**, A platinum resistance thermometer for use at high temperatures, *J. Sci. Instrum.*, 38, 17, 1961.
66. A thorough discussion on modern PRTs is given by **Curtis, D. J.**, Platinum resistance interpolation standards, in *Temperature, Its Measurement and Control in Science and Industry*, Vol. 4, Plumb, H. H., Ed.-in-Chief, Instrument Society of America, Pittsburgh, 1972, 951.
67. **Rosebury, F.**, *Handbook of Electron Tube and Vacuum Techniques*, Addison-Wesley, Reading, Mass., 1965, 371.
68. **Riddle, J. L., Furukawa, G. T., and Plumb, H. H.**, *Platinum Resistance Thermometry*, U.S. National Bureau of Standards Monogr. 126, 1973, sect. 3.
69. **Berry, R. J.**, Effect of oxidation on Pt resistance thermometry, *Metrologia*, 16, 117, 1980.
70. **Berry, R. J.**, Evaluation and control of platinum oxidation errors in standard platinum resistance thermometers, in *Temperature, Its Measurement and Control in Science and Industry*, Vol. 5, Schooley, J. F., Ed.-in-Chief, American Institute of Physics, New York, 1982, 743.
71. **Moser, H.**, Temperature measurement with platinum resistance thermometers up to 1100°C, *Ann. Physik (5)*, 6, 582, 1930.
72. Note that this construction technique is less effective than those of Figure 21 in tempering the sensor coils.
73. **Evans, J. P. and Burns, G. W.**, A study of stability of high temperature platinum resistance thermometers, in *Temperature, Its Measurement and Control in Science and Industry*, Vol. 3, Part I, Herzfeld, C. M., Ed.-in-Chief, Reinhold, New York, 1962, 313.
74. **Curtis, D. J. and Thomas, G. J.**, Long term stability of platinum resistance thermometers for use to 1063°C, *Metrologia*, 4, 184, 1968.
75. **Evans, J. P. and Wood, S. D.**, An intercomparison of high temperature platinum resistance thermometers and standard thermocouples, *Metrologia*, 7, 108, 1971.
75a. **Evans, J. P.**, High temperature platinum resistance thermometry, in *Temperature, Its Measurement and Control in Science and Industry*, Vol. 4, Plumb, H. H., Ed.-in-Chief, Instrument Society of America, Pittsburgh, 1972, 899.
76. **Jones, T. P.**, The accuracies of calibration and use of IPTS thermocouples, *Metrologia*, 4, 80, 1968.
77. **Chattle, M. V.**, Platinum resistance thermometry up to the gold point, in *Temperature, Its Measurement and Control in Science and Industry*, Vol. 4, Plumb, H. H., Ed.-in-Chief, Instrument Society of America, Pittsburgh, 1972, 907.
78. **Sawada, S. and Mochizuki, T.**, Stability of 25 ohm platinum thermometer up to 1100°C, in *Temperature, Its Measurement and Control in Science and Industry*, Vol. 4, Plumb, H. H., Ed.-in-Chief, Instrument Society of America, Pittsburgh, 1972, 919.
79. **Anderson, R. L.**, The high temperature stability of platinum resistance thermometers, in *Temperature, Its Measurement and Control in Science and Industry*, Vol. 4, Plumb, H. H., Ed.-in-Chief, Instrument Society of America, Pittsburgh, 1972, 927.
80. **Jung, H. J. and Nubbemeyer, H.**, The stability of commercially available high temperature platinum resistance thermometers of a 5 ohm silica cross type up to 961.93°C, in *Temperature, Its Measurement and Control in Science and Industry*, Vol. 5, Schooley, J. F., Ed.-in-Chief, American Institute of Physics, New York, 1982, 763.
81. **Evans, J. P.**, Experiences with high-temperature platinum resistance thermometers, in *Temperature, Its Measurement and Control in Science and Industry*, Vol. 5, Schooley, J. F., Ed.-in-Chief, American Institute of Physics, New York, 1982, 771.
82. **Long Guang and Tao Hongtu**, Stability of precision high temperature platinum resistance thermometers, in *Temperature, Its Measurement and Control in Science and Industry*, Vol. 5, Schooley, J. F., Ed.-in-Chief, American Institute of Physics, New York, 1982, 783.

83. **Ling Shankang, Zhang Guoquan, Li Ruisheng, Wang Zilin, Li Zhiran, Zhao Qi, and Li Xumo,** The development of temperature standards at NIM of China, in *Temperature, Its Measurement and Control in Science and Industry,* Vol. 5, Schooley, J. F., Ed.-in-Chief, American Institute of Physics, New York, 1982, 191.
84. **Li Xumo, Zhang Jinde, Su Jinrong, and Chen Deming,** A new high-temperature platinum resistance thermometer, *Metrologia,* 18, 203, 1984.
85. **Bass, N. and Evans, J. P.,** *Techniques in High-Temperature Resistance Thermometry,* U.S. National Bureau of Standards Tech. Note 1183, January 1984.
86. **Evans, J. P.,** Report to the 15th Meeting of the Consultative Committee for Thermometry, Paper CCT 84/17 Bureau International des Poids et Mesures, Paris, June 5 to 7, 1984.
86a. **Evans, J. P.,** Evaluation of some high-temperature platinum resistance thermometers, *J. Res. Natl. Bur. Stand.,* 89, 349, 1984.
87. **Lucas, D. A.,** Miniature precision platinum resistance thermometers, in *Temperature, Its Measurement and Control in Science and Industry,* Vol. 4, Plumb, H. H., Ed.-in-Chief, Instrument Scoiety of America, Pittsburgh, 1972, 963.
88. **Curtis,, D. J.,** Thermal hysteresis and stress effects in platinum resistance thermometers, in *Temperature, Its Measurement and Control in Science and Industry,* Vol. 5, Schooley, James F., Ed.-in-Chief, American Institute of Physics, New York, 1982, 803.
89. **Sinclair, D. H., Terbeek, H. G., and Malone, J. H.,** Calibration of platinum resistance thermometers, in *Temperature, Its Measurement and Control in Science and Industry,* Vol. 4, Plumb, H. H., Ed.-in-Chief, Instrument Society of America, Pittsburgh, 1972, 983.
90. **Dutt, M.,** Practical applications of platinum resistance sensors, in *Temperature, Its Measurement and Control in Science and Industry,* Vol. 4, Plumb, H. H., Ed.-in-Chief, Instrument Society of America, Pittsburgh, 1972, 1013.
91. **Kleven, L. and Lofgren, L.,** Unique platinum resistance temperature sensors for lunar heat flow measurements, in *Temperature, Its Measurement and Control in Science and Industry,* Vol. 4, Plumb, H. H., Ed.-in-Chief, Instrument Society of America, Pittsburgh, 1972, 1021.
92. **Kennedy, W. K. and Brown, J. L.,** High speed temperature sensor, in *Temperature, Its Measurement and Control in Science and Industry,* Vol. 4, Plumb, H. H., Ed.-in-Chief, Instrument Society of America, Pittsburgh, 1972, 1035.
93. **Moeller, C. E.,** Gold film resistance thermometers for surface temperature measurements, in *Temperature, Its Measurement and Control in Science and Industry,* Vol. 4, Plumb, H. H., Ed.-in-Chief, Instrument Society of America, Pittsburgh, 1972, 1049.
94. **Carr, K. R.,** An evaluation of industrial platinum resistance thermometers, in *Temperature, Its Measurement and Control in Science and Industry,* Vol. 4, Plumb, H. H., Ed.-in-Chief, Instrument Society of America, Pittsburgh, 1972, 971.
95. **Saffell, J. R.,** Designing accurate platinum RTD measuring systems for industry, in *Temperature, Its Measurement and Control in Science and Industry,* Vol. 5, Schooley, J. F., Ed.-in-Chief, American Institute of Physics, New York, 1982, 733.
96. **Fritschen, L. J. and Simpson, J. R.,** An automatic system for measuring Bowen ratio gradients using platinum resistance elements, in *Temperature, Its Measurement and Control in Science and Industry,* Vol. 5, Schooley, J. F., Ed.-in-Chief, American Institute of Physics, New York, 1982, 739.
97. **McAllan, J. V.,** Practical high temperature resistance thermometry, in *Temperature, Its Measurement and Control in Science and Industry,* Vol. 5, Schooley, J. F., Ed.-in-Chief, American Institute of Physics, New York, 1982, 789.
98. **Mangum, B. W. and Evans, G. A., Jr.,** Investigation of the stability of small platinum resistance thermometers, in *Temperature, Its Measurement and Control in Science and Industry,* Vol. 5, Schooley, J. F., Ed.-in-Chief, American Institute of Physics, New York, 1982, 795.
99. **Bass, N. M.,** Construction of a laboratory working thermometer using industrial platinum resistance sensors, in *Temperature, Its Measurement and Control in Science and Industry,* Vol. 5, Schooley, J. F., Ed.-in-Chief, American Institute of Physics, New York, 1982, 813.
100. **Connolly, J. J.,** The calibration characteristics of industrial platinum resistance thermometers, in *Temperature, Its Measurement and Control in Science and Industry,* Vol. 5, Schooley, J. F., Ed.-in-Chief, American Institute of Physics, New York, 1982, 815.
101. **Actis, A. and Crovini, L.,** Interpolating equations for industrial platinum resistance thermometers in the temperature range from -200 to $+420°C$, in *Temperature, Its Measurement and Control in Science and Industry,* Vol. 5, Schooley, J. F., Ed.-in-Chief, American Institute of Physics, New York, 1982, 819.
102. **Giarratano, P. J., Lloyd, F. L., Mullen, L. O., and Chen, G. B.,** A thin platinum film for transient heat transfer studies, in *Temperature, Its Measurement and Control in Science and Industry,* Vol. 5, Schooley, J. F., Ed.-in-Chief, American Institute of Physics, New York, 1982, 859.

103. **Dauphinee, T. M.**, Deep-ocean temperature measurement, in *Temperature, Its Measurement and Control in Science and Industry*, Vol. 5, Schooley, J. F., Ed.-in-Chief, American Institute of Physics, New York, 1982, 1317.
104. **Stickney, T. M. and Stiles, M. T.**, Down-to-earth air temperature measurements during space shuttle earth atmosphere re-entry, in *Temperature, Its Measurement and Control in Science and Industry*, Vol. 5, Schooley, J. F., Ed.-in-Chief, American Institute of Physics, New York, 1982, 1327.
105. **Kerlin, T. W., Shepard, R. L., Hashemian, H. M., and Petersen, K. M.**, Response of installed temperature sensors, in *Temperature, Its Measurement and Control in Science and Industry*, Vol. 5, Schooley, J. F., Ed.-in-Chief, American Institute of Physics, New York, 1982, 1367.
106. **Mayer, E.**, Thermal environments and thermal comfort: new instruments and methods, in *Temperature, Its Measurement and Control in Science and Industry*, Vol. 5, Schooley, J. F., Ed.-in-Chief, American Institute of Physics, New York, 1982, 1381.
107. See, for example, *The Thomas Register*, Thomas Publishing, New York, for information on commercial products.
108. **Sachse, H. B.**, *Semiconducting Temperature Sensors and Their Applications*, John Wiley & Sons, Inc., New York, 1975.
109. **Sapoff, M.**, Thermistors for biomedical use, in *Temperature, Its Measurement and Control in Science and Industry*, Vol. 4, Plumb, H. H., Ed.-in-Chief, Instrument Society of America, Pittsburgh, 1972, 2109.
110. **Trolander, H. W., Case, D. A., and Harruff, R. W.**, Reproducibility, stability, and linearization of thermistor resistance thermometers, in *Temperature, Its Measurement and Control in Science and Industry*, Vol. 4, Plumb, H. H., Ed.-in-Chief, Instrument Society of America, Pittsburgh, 1972, 997.
111. **Sapoff, M.**, *Measurements and Control*, 14(4), 144, 1980.
112. **Sapoff, M., Siwek, W. R., Johnson, H. C., Slepian, J., and Weber, S.**, The exactness of fit of resistance-temperature data of thermistors with third-degree polynomials, in *Temperature, Its Measurement and Control in Science and Industry*, Vol. 5, Schooley, J. F., Ed.-in-Chief, American Institute of Physics, New York, 1982, 875.
113. **Steinhart, J. S. and Hart, S. R.**, *Deep-Sea Res.*, 15, 497, 1968.
114. **Wood, S. D., Mangum, B. W., Filliben, J. J., and Tillett, S. B.**, An investigation of the stability of thermistors, *J. Res. Natl. Bur. Stand.*, 83, 247, 1978.
115. **LaMers, T. H., Zurbuchen, J. M., and Trolander, H.**, Enhanced stability in precision interchangeable thermistors, in *Temperature, Its Measurement and Control in Science and Industry*, Vol. 5, Schooley, J. F., Ed.-in-Chief, American Institute of Physics, New York, 1982, 865.
116. **Mangum, B. W. and Thornton, D. D., Eds.**, *Gallium Melting-Point Standard*, U.S. National Bureau of Standards, Spec. Publ. 481, 1977.
116a. **Mangum, B. W. and Thornton, D. D.**, Determination of the triple-point temperature of gallium, *Metrologia*, 15, 201, 1979.
117. **Mangum, B. W.**, Triple point of succinonitrile and its use in the calibration of thermistor thermometers, *Rev. Sci. Instrum.*, 54, 1687, 1983.
118. **Rusby, R. L.**, A rhodium-iron resistance thermometer for use below 20 K, in *Temperature, Its Measurement and Control in Science and Industry*, Vol. 4, Plumb, H. H., Ed.-in-Chief, Instrument Society of America, Pittsburgh, 1972, 865.
119. **Coles, B. R.**, *Phys. Lett.*, 8, 243, 1964.
120. **Oliveira, N. F. and Foner, S.**, *Phys. Lett.*, 34A, 15, 1971.
121. **Rusby, R. L.**, Resistance thermometry using rhodium-iron, 0.1 K to 273 K, in *Temperature Measurement, 1975*, Billing, B. F. and Quinn, T. J., Eds., Conference Series No. 26, The Institute of Physics, London, 1975, 125.
122. **Rusby, R. L.**, The rhodium-iron resistance thermometer: ten years on, in *Temperature, Its Measurement and Control in Science and Industry*, Vol. 5, Schooley, J. F., Ed.-in-Chief, American Institute of Physics, New York, 1982, 829.
123. See, for example, **Tiggelman, J. L. and Durieux, M.**, Platinum resistance thermometry below 13.81 K, in *Temperature, Its Measurement and Control in Science and Industry*, Vol. 4, Plumb, H. H., Ed.-in-Chief, Instrument Society of America, Pittsburgh, 1972, 849.
124. **Besley, L. M.**, Stability characteristics of rhodium-iron alloy resistance thermometers, *J. Phys. E: Sci. Instrum.*, 15, 824, 1982.
125. **Berry, K. H.**, NPL-75: a low temperature gas thermometry scale from 2.6 K to 27.1 K, *Metrologia*, 15, 89, 1979.
126. **Soulen, R. J., Rusby, R. L., and Van Vechten, D.**, A self-calibrating rhodium-iron resistive SQUID thermometer for the range below 0.5 K, *J. Low Temp. Phys.*, 40, 553, 1980.
127. **Rubin, L. G.**, Cyrogenic thermometry: a review of recent progress, *Cryogenics*, 10, 14, 1970.
128. **Hudson, R. P., Marshak, H., Soulen, R. J., Jr., and Utton, D. B.**, Recent advances in thermometry below 300 mK, *J. Low Temp. Phys.*, 20, 1, 1975.

129. **Betts, D. S.**, *Refrigeration and Thermometry Below One Kelvin*, Crane-Russak, New York, 1976.
130. Fourteen papers on resistance thermometry in cryogenics are collected in *Temperature, Its Measurement and Control in Science and Industry*, Vol. 4, Section IV, Plumb, H. H., Ed.-in-Chief, Instrument Society of America, Pittsburgh, 1972, 773—874.
131. **Durieux, M.**, Cryogenic thermometry between 0.1 K and 100 K, in *Temperature Measurement, 1975*, Billing, B. F. and Quinn, T. J., Eds., Conference Series No. 26, The Institute of Physics, London, 1975, 17.
132. **White, G. K.**, *Experimental Techniques in Low Temperature Physics*, Clarendon, Oxford University Press, 1979, chap. 4.
133. **Rubin, L. G., Brandt, B. L., and Sample, H. H.**, Cryogenic thermometry: a review of recent progress. II, in *Temperature, Its Measurement and Control in Science and Industry*, Vol. 5, Schooley, J. F., Ed.-in-Chief, American Institute of Physics, New York, 1982, 1333.
134. **Estermann, I., Foner, A., and Randall, J. A.**, *Phys. Rev.*, 71, 484, 1947.
134a. **Estermann, I.**, *Phys. Rev.*, 78, 83, 1950.
135. **Gerritson, A. N.**, *Physica*, 15, 427, 1949.
136. **Friedberg, S. A.**, Semiconductors as thermometers, in *Temperature, Its Measurement and Control in Science and Industry*, Vol. 2, Wolfe, H. D., Ed., Reinhold, New York, 1955, 359.
137. **Kunzler, J. E., Geballe, T. H., and Hull, G. W., Jr.**, Germanium resistance thermometers, in *Temperature, Its Measurement and Control in Science and Industry*, Vol. 3, Part 1, Herzfeld, C. M., Ed.-in-Chief, Reinhold, New York, 1962, 391.
138. **Besley, L. M. and Kemp, W. R. G.**, An intercomparison of temperature scales in the range 1 to 30 K using germanium resistance thermometry, *Metrologia*, 13, 35, 1977.
139. **Besley, L. M. and Plumb, H. H.**, Stability of germanium resistance thermometers at 20 K, *Rev. Sci. Instrum.*, 49, 68, 1978.
140. **Besley, L. M.**, Further stability studies on germanium resistance thermometers at 20 K, *Rev. Sci. Instrum.*, 49, 1041, 1978.
141. **Besley, L. M.**, Stability studies for germanium resistance thermometers at three temperatures, *Rev. Sci. Instrum.*, 51, 972, 1980.
142. **Swinehart, P. R.**, The state of development of planar germanium cyrogenic thermometers, in *Temperature, Its Measurement and Control in Science and Industry*, Vol. 5, Schooley, J. F., Ed.-in-Chief, American Institute of Physics, New York, 1982, 835.
143. See for example, **Kittel, C.**, *Introduction to Solid State Physics*, 2nd ed., John Wiley & Sons, New York, 1956, chap. 13.
144. **Ganapati Rao, M.**, Semiconductor junctions as cryogenic temperature sensors, in *Temperature, Its Measurement and Control in Science and Industry*, Vol. 5, Schooley, J. F., Ed.-in-Chief, American Institute of Physics, New York, 1982, 1205.
145. **Pavese, F. and Limbarinu, S.**, Accuracy of gallium arsenide diode thermometers in the range 4—300 K, in *Temperature, Its Measurement and Control in Science and Industry*, Vol. 4, Plumb, H. H., Ed.-in-Chief, Instrument Society of America, Pittsburgh, 1972, 1103.
146. **Swartz, J. M. and Gaines, J. R.**, Wide range thermometry using gallium arsenide sensors, in *Temperature, Its Measurement and Control in Science and Industry*, Vol. 4, Plumb, H. H., Ed.-in-Chief, Instrument Society of America, Pittsburgh, 1972, 1117.
147. **Ohte, A., Yamagata, M., and Akiyama, K.**, Precision silicon transistor thermometer, in *Temperature, Its Measurement and Control in Science and Industry*, Vol. 5, Schooley, J. F., Ed.-in-Chief, American Institute of Physics, New York, 1982, 1197.
148. **Verster, T. C.**, The silicon transistor as a temperature sensor, in *Temperature, Its Measurement and Control in Science and Industry*, Vol. 4, Plumb, H. H., Ed.-in-Chief, Instrument Society of America, Pittsburgh, 1972, 1125.
149. **Clement, J. R. and Quinnell, E. H.**, The low temperature characteristics of carbon-composition thermometers, *Rev. Sci. Instrum.*, 23, 213, 1952.
150. **Lindenfeld, P.**, Carbon and semiconductor thermometers for low temperatures, in *Temperature, Its Measurement and Control in Science and Industry*, Vol. 3, Part 1, Herzfeld, C. M., Ed.-in-Chief, Reinhold, New York, 1962, 399.
151. **Anderson, A. C.**, Instrumentation at temperatures below 1 K, *Rev. Sci. Instrum.*, 51, 1603, 1980.
152. **Johnson, W. L. and Anderson, A. C.**, The stability of carbon resistance thermometers, *Rev. Sci. Instrum.*, 42, 1296, 1971.
153. **Lawless, W. N.**, Thermometric properties of carbon-impregnated porous glass at low temperatures, *Rev. Sci. Instrum.*, 43, 1743, 1972.
154. Temperature cycling of carbon-glass thermometers was studied by **Besley, L. M.**, *Rev. Sci. Instrum.*, 50, 1626, 1979. Their performance in magnetic fields was reported by **Sample, H. H. and Rubin, L. G.**, Instrumentation and methods for low temperature measurements in high magnetic fields, *Cryogenics*, 17, 597, 1977.

155. **Ricketson, B. W. and Grinter, R.**, Carbon-glass sensors: reproducibility and polynomial fitting of temperature vs. resistance, in *Temperature, Its Measurement and Control in Science and Industry*, Vol. 5, Schooley, J. F., Ed.-in-Chief, American Institute of Physics, New York, 1982, 845.
156. **Yao Quanfa, Deng Daren, Ma Hongqi, Jiang Dehua, Ji Yunsong, and Huang Xihuai**, Carbon-glass thermometry in China, in *Temperature, Its Measurement and Control in Science and Industry*, Vol. 5, Schooley, J. F., Ed.-in-Chief, American Institute of Physics, New York, 1982, 853.
157. See, for example, **Kittel, C.**, *Introduction to Solid State Physics*, 2nd ed., John Wiley & Sons, New York, 1956, chap. 5.
158. See, for example, **Halliday, D. and Resnick, R.**, *Physics for Students of Science and Engineering*, combined ed., John Wiley & Sons, New York, 1965, sect. 47-2.
159. **Wien, W.**, *Ann. Physik*, 58, 662, 1896.
160. **Planck, M.**, *Ann. Physik*, 4, 553, 1901.
161. **Richmond, J. C. and Nicodemus, F. E.**, Blackbodies, blackbody radiation, and temperature scales, in *Self-Study Manual on Optical Radiation Measurements: Part I, Concepts*, NBS Tech. Note 910-8, 1985, chap. 12.
161a. **Cohen, E. R. and Taylor, B. N.**, The 1973 least-squares adjustment of the fundamental constants, *J. Phys. Chem. Ref. Data*, 2, 663, 1973.
162. See, for example, **Brown, E. B.**, *Modern Optics*, Reinhold, New York, 1965, chap. 3.
163. See, for example, *American Institute of Physics Handbook*, 3rd ed., Gray, D. E., Ed., McGraw-Hill, New York, 1972, chap. 6.
164. **Kirchhoff, G.-R.**, [Gesammelte Abhandlungen, Leipzig, Barth, J. A., 1882], p. 594. See also **Grum, F. and Becherer, R. J.**, *Optical Radiation Measurements: Vol. 1, Radiometry*, Academic Press, New York, 1979, sect. 4.5.
165. These values were obtained using solar radiation. See **Gray, D. E.**, Ed., *American Institute of Physics Handbook*, 3rd ed., 1972, McGraw-Hill, New York, sect. 6.
166. See, for example, **Herzberg, G.**, *Molecular Spectra and Molecular Structure*, 2nd ed., Vol. 1, D. Van Nostrand, Princeton, 1951; 2nd ed., Vol. 2, 1959.
167. **Blevin, W. R.**, Corrections in optical pyrometry and photometry for the refractive index of air, *Metrologia*, 8, 146, 1972.
168. **Lovejoy, D. R.**, Recent advances in optical pyrometry, in *Temperature, Its Measurement and Control in Science and Industry*, Vol. 3, Part 1, Herzfeld, C. M., Ed.-in-Chief, Reinhold, New York, 1962, 487.
169. **Quinn, T. J. and Martin, J. E.**, Radiometric measurement of thermodynamic temperature between 327 and 365 K, in *Temperature, Its Measurement and Control in Science and Industry*, Vol. 5, Schooley, J. F., Ed.-in-Chief, American Institute of Physics, New York, 1982, 103. See also **Quinn, T. J. and Martin, J. E.**, Radiometric measurements of the Stefan-Boltzmann constant and thermodynamic temperature between $-40°C$ and $+100°C$, *Metrologia*, 20, 163, 1984.
170. **Ginnings, D. C. and Reilly, M. L.**, Calorimetric measurement of thermodynamic temperatures above 0°C using total blackbody radiation, in *Temperature, Its Measurement and Control in Science and Industry*, Vol. 4, Plumb, H. H., Ed.-in-Chief, Instrument Society of America, Pittsburgh, 1972, 339.
171. **Jung, H. J.**, An optical measurement of the deviation of International Practical Temperatures T68 from thermodynamic temperatures in the range from 730 K to 903 K, *Metrologia*, 20, 67, 1984.
172. **Bonhoure, J. and Pello, R.**, Étude pyrométrique des températures comprises entre 420 et 630°C, Informal communication no. CCT/84-21 to the 15th Session of the Consultative Committee for Thermometry, International Bureau of Weights and Measures, Sèvres, France, 5—7 June, 1984. See also **Bonhoure, J.**, Determination radiométrique des températures thermodynamiques comprises entre 904 et 1338 K, *Metrologia*, 11, 141, 1975.
173. **Hall, J. A.**, The radiation scale of temperature between 175°C and 1063°C, *Metrologia*, 1, 140, 1965.
174. **Heusinkveld, W. A.**, Determination of the differences between the thermodynamic and the practical temperature scale in the range 630 to 1063°C from radiation measurements, *Metrologia*, 2, 61, 1966.
175. **Guildner, L. A. and Edsinger, R. E.**, Deviation of international practical temperatures from thermodynamic temperatures in the temperature range from 273.16 K to 730 K, *J. Res. Natl. Bur. Stand.*, 80A, 703, 1976.
175a. **Dils, R. R., Geist, J., and Reilly, M. L.**, Measurement of the silver freezing point with an optical fiber thermometer: proof of concept, *J. Appl. Phys.*, in press.
176. **Smith, R. A., Jones, F. E., and Chasmar, R. P.**, *The Detection and Measurement of Infra-Red Radiation*, Clarendon Press, Oxford, 1957.
177. **Worthing, A. G. and Halliday, D.**, *Heat*, John Wiley & Sons, New York, 1948.
178. **Campbell, C. H.**, *Modern Pyrometry*, Chemical Publishing, New York, 1951.
179. A thorough discussion of this type of temperature measurement is given by **Kostkowski, H. J. and Lee, R. D.**, Theory and methods of optical pyrometry, in *Temperature, Its Measurement and Control in Science and Industry*, Vol. 3, Part 1, Herzfeld, C. M., Ed.-in-Chief, Reinhold, New York, 1962, 449.

180. **Ono, A.**, Apparent emissivities of cylindrical cavities with partially specular conical bottoms, in *Temperature, Its Measurement and Control in Science and Industry*, Vol. 5, Schooley, J. F., Ed.-in-Chief, American Institute of Physics, New York, 1982, 513.
181. **Ono, A., Trusty, R. M., and DeWitt, D. P.**, Experimental and theoretical study on the quality of reference blackbodies formed by lateral holes on a metallic tube, in *Temperature, Its Measurement and Control in Science and Industry*, Vol. 5, Schooley, J. F., Ed.-in-Chief, American Institute of Physics, New York, 1982, 541.
182. **Beynon, T. G. R.**, Radiation thermometry applied to the development and control of gas turbine engines, in *Temperature, Its Measurement and Control in Science and Industry*, Vol. 5, Schooley, J. F., Ed.-in-Chief, American Institute of Physics, New York, 1982, 471.
183. **Iuchi, T. and Kusaka, R.**, Two methods for simultaneous measurement of temperature and emittance using multiple reflection and specular reflection, and their applications to industrial processes, in *Temperature, Its Measurement and Control in Science and Industry*, Vol. 5, Schooley, J. F., Ed.-in-Chief, American Institute of Physics, New York, 1982, 491.
184. **Warnke, G. F.**, Commercial pyrometers, in *Temperature, Its Measurement and Control in Science and Industry*, Vol. 4, Plumb, H. H., Ed.-in-Chief, Instrument Society of America, Pittsburgh, 1972, 503.
185. **Kwap, T. W., O'Kane, D. F., and Gulitz, L.**, Temperature measurements with an infrared television system, in *Temperature, Its Measurement and Control in Science and Industry*, Vol. 4, Plumb, H. H., Ed.-in-Chief, Instrument Society of America, Pittsburgh, 1972, 541.
186. **Toyota, H., Itoh, T., Takumi, Y., and Nezu, K.**, Measurement and evaluation of temperature patterns in cement industry, in *Temperature, Its Measurement and Control in Science and Industry*, Vol. 4, Plumb, H. H., Ed.-in-Chief, Instrument Society of America, Pittsburgh, 1972, 545.
187. **Kaplan, H. and Leftwich, R. F.**, Real-time thermal mapping of microscopic targets, in *Temperature, Its Measurement and Control in Science and Industry*, Vol. 4, Plumb, H. H., Ed.-in-Chief, Instrument Society of America, Pittsburgh, 1972, 557.
188. **White, E. L.**, Development of an infrared scanning system for the empirical evaluation of aerodynamic heating, in *Temperature, Its Measurement and Control in Science and Industry*, Vol. 4, Plumb, H. H., Ed.-in-Chief, Instrument Society of America, Pittsburgh, 1972, 565.
189. **Agerskans, J.**, Thermal imaging, A technical review, in *Temperature Measurement, 1975*, Billing, B. F. and Quinn, T. J., Eds., Conference Series No. 26, The Institute of Physics, London, 1975, 375.
190. **Wallace, J. D. and Cade, C. M.**, *Clinical Thermography*, CRC Press, Cleveland, 1975.
191. **Edrich, J. and Jobe, W. E.**, Imaging microwave thermography, in *Temperature, Its Measurement and Control in Science and Industry*, Vol. 5, Schooley, J. F., Ed.-in-Chief, American Institute of Physics, New York, 1982, 1379.
192. **Hurley, C. W. and Kreider, K. G.**, *Applications of Thermography for Energy Conservation in Industry*, U.S. National Bureau of Standards Tech. Note 923, October 1976. See also **Kreider, K. G. and Sheahen, T. P.**, *Use of Infrared Thermography for Industrial Heat Balance Calculations*, U.S. National Bureau of Standards Tech. Note 1129, July 1980.
193. **Cheng, A. F.**, Fluoroptic thermometry, *Measurements and Control*, 86, 115, April 1981.
194. **Sholes, R. R. and Small, J. G.**, Fluorescent decay thermometer with biological applications, *Rev. Sci. Instrum.*, 51, 882, 1980.
195. **Vaguine, V. A. et al.**, Multiple sensor optical thermometry system for application in clinical hyperthermia, *IEEE Transactions on Biomedical Engineering*, BME-31, 168, 1984.
196. **Cetas, T. C. and Connor, W. G.**, *Med. Phys.*, 5, 79, 1978.
197. **Wickersheim, K. A. and Alves, R. B.**, Optical temperature measurement, *Industrial Res. and Devel.*, 21, 82, 1979. See also **Wickersheim, K. A. and Alves, R. B.**, Fluoptic thermometry: a new RF-immune technology, in *Biomedical Thermology*, Alan R. Liss, New York, 1982.
198. **Morgan, F. H. and Danforth, W. E.**, *J. Appl. Phys.*, 21, 112, 1950.
199. **Payne-Gaposchkin, C.**, Astrophysical temperatures, in *Temperature, Its Measurement and Control in Science and Industry*, Vol. 2, Wolfe, H. D., Ed., Reinhold, New York, 1955, 31.
200. **Russell, H. N., Dugan, R. S., and Stewart, J. Q.**, *Astronomy*, Ginn & Co., 1927, See also **Kuiper, G. P.**, *Astrophys. J.*, 88, 429, 1938.
201. **Smith, E. P.**, Temperature in solar flares, in *Temperature, Its Measurement and Control in Science and Industry*, Vol. 4, Plumb, H. H., Ed.-in-Chief, Instrument Society of America, Pittsburgh, 1972, 2359.
202. **Stratton, T. F.**, Plasma temperatures by spectroscopy in the X-ray region, in *Temperature, Its Measurement and Control in Science and Industry*, Vol. 3, Part 1, Herzfeld, C. M., Ed.-in-Chief, Reinhold, New York, 1962, 663.
203. **Herzfeld, C. M., Ed.-in-Chief**, *Temperature, Its Measurement and Control in Science and Industry*, Vol. 3, Part 1, Reinhold, New York, 1962, sect. X.
203a. **Herzfeld, C. M., Ed.-in-Chief**, *Temperature, Its Measurement and Control in Science and Industry*, Vol. 3, Part 2, Reinhold, New York, 1962, sect. III. C.

203b. **Stuck, D. and Wende, B.**, Basic method for realization of temperature scale at 10^4 K by photometric comparison between vacuum-UV-blackbody radiation of a plasma and synchrotron radiation, in *Temperature, Its Measurement and Control in Science and Industry*, Vol. 4, Plumb, H. H., Ed.-in-Chief, Instrument Society of America, Pittsburgh, 1972, 105.
204. **Efthimion, P. C., Arunasalam, V., Bitzer, R. A., and Hosea, J. C.**, Measurement of the time evolution of the electron temperature profile of reactor-like plasmas from the measurement of blackbody electron cyclotron emission, in *Temperature, Its Measurement and Control in Science and Industry*, Vol. 5, Schooley, J. F., Ed.-in-Chief, American Institute of Physics, New York, 1982, 677.
205. **Stauffer, F. J.**, Measurement of the electron temperature profile in a Tokomak by observation of the electron cyclotron emission using a Fourier transform spectrometer, in *Temperature, Its Measurement and Control in Science and Industry*, Vol. 5, Schooley, J. F., Ed.-in-Chief, American Institute of Physics, New York, 1982, 687.
206. **Bitter, M., von Goeler, S., Goldman, M., Hill, K. W., Horton, R., Roney, W., Sauthoff, N., and Stodiek, W.**, Measurement of the central ion and electron temperature of Tokomak plasmas from the X-ray line radiation of high-Z impurity ions, in *Temperature, Its Measurement and Control in Science and Industry*, Vol. 5, Schooley, J. F., Ed.-in-Chief, American Institute of Physics, New York, 1982, 693.
207. **Danielewicz, E. J., Luhmann, N. C., Jr., and Peebles, W. A.**, Applications of lasers to magnetic confinement fusion plasmas, *Optical Eng.*, 23, 475, 1984.
208. **Bechtel, J. H., Dasch, C. J., and Teets, R. E.**, Combustion research with lasers, *Laser Appl.*, 5, 129, 1984.
209. **Taran, J. P. and Péalat, M.**, Practical CARS temperature measurements, in *Temperature, Its Measurement and Control in Science and Industry*, Vol. 5, Schooley, J. F., Ed.-in-Chief, American Institute of Physics, New York, 1982, 575.
210. **Hall, R. J. and Eckbreth, A. C.**, Coherent anti-Stokes Raman spectroscopy (CARS): application to combustion diagnostics, in *Laser Appl.*, 5, 213, 1984.
211. See, for example, **Lempert, W., Rosasco, G. J., and Hurst, W. S.**, Rotational collisional narrowing in the NO fundamental Q branch, studied with CW stimulated Raman spectroscopy, *J. Chem. Phys.*, 81, 4241, 1984.
212. **Semerjian, H. G., Santoro, R. J., Emmerman, P. J., and Goulard, R.**, Laser tomography for temperature measurements in flames, in *Temperature, Its Measurement and Control in Science and Industry*, Vol. 5, Schooley, J. F., Ed.-in-Chief, American Institute of Physics, New York, 1982, 649.
213. See, for example, **Tourin, R. H.**, *Spectroscopic Gas Temperature Measurements*, Elsevier, New York, 1966.
214. **Ray, S. R. and Semerjian, H. G.**, Laser tomography for simultaneous concentration and temperature measurement in reacting flows, in *Combustion Diagnostics by Nonintrusive Methods*, McCoy, T. D. and Roux, J. A., Eds., Vol. 92 in the series *Progress in Astronautics and Aeronautics*, Summerfield, M., Series Ed.-in-Chief, American Institute of Aeronautics and Astronautics, New York, 1984, 300.
215. **Zizak, G., Omenetto, N., and Winefordner, J. D.**, Laser-excited atomic fluorescence techniques for temperature measurements in flames: a summary, *Optical Eng.*, 23, 749, 1984.
216. **Rusby, R. L. and Swenson, C. A.**, A new determination of the helium vapour pressure scales using a CMN magnetic thermometer and the NPL-75 gas thermometer scale, *Metrologia*, 16, 73, 1980.
217. **Durieux, M., van Dijk, J. E., ter Harmsel, H., Rem, P. C., and Rusby, R. L.**, Helium vapor pressure equation on the EPT-76, in *Temperature, Its Measurement and Control in Science and Industry*, Vol. 5, Schooley, J. F., Ed.-in-Chief, American Institute of Physics, New York, 1982, 145—153.
218. **Durieux, M. and Rusby, R. L.**, Helium vapor pressure equations on the EPT-76, *Metrologia*, 19, 67, 1983.
219. Equations for H_2, N_2, CO, and Ne are given by **van Dijk, H.**, Selected values for the thermodynamic temperatures of thermometric fixed points below 0°C, in *Temperature, Its Measurement and Control in Science and Industry*, Vol. 3, Part 1, Herzfeld, C. M., Ed.-in-Chief, Reinhold, New York, 1962, 173.
220. See, for example, *Handbook of Chemistry and Physics*, Weast, R. C., Ed., CRC Press, Boca Raton, Fla.
221. **Bayer, H.**, *Z. Physik*, 130, 227, 1951.
222. **Kushida, T.**, *J. Sci. Hiroshima Univ.*, A19, 327, 1955.
223. **Vanier, J.**, *Can. J. Phys.*, 38, 1397, 1960.
224. **Utton, D. B.**, Nuclear quadrupole resonance thermometry, *Metrologia*, 3, 98, 1967.
225. **Ohte, A. and Iwaoka, H.**, A new nuclear quadrupole resonance standard thermometer, in *Temperature, Its Measurement and Control in Science and Industry*, Vol. 5, Schooley, J. F., Ed.-in-Chief, American Institute of Physics, New York, 1982, 1173.
226. **Wade, W. H. and Slutsky, L. J.**, *Rev. Sci. Instrum.*, 33, 212, 1962.
227. **Hammond, D. L., Adams, C. A., and Schmidt, P.**, *Trans. Instrum. Sci. Am.*, 4, 349, 1965.
228. **Benjaminson, A. and Rowland, F.**, The development of the quartz resonator as a digital temperature sensor with a precision of 1×10^{-4}, in *Temperature, Its Measurement and Control in Science and Industry*, Vol. 4, Plumb, H. H., Ed.-in-Chief, Instrument Society of America, Pittsburgh, 1972, 701.

229. **Van Vleck, J. H.**, *Electric and Magnetic Susceptibilities,* Clarendon Press, Oxford, 1932.
230. See, for example, **de Klerk, D. and Hudson, R. P.**, *J. Res. Natl. Bur. Stand.* 53, 173, 1954. See also **Erickson, R. A., Roberts, L. D., and Dabbs, J. W. T.**, *Rev. Sci. Instrum.,* 25, 1178, 1954.
231. See, for example, **Lounasmaa, O. V.**, *Experimental Principles and Methods Below 1 K,* Academic Press, London, 1976, sect. 7.9 and 8.9. See also **Giffard, R. P., Webb, R. A., and Wheatley, J. C.**, *J. Low Temp. Phys.,* 6, 533, 1972.
232. See, for example, **van den Handel, J.**, Low temperature magnetism, and **de Klerk, D.**, Adiabatic demagnetization, in *Encyclopedia of Physics,* Flügge, S., Ed., Vol. 15, Springer-Verlag, Heidelberg, 1956.
233. **Bowers, K. D. and Owen, J.**, Paramagnetic resonance: II, *Rep. Prog. Phys.,* 18, 304, 1955.
234. **Hudson, R. P.**, *Principles and Application of Magnetic Cooling,* North-Holland, Amsterdam, 1972, section 3.
235. **Cetas, T. C. and Swenson, C. A.**, A paramagnetic salt temperature scale from 0.9 to 18 K, *Metrologia,* 8, 46, 1972.
236. **Cetas, T. C.**, A magnetic temperature scale from 1 to 83 K, *Metrologia,* 12, 27, 1976.
237. **Blalock, T. V. and Shepard, R. L.**, A decade of progress in high temperature Johnson noise thermometry, in *Temperature, Its Measurement and Control in Science and Industry,* Vol. 5, Schooley, J. F., Ed.-in-Chief, American Institute of Physics, New York, 1982, 1219.
238. **Brixy, H., Hecker, R., Rittinghaus, K. F., and Höwener, H.**, Application of noise thermometry in industry under plant conditions, in *Temperature, Its Measurement and Control in Science and Industry,* Vol. 5, Schooley, J. F., Ed.-in-Chief, American Institute of Physics, New York, 1982, 1225.
239. **Borkowski, C. J. and Blalock, T. V.**, A new method of noise thermometry, *Rev. Sci. Instrum.* 45, 151, 1974.
240. **Blalock, T. V., Horton, J. L., and Shepard, R. L.**, Johnson noise power thermometer and its application in process temperature measurement, in *Temperature, Its Measurement and Control in Science and Industry,* Vol. 5, Schooley, J. F., Ed.-in-Chief, American Institute of Physics, New York, 1982, 1249.
241. **Blalock, T. V.**, private communication.

INDEX

A

Absolute P-V isotherm thermometry, 131—132
Absolute scale, 77
Absolute zero of temperature, 10—16
AC measurements, 190—191
Acoustic thermometry, 134, 136—137
Active devices, 213
Active techniques, 221
Adsorbed impurities, 124—125
Air thermoscope, 8
Alcohol thermometers, 9—10
Aligned spin system, 145
Aluminum fixed-point cell, 64—65
Aluminum fixed points, 62
Aluminum freezing point, 63
American Society for Testing and Materials (ASTM), 167
Amonton, 77
Analog meter, 164
Annealing, 180
Artificial temperature scale, platinum resistance thermometers, 196
Automatic resistance thermometer bridges, 190
Avogadro's constant, 88, 116
Avogadro's law, 28—29

B

Basic thermometry terms defined, 7
Berthelot's equation, 119
Bimetallic thermometers, 165—166
Biomedical microwave installations, 219
Bird cage platinum thermometers, 195
Black body, 149—150, 152, 209—212
Boiling-point elevation, 46
Boltzmann distribution function, 145
Boltzmann equation, 220
Boltzmann's constant, 28—29, 88, 116
Boyle temperature, 117
Boyle's law, 11—13, 16, 28
Bridge measurements, 189—190
Btu, 20—21

C

Cadmium, 72
Calibration, 10, 33, 81, 169
Callendar equation, 89, 92, 105
Callendar-Van Dusen equation, 89
15°C Calorie, 20—21
20°C Calorie, 20
Capsule PRT, 96, 98—100, 102—104, 193
Carbon radio resistor, 208
Carnot cycle, 16—18
Carnot-cycle efficiencies, 77

Carnot's heat engine, 16—18
Cathetometer, 171
Cavity radiators, 209—210, 214—215
Celsius, 91, 94
Celsius temperature scale, 2—4, 10
Centigrade, 91, 94
Centigrade scale, 89, 93
Charles' law, 12—14, 16
Clapeyron equation, 37—38
Coherent Anti-Stokes Raman Spectroscopy (CARS), 221
Commercial radiation thermometers, 215—217
Comparison thermocouple calibration, 181
Comparisons of national temperature scales in cryogenic range, 108
Component of phase, 34
Compressibility factor, 117—118
Concept of temperature, 7—32
Consensus scales, 79—80
Constant bulb temperature gas thermometry, 117, 132
Constant bulb volume gas thermometry, 122
Constant pressure gas thermometry, 116—117
Constant volume gas thermometry, 116, 120—121, 131, 134—138
    KOL, 138
    NBS, 125—131
    NML, 137—138
Constant volume valve, 121
Consultative Committee for Thermometry (CCT), 87, 91, 107—108
Convention of the Meter, 86—87
Cooling curves, 65
Correlation noise thermometer, 140
Critical magnetic fields, 69, 71
Critical point, 39
Cross-correlation technique of noise thermometry, 140—141
Cryogenics
    semiconducting thermometers in, 206—208
    thermocouples for use in, 183
Cryoscopic constants, 46—47
Current practical temperature scales, 135
Cyclotron, 220

D

Dalton's law, 28—29
Dead space corrections, 120—121
Debye temperature, 30—31, 69
Decigrade scale, 77
Defining fixed points, 86, 89
    comparison of temperature scale fixed-point assignments, 110
    IPTS-68, 97
Definitions, see specific terms
Degrees of freedom, 37

Dectectors, 217
Deviation functions, 98
Dielectric constant gas thermometry, 136—137
Diffusion of impurities, 182
Dilute solutions, 45—48
Diode thermometers, 163, 207
Diplomatic Conference on the Meter, 86—87
Disappearing filament optical pyrometer, 216
D_2O, triple points of, 43—44
Doped germanium resistors, 207
Dulong-Petit value, 30

E

Early thermometers, 7—10
Ebullioscopic constants, 46—47
Einstein temperature, 30—31
Electrical leakage, 182
Electromagnetic interference, 182
Electromagnetic radiation, 218
Electromagnetic spectrum, 209
Electronic temperature, 220
emf measurement, 181
emf-temperature relations, 174
Emissivity, 212—214
Energy units, 20
Enthalpy, 20—24
Entropy, 23—25
EPT-76, see 1976 Provisional 0.5 to 30 K Temperature Scale
Equilibrium hydrogen, 55
Equilibrium state, 29
Everyday scales, 2—4
Experimental imprecision, 78, 80, 84—85
Extension wires, 182

F

Fahrenheit terperature scale, 2—4, 9—10
Filled-system thermometers, 164—165
First law of thermodynamics, 19—23
First-order transitions, 35, 67
First thermometers, 8—10
Fixed-point device, superconductive transitions, 68
Fixed points, see also Triple point, 4, 7, 33—75
   defining, see Defining fixed points
   EPT-76, 109
   freezing, 56
   high-pressure cells for low-temperature triple points, 49—55
   Kelvin thermodynamic temperature, 149, 156
   life sciences, 63, 65—66
   melting, 56
   metal, 66
   metallic liquid-solid, 55
   phase equilibria, 33—35
   phases, 33—35
   phase transitions, 33—35
   states of matter, 33—34
   superconductive, 71
   types, 35—37
   water, 37—49
Fixed point temperature ratios, 155
Flames, temperature measurement in, 219—222
Fluorescent decay, 219
Free energy, 26
Freezing fixed point, 56
Freezing methods, 61
Freezing-point depressions in water, 45—46
Freezing point of metals, 55—63

G

Gas constant, 15, 22, 29, 88, 116—117
Gases, 33
   temperature measurement in, 219—222
Gas-expansion thermometers, 10
Gas thermometer, 84, 89, 137, 163
Gas thermometer scale, 88
Gas thermometry, 115—138
   adsorbed impurities, 124—125
   constant bulb temperature, 117, 132
   constant pressure, 116—117
   constant volume, 116, 131, 134—138
      KOL, 138
      NBS, 125—131
      NPL, 137—138
   current experiments, 125—138
   dead space corrections, 120—121
   dielectric constant, 136—137
   gas constant, 116—117
   hydrostatic pressure head corrections, 123, 128
   ideal gas law, 115
      deviations from, 117—120
   KOL determinations, 138
   manometer system, 127—128
   methods, 116—117
   normal hydrogen scale, 115
   NPL, 131—136
   ratio technique, 116
   second virial coefficients, 129, 133
   thermal expansion, corrections for, 122—123
   thermomolecular pressure corrections, 123—124
   virial coefficients, 129, 133
Gay-Lussac's law, 12—13
General Conference on Weights and Measures, 86—88
Gibbs free energy, 26, 34
Gibbs phase rule, 38
Gold fixed-point cells, 84
Graduation intervals, 169
Gray body, 217

H

^{3}He
   vapor pressure/temperature equations, 224
   vapor pressure/temperature relations, 223

^{4}He
    second virial coefficients, 129, 133
    vapor pressure/temperature equations, 224
    vapor pressure/temperature relations, 223
    volume virial coefficients, 135
Heat, 16—18
Heat capacity, 20—22, 30, 70
Heat pump, 17
Heats of fusion, 23
Helmholtz free energy, 26
Heterogeneous system, 34—35
High-pressure cells for low-temperature triple points, 49—55
High-temperature furnace, 56, 60—61
High-temperature platinum resistance thermometers, 194—199
High temperature thermocouples, 183—186
Homogeneous metals, law of, 176
Homogeneous system, 34—35
Hot gaseous systems, 221
Hotness, 7, 33, 77, 79, 83
Hydrostatic pressure head corrections, 40, 123, 128

I

Ice mantle, 40
Ice phases, 37—38
Ice point, 37—38
Ideal gas, 10—16
Ideal gas law, 18, 27—29, 33, 115
    compressibility factor, 117—118
    deviations from, 117—120
    second virial coefficient, 119
    second virial correction, 128
    thermal expansion, 122
    virial coefficients, 119
    virial equation, 117
Ideal heat engine, 17
Imprecision, 164
    platinum resistance thermometer, 106
    standard thermocouple thermometer, 196
    thermocouple thermometer, 107
Impurity effects, 44—45
Indium, 70
Induction-heating apparatus, 57
Industrial Johnson noise thermometry, 227—228
Inner melt, 65
Intermediate metals at different temperatures, law of, 178—179
Intermediate metals at a single temperature, law of, 176—178
Intermediate temperatures, law of, 178—179
Internal energy, 19, 24
International Bureau of Weights and Measures (BIPM), 86—87
International Committee for Weights and Measures, 86—87
International Practical Temperature Scale of 1948 (IPTS-48), 91—94
International Practical Temperature Scale of 1968 (IPTS-68), 94—108, 135
    comparison with thermodynamic temperature, 129—130
    Kelvin thermodynamic temperatures, 157
    nonuniqueness of, 106
    thermodynamic accuracy of, 198
International scales of temperature, 77—114
    future, 109—111
    normal hydrogen scale, 88
International Steam Table calorie, 20
International Temperature Scale of 1927 (ITS-27), 88—90
International Temperature Scale of 1948 (ITS-48), 91—94
Interpolating equation, 92, 98
Iron-doped rhodium resistance thermometers, 108
Iron scale, 77
ISA standard thermocouples, 180
Isotopic effects, 43—44

J

Johnson noise power thermometer, 227—228
Johnson noise thermometers, 163
Johnson noise thermometry, see also Noise thermometry, 140—143

K

Kamerlingh Onnes Laboratory (KOL), 138
Kelvin temperature scale, 2, 4, 18
Kelvin thermodynamic scale, 93, 130
Kelvin thermodynamic temperatures, 131, 136, 144, 148, 152—153
    above 273 K, 156—157
    acoustic thermometer determinations, 137
    below 273 K, 147—148
    compared with current practical temperature scales, 135
    compared wtih NPL-75 scale, 135
    fixed points, 149, 156
    gas thermometer determinations, 137
    gas thermometry, 151
    hypothetical example of, 81—86
    IPTS-68, 157
    NBS gas thermometry determination, 130—131
    noise thermometry determination, 139—141
    nuclear orientation thermometer compared with noise thermometer, 147
    spectral radiation thermometry, 153—155
    total radiation thermometry, 151
Kelvin thermodynamic temperature scale (KTTS), 14, 18, 78—79
    radiation thermometers, 214
Kinetic theory, 27—29
Kirchhoff, law of, 212

Kirchhoff relation, 219

L

Laboratory scales, 79
Laboratory temperature scale, 7, 147
Lambert law, 212
Laser-based measurements, 221
Laser-excited atomic fluorescence, 222
Laser tomography, 221
Latent heat of fusion, 23
Latent heats of phase transitions, 35—36
Laws of thermodynamics, see also Thermodynamics, 18—26
Least squares fitting, 206
Legendre polynomials, 146
Life sciences, fixed points for, 63, 65—66
Liquid-expansion thermometers, 10
Liquid-in-glass thermometers, 9—10, 166—171
Liquids, 33
Liquidus curve, 47—48
Long-stem PRT, 95—96
  construction, 193
Loop current errors, 182
Low-temperature triple points
  high-pressure cells for, 49—55
  sealed high-pressure cells, 51—53

M

Magnetic field sensitivities, 71
Manometer system, 127—128
Melting fixed point, 56
Melting point of metals, 55—63
Mercury-in-glass thermometers, 88
Mercury thermometers, 9—10, 171
Metal fixed-point cells, 55
Metallic liquid-solid fixed points, 55
Metals
  fixed points, 66
  freezing point, 55—63
  melting point, 55—63
  superconductive-normal phase equilibria in, 68—72
  triple points, 55—63
Method of realization, 78—79
  EPT-76, 110
Metrologia, 87, 95
Microprocessor-controlled resistance bridges, 190—191
Microwave compatible thermometers, 219
Modern thermometers, see also specific types, 163—238
  block diagram illustrating components, 164
  radiation thermometers, 208—222
  resistance thermometers, 186—208
  thermal expansion thermometers, 164—171
  thermocouple thermometry, 172—186

N

National Bureau of Standards (NBS), constant volume gas thermometry, 125—131
  freezing-point cell, 56, 59
  high-temperature fixed-point furnace, 56, 60—61
National Physical Laboratory (NPL), 131—136
Nicrosil vs. nisil thermocouple, 184—185
1976 Provisional 0.5 to 30 K Temperature Scale (EPT-76) 108—110, 135
Nitrogen, compressibility factor, 118
Noise thermometry, see also Johnson noise thermometry, 134, 138—143, 227
Non-uniqueness, 78—80, 106
Normal boiling point of water, 49
Normal distribution of a measurement process, 81
Normal hydrogen, 54
Normal hydrogen scale, 88, 115
NPL-75 gas thermometer scale, 135
Nuclear orientation thermometry, 143—147
Nuclear quadrupole resonance thermometry, 223, 225
Nyquist equation, 139
Nyquist theory of Johnson noise, 79

O

Optical fiber, 215, 218
Ortho hydrogen, 54—55
Overspecified scale, 78
Overspecified thermodynamic scale, 91
Oxygen transitions, 67

P

Para hydrogen, 54—55
Paramagnetic salt thermometry, 108
Paramagnetic thermometers, 163, 225—227
Passive devices, 213
Peltier effect, 175—176
Phase diagram of ice, 37—38
Phase diagram of water, 40
Phase equilibrium, 33—35
Phase equilibrium temperatures, 35—36
Phases, 33—35
Phase transitions, 33—35
Photoelectric pyrometer, 152
Photomultiplier, 153
Planck equation, 211, 219
Planck law, 79, 92—93, 107, 148—149, 151—152
Planck's constant, 211
Plasmas, temperature measurement in, 219—222
Plasma thermometry, 220
Platinum vs. platinum-rhodium thermocouple thermometer, 90
Platinum vs. platinum-10 wt% rhodium thermocouple thermometer, 92
Platinum resistance thermometers, see also Standard platinum resistance thermometers; other types, 89, 101, 105, 193, 215

imprecision, 106
rhodium-iron resistance thermometers compared, 205
thermistor thermometer compared, 203
thermocouple thermometers distinguished, 197
Polarized spin system, 145
Potentiometer measurements, 187—189
Practical scales, 79—80
Practical temperature scale, 7, 79, 89
construction, 86
Precision gap, 88
Probability, 29
Probe crystals, 225
Pulse-heating techniques, 51, 54

Q

Quantization of light and heat, 29—31
Quantized energy levels, 29
Quartz resonance thermometers, 225

R

Radiation detectors
characteristics, 217
properties, 214
Radiation thermometers, 163, 208—222
absorption edge, shift in, 219
active devices, 213
active techniques, 221
black body, 209—212
commercial, 215—217
fluorescent decay, 219
general features of, 213—219
optical fiber, 218
passive devices, 213
presence of electromagnetic radiation, 218
real materials, 212—213
reference sources, 214
temperature measurement in gases, flames, plasmas and stars, 219—222
thermometry research, 214—215
Radiation thermometry, 148—155
spectral, 151—156
total, 149—151
Radiometer, 215—216
Range, 78—79
Rankine temperature scale, 2, 4
Raoult's law, 45—48, 59
Ratio technique, 116
Real materials, radiation properties of, 212—213
Reference function, 98, 101
IPTS-68, 101
platinum resistance thermometer, 104
Reference tables, 102, 180, 182, 184
Reference temperatures, 4, 7
Relative P-V isotherm thermometry, 131, 133—134, 138
Resistance bridges, 189—191

Resistance temperature detectors (RTDs), 199—202
Resistance thermometer measurements, 186—191
Resistance thermometers, see also specific types, 81, 84—86, 163, 186—208
high-temperature platinum, 194—199
measurement, see also Resistance thermometer measurements, 186—191
porous glass with carbon containing material, 208
pure metal, 192—202
resistance temperature detectors, 199—202
rhodium-iron, 204—206
semiconducting thermometers in cryogenics, 206—208
standard platinum, 192—194
thermistor thermometers, 202—204
Rhodium-iron resistance thermometers, 204—206
R-SQUID, 141—143

S

Saha equation, 220
Scales, see specific types
Scientific scales, 2—3
Sealed high-pressure cells, 51—53
Second law of thermodynamics, 23—25
Second order transitions, 68
Second radiation constant, 92, 107, 149
Second virial coefficients, 119
^{4}He, 129, 133
Second virial correction, 128
Seebeck coefficients, 173—174
Seebeck effect, 172
Seebeck voltage, 172—173
Semiconducting thermometers in cryogenics, 206—208
Semiconductor diode, 207
Sensitivity function, 146
Signal conditioner, 163
Silicon photodiode radiation detector, 152
Silicon transistor thermometers, 207
Solid oxygen, 67
Solids, 33—34
Solid-solid phase equilibria, 65, 67—72
oxygen, 67—68
Solidus curve, 47
Solubilities in water, 45
Specific heat, 21
Spectral distribution of radiation, 219
Spectral emittance, 212—214
Spectral radiances, of black body, 210
Spectral radiation thermometers, 215
Spectral radiation thermometry, 151—156
Standard deviation of the fit, 206
Standard deviation of the mean, 80—81, 85, 130
Standard instruments for the IPTS-48, 94
Standard noise circuitry, 139—140
Standard platinum resistance thermometer (SPRT), 56, 61, 91, 95—98, 192—194, 197
Standard platinum-10 wt% rhodium vs. platinum thermocouple thermometer, specifications, 107

Standard PRT, see Standard platinum resistance thermometer
Standard thermocouple thermometers, 179—182, 196
Star, 219—222
States of matter, 33—34
State variables, 10
Statistical mechanics, 26—31
Steam point, 48—49
Stefan-Boltzmann constant, 211
Stefan-Boltzmann equation, 211
Stefan-Boltzmann law, 79, 82, 84, 148
Steinhart-Hart equation, 202, 204
Stellar line profiles, 220
Superconducting transition point, 109—110
Superconducting transition temperature, 69
Superconductive fixed points, 71
Superconductive-normal phase equilibria in metals, 68—72
Superconductive transitions, 35, 68, 70—71, 108
Superconductive transition temperature, 72
Superconductors, 71
Superfluid transitions, 35
Supplementary information, 107—108
Systematic errors, 84—86, 164
  total radiation thermometry, 150—151

T

Temperature, see also specific topics
  defined, 7
  distribution, 42
  fixed points, see Fixed points
  gradients, 42—43
  reference points
    solid-solid phase equilibria, 65, 67—72
    superconductive normal phase equilibria in metals, 68—72
Temperature-averaging thermocouple, 176
Temperature scales, see also specific types, 2—4, 7—10
  characteristics of, 77—81
  construction, 94
  defined, 7
  elements of, 77—86
  experimental imprecision, 78, 80, 84—85
  method of realization, 78—79
  multiply realizable, 79
  NBS-CTS-1, 147
  1958 ^{4}He vapor pressure scale, 135
  1962 ^{3}He vapor pressure scale, 135
  non-uniqueness, 78—80
  range, 78—79
  singly realizable, 79
  systematic errors, 84—86
  thermometry, 81—86
  uncertainty of fixed-point realization, 78, 80
  uncertainty of thermodynamic temperature, 84—85
  uncertainty with respect to the KTTS, 78, 80

Temperature sensor, 163
Temperature-time curve, 48, 62—63
Terrestrial water samples, 44
Test junction assembly, 181
Thermal diffusivity, 55—56
Thermal expansion coefficients, 122
Thermal expansion corrections, 122—123
Thermal expansion thermometers, 163—171
Thermal imagers, 217—218
Thermistor thermometers, 202—204
Thermochemical calorie, 20
Thermocouple circuits, laws of, 176—179
Thermocouple reference wire, 175
Thermocouples for special applications, 183—186
Thermocouple thermometer circuit, 172
Thermocouple thermometers, 163, 172
  imprecision, 107
  platinum resistance thermometer compared, 197
Thermocouple thermometry, 172—186
  calibration, standard thermocouple thermometers, 179—182
  diffusion of impurities, 182
  electrical leakage, 182
  electromagnetic interference, 182
  emf-temperature reference, 173—174
  extension wires, 182
  homogeneous metals, law of, 176
  intermediate metals at different temperatures, law of, 178—179
  intermediate metals at a single temperature, law of, 176—178
  intermediate temperatures, law of, 178—179
  loop current errors, 182
  Peltier effect, 175—176
  reference tables, 184
  Seebeck coefficients, 173—174
  Seebeck effect, 172
  Seebeck voltage, 172—173
  special applications, 183—186
  special problems, 182—183
  standard thermocouple thermometers, 179—182
  thermocouple circuits, laws of, 176—179
  Type E, adjustment for different reference temperature, 179
Thermodynamic determinations of temperature, 148
Thermodynamic equilibrium, 34
Thermodynamic ice-point temperature, 124
Thermodynamics, 10—26
  absolute zero of temperature, 10—16
  Boyle's law, 11—13, 16
  Carnot cycle, 16—18
  Carnot's heat engine, 16—18
  Charles' law, 12—14, 16
  enthalpy, 20—24
  entropy, 23—25

first law of, 19—23
free energy, 26
gas constant, 15, 22
Gay-Lussac's law, 12—14
heat, 16—18
heat capacity, 20—22
heat of fusion, 23
ideal gas, 10—16
ideal gas law, 18
ideal heat engine, 17
internal energy, 19, 24
laws of, 18—26
second law of, 23—25
specific heat, 21
temperature, role of, 10
third law of, 26
vaporization, 23
work, 16—18
zeroth law of, 18—19
Thermodynamic temperatures, 109, 135, 143
 comparison with IPTS-68, 129—130
 gas thermometry, see also Gas thermometry, 115—138
 Kelvin thermodynamic temperatures above 273 K, 156—157
 Kelvin thermodynamic temperatures below 273 K, 147—148
 measurement of, 115—161
 noise thermometry, see also Noise thermometry, 138—143
 nuclear orientation thermometry, see also Nuclear orientation thermometry, 143—147
 radiation thermometry, see also Radiation thermometry, 148—155
Thermodynamic temperature scale, 7, 77—78, 134
Thermodynamic triple-point temperature, 124
Thermometer circuit, 178
Thermometer defined, 7, 163
Thermometer system, 163
Thermometry, see also specific topics
 elements of, 77—86
 research, radiation thermometers for, 214—215
Thermomolecular pressure corrections, 123—124
Thermopile, 173, 175
Third law of thermodynamics, 26
Tin freezing point, 61—62
Total radiation thermometer, 150, 214
Total radiation thermometry, 149—151
Transducer, 163
Transition temperatures, 71
Transition width, 71—72
Transmitter, 164
"Triple-point cell", 39—43
Triple points
 defined, 38
 D_2O, 43—44
 low-temperature, see Low-temperature triple points
 metals, 55—63
 oxygen, 53
 water, 37—43, 79, 124—125
 isotopic effects on, 43—44
Triple point temperatures of terrestrial water samples, 44
Two-color pyrometry, 216—217
Type K thermocouples, emf drift, 185

U

Uncertainty of fixed-point realization, 78, 80
Uncertainty of thermodynamic temperature, 84—85
Uncertainty with respect to the KTTS, 78, 80
Undercooling, 61
Universal scale of temperature, 18

V

Van der Waals equation, 119
Van Dusen equation, 91
Vaporization, 23
Vapor pressure/temperature, 223—224
Vapor pressure thermometers, 163
Vapor pressure thermometry, 222—224
Virial coefficients, 119, 129, 133
Virial equation, 117
Volume virial coefficients, see also Second virial coefficient, 135

W

Water
 boiling-point elevation, 46
 dilute solutions, 45—48
 fixed points in, 37—49
 freezing-point depressions, 45—46
 ice points, 37—38
 impurity effects, 44—45
 normal boiling point, 49
 phase diagram, 40
 Raoult's law, 45—49
 solubilities, 45
 steam point, 48—49
 triple point, see Triple point
 triple-point cell, see "Triple-point cell"
Wien displacement law, 211
Wien law of radiation, 90
Work, 16—18, 20

X

X-ray emission, 220

Z

Zero-current resistance value, 43
Zeroth law of thermodynamics, 18—19
Zinc freezing point, 61, 63